Springer Geology

More information about this series at http://www.springer.com/series/10172

M.N. Rajeevan · Shailesh Nayak
Editors

Observed Climate Variability and Change over the Indian Region

Springer

Editors
M.N. Rajeevan
Ministry of Earth Sciences
Government of India
New Delhi, Delhi
India

Shailesh Nayak
Ministry of Earth Sciences
Government of India
New Delhi, Delhi
India

ISSN 2197-9545 ISSN 2197-9553 (electronic)
Springer Geology
ISBN 978-981-10-9641-9 ISBN 978-981-10-2531-0 (eBook)
DOI 10.1007/978-981-10-2531-0

Softcover reprint of the hardcover 1st edition 2016

Printed on acid-free paper

This Springer imprint is published by Springer Nature
The registered company is Springer Nature Singapore Pte Ltd.
The registered company address is: 152 Beach Road, #22-06/08 Gateway East, Singapore 189721, Singapore

Foreword

India is one of the few countries in the tropics that have a long record of observations that are necessary to understand tropical climate variability. These long records are essential to understand changes in climate that occur over both short and long time scales. During the past 50 years, there has been a concern about the influence of human beings on the earth's climate. Climate can change over a long period on account of natural causes. Major ice ages have occurred on our Earth every 100,000 years during the past few million years. These were caused by small changes in the Earth's orbital parameters. In the twentieth century, the changes in climate have occurred both on account of human activities and natural variability. To unravel the relative contribution of anthropogenic and natural causes is a great challenge. We need to differentiate between modes of natural variability from changes induced by human actions.

This book documents the changes in climate that have occurred in India during the past 150 years. There is a detailed discussion on the variability of temperature, humidity and rainfall in the India. An exhaustive study of extreme events such as droughts, floods, heat and cold waves has been made. The variations in sea surface temperature in seas around India influence the Indian monsoon. Hence, one chapter is devoted to the documentation of the variability of sea surface temperature in this region. One of the important consequences of global warming is the rise in sea level and hence the rise in sea level in various coastal stations in India has been discussed. Global warming has caused the retreat of many glaciers in the Himalayas and these changes have been delineated comprehensively. Air pollution has become an important issue in India today. Hence, there is a thorough discussion on the changes in aerosols, trace gases and ozone and their impact on radiation budget at the surface. The changes that occurred in the monsoon in the past have been highlighted through proxy indicators such as tree rings and stalagmites. There is a discussion on change in regional climate in the future under various scenarios. This book will serve as an excellent reference book for those who want to understand climate variability and change in India. I congratulate the editors and the authors for bringing out such an excellent compilation.

Prof. J. Srinivasan
Divecha Centre for Climate Change
Indian Institute of Science
Bengaluru, India

Preface

The Earth's climate system includes the land surface, atmosphere, rivers and oceans and cryosphere. Many aspects of the global climate are changing as evident in the long-term observations from the top of the atmosphere to the depths of the oceans. All these observations provide the unambiguous evidence of global warming. However, the changes in climate are not expected to be uniform across the Earth. There are significant regional variations in terms of magnitude of changes in the climate system. In some cases, regional changes may not necessarily follow the global trends. Climate change impacts are already evident and are expected to become increasingly disruptive across the globe. However, the nature of the impacts and associated vulnerability varies geographically.

To make an assessment of regional climate change and the possible impacts on agriculture, water resources, health, etc., it is important to have a clear comprehensive view on the observed climate variability and change over the region. There are many research studies documenting changes in the climate system over the Indian region using different instrumental and re-analysis data sets. However, these studies used data sets of varying lengths and the results differ both quantitatively and qualitatively. There is also a strong evidence of multi-decadal variability in the climate system over the region. Therefore, it is important to make a comprehensive assessment of the changes in the regional climate system using data sets of longer period and preferably for the common period.

This book compiles articles that review observed changes in the regional climate system. Using various long-term instrumental data sets starting from 1901, the contributing authors describe the observed changes in different components of the regional climate system, atmosphere, ocean and the cryosphere. The last chapter, however, deals with the future climate change scenarios over the region derived from various coupled climate models.

It is hoped that this book will immensely contribute in better understanding of the regional climate change and its possible attribution to natural and anthropogenic activities. The book may be used as reference material for research students and policymakers.

New Delhi, India
July 2016

M.N. Rajeevan
Shailesh Nayak

Acknowledgements

We would like to thank, Drs. D.R. Sikka, J. Srinivasan, M. Sarin, K. Krishnamoorthy and G.B. Pant in reviewing the first draft of some of the articles published in this book and Mr. Abhay Rajput for helping us in the editorial work.

M.N. Rajeevan
Shailesh Nayak

Contents

Editors and Contributors

About the Editors

Dr. Madhavan Nair Rajeevan is at present working as the Secretary, Ministry of Earth Sciences, Government of India at New Delhi. He has more than 30 years of research and operational experience in many aspects of Tropical Meteorology. His main research contributions include understanding monsoon variability and development of monsoon prediction models, understanding extreme weather events like heavy rainfall and heat waves and cloud–radiation interaction. He had served in many international expert panels of WMO/WCRP including as a member of the Asian-Australian Monsoon Panel. He briefly served as the High-Level Adviser to the Commission for Climatology, WMO. He is a Fellow of the Indian Academy of Sciences and National Academy of Sciences, India.

Dr. Shailesh Nayak is the former secretary, Ministry of Earth Sciences. He did his Ph.D. in Geology in 1980 from the M.S. University of Baroda, and specialized in Oceanography and remote sensing. Initially he has served in Indian Space Research Organization (ISRO) and worked on research areas of Coastal and Ocean processes and ocean-atmosphere interaction, coastal geomorphology, hazards. He has also served as the Director, Indian National Centre for Ocean Information Services, INCOIS, Hyderabad, where he has set up a state-of-the-art Early Warning System for Tsunami and Storm Surges in the Indian Ocean. He has been providing leadership for the programs related to science of climate change, weather services, polar science, ocean science and modelling, ocean survey, resources, and technology. He has published more than 100 papers in Journals, conferences & reports. He was honoured with an Honorary Doctorate of Science by Andhra University in 2011 and Assam University in 2013.

Contributors

Dr. S. Suresh Babu is the Head of the Aerosols and Radiative Forcing Section, Space Physics Laboratory (SPL), Vikram Sarabhai Space Centre (VSSC), Thiruvananthapuram. He has made pioneering scientific contributions in the area of atmospheric aerosols and its climate forcing. He was instrumental in setting up the national network of aerosol observatories all over India under ISRO-GBP, designed and executed thematic field experiments using high-altitude balloons, aircrafts and research ships and established aerosol observatories at Himalayas and Polar regions. He has published more than 100 papers in internationally reputed journals and edited books and has several awards to his credit which include the Swarna Jayanti Fellowship Award from Department of Science and Technology, Government of India; NASI-SCOPUS Young Scientist Award; Indian National Science Academy (INSA) Young Scientist Medal Award and National Academy of Sciences India (NASI) platinum jubilee Young Scientist Award.

Dr. S. Balachandran is serving as a Scientist and the Head of Area Cyclone Warning Centre and Cyclone Research Centre at India Meteorological Department (IMD), Chennai . He joined IMD in 1992 and since then is involved in operational weather forecasting, training and research on tropical cyclones and northeast monsoon. He has published over 25 research papers in National and International journals. His other major contributions include development of web based atlas of tracks and statistics of cyclones and depressions over north Indian Ocean, *Cyclone eAtlas—IMD* and Tropical Cyclone Rainfall Atlas for north Indian Ocean—*TCRAIN*.

Ms. Shraddha Band is Senior Research Fellow in the Geosciences Division, Physical Research Laboratory, Ahmedabad. She is interested in Multiproxy paleoclimate studies in the Indian region. Her work is focused on reconstructing Indian Summer Monsoon variability from late Pleistocene to present based on speleothems from caves located in Core Monsoon Zone of India, and to compare these records with the Indian Ocean core data to check for the coherence between different proxies. She has reconstructed a continuous high resolution record of Indian Summer Monsoon covering entire Holocene and inception of last Glacial period in the late Pleistocene. In her work, she has addressed the effect of major climate events like Little Ice Age, Medieval Warming on the strength of Indian Monsoon.

Dr. Hemant Boragaonkar is Senior Scientist at Indian Institute of Tropical Meteorology, Pune. He has more than three decades of vast experience in Long-Term Climate Change and Variability research with palaeoclimatic aspect, particularly with dendroclimatic approach. Over the Indian region Dr. Borgaonkar developed wide network of tree-ring chronologies of Himalayan conifers and teak (*Tectona grandis*) chronologies from central and peninsular India to understand the past climate changes since past 2000 years. Dendroclimatic studies of Himalayan conifers provide reliable reconstructions of pre-monsoon climate of the region since the past millennium. More than 40 research publications in national and international journals are in his credit. He is member of various national and international Societies including the Indian Meteorological Society, International Tree-ring Society, USA, Asia Pacific Network (APN) for Global Change, Japan, Working Group of PAGES ASIA-2K activity, Switzerland, Adjunct Scientist, University of Columbia, USA, Indo-German HIMPAC Collaborative project etc

Dr. Jasti S. Chowdary received the Ph.D. degree from University of Pune, India. He is working as a Scientist with Indian Institute of Tropical Meteorology, Pune, India. His research focus is on understanding tropical Indian Ocean climate variability and impact on summer monsoon. He has published 40 papers on various topics in peer-reviewed journals. He has been actively involved in teaching physical Oceanography and ocean–atmosphere interaction courses.

Ms. J.S. Deepa is a Scientist at the Indian Institute of Tropical Meteorology (IITM), Pune. She obtained her M.Sc. in Mathematics from IIT Madras in 2011. Her current research interests are in the area of understanding climate variability. Her research includes the analysis of Indian Ocean sea level trends and variability and the underlying mechanisms using observations and model with a focus on warming of Indian Ocean and impacts of Indonesian through flow. She has also worked on atmospheric data assimilation using the GFS-LETKF data assimilation system.

Dr. Aditi Deshpande has completed her Ph.D. in 2016 from the Savitribai Phule Pune University on Equatorial Indian Ocean current variability and its regional impacts. She has carried out her Ph.D. work at Indian Institute of Tropical Meteorology, Pune. She has published two papers in peer-reviewed international journals. She is currently working as Assistant Professor at Savitribai Phule Pune University. She is interested in Indian Ocean circulation and changes and sustainability.

Dr. Medha Deshpande graduated and postgraduated in Physics from Shivaji University, Kolhapur (Maharashtra) and Ph.D. degree in Atmospheric & Space Sciences from S. P. Pune University, Pune (Maharashtra) awarded in July 2010. She is working as a Scientist at Indian Institute of Tropical Meteorology from November 2011. Her research interests include tropical cyclones, Indian summer monsoon, numerical weather prediction and cloud parameterization schemes. She has published number of research papers in peer-reviewed journals.

Dr. B. Geetha is serving as a Assistant Meteorologist at the Cyclone Warning Division of IMD, New Delhi. She has over 20 years of experience in meteorological services. Her main research interests are Northeast Monsoon and Tropical Cyclones (TCs). She has authored 11 research papers published in peer-reviewed journals. She has made significant contributions towards development of various data sets for TC research including electronic version of the TC best track data sets, entitled, *Cyclone eAtlas*, its web version and Tropical Cyclone Rainfall Atlas for North Indian Ocean, *TCRAIN*.

Dr. C. Gnanaseelan is a Senior Scientist at Indian Institute of Tropical Meteorology, Pune, India. He is also an Adjunct Professor at the Savitribai Phule Pune University. He received Ph.D. degree in Mathematics and M.Tech. degree in Atmospheric Science and Technology from the Indian Institute of Technology, Kharagpur, India. He was a visiting fellow in the Florida State University during August 2005 to February 2007. He is currently leading a research group on climate variability and ocean data assimilation at IITM. His research interest includes climate variability, Indian Ocean modelling, air–sea interaction, monsoon variability, Indian Ocean dynamics and variability and data assimilation. He has published more than 75 papers in peer-reviewed journals. He has successfully guided 11 PhDs and several M.Tech. graduate students.

Dr. Pulak Guhathakurta has a Ph.D. degree in Applied Mathematics from the University of Calcutta. Before joining the India Meteorological Department (IMD) in 1992 he was a CSIR Research fellow at the University of Calcutta. At IMD, his field of work includes climate data processing, long range forecasting, operational weather forecasting, Hydrometeorology, drought monitoring and prediction etc. His research works is mainly in the field of climate change and variability, climate prediction based on empirical models, hydrometeorology and drought analysis and prediction, application of neural network model in weather analysis, prediction and spatial interpolation. He has more than 37 research papers published in various international and national journals and attended several national and international seminars. He has also worked in the WMO expert groups on climate extremes.

Dr. A.K. Jaswal obtained his Master's degree in Mathematics from Jammu University in 1978 and joined India Meteorological Department (IMD) in 1980. After putting in 35 years of service, he retired from IMD in July 2015 from the National Data Centre, Pune. He has expertise in computer programming, data analysis, data processing & archival and climatological data products generation. He has significantly contributed in bringing out many climatological publications and reports. His research work in the field of Climatology and Climate Change has been published in referred national and international journals. He worked as core-member of WMO CBS Expert Team on Metadata Interoperability IPET(MDI) during 2009–2011. He is a co-author of a chapter in "Global Warming Challenges and Opportunities for Policy and Practice", published by InTech. He is the recipient of 26th Biennial MAUSAM Award for best research paper.

Ms. Ananya Karmakar received the M.Sc. degree in Applied Mathematics from IIEST, Shibpur Howrah, India in 2013. Currently, she is working as Research Fellow at IITM, Pune and pursuing her Ph.D. degree at the Savitribai Phule Pune University, India. She is working on the role of Forcing, Resolution and Physics on the Interannual variability of Indo-Pacific Ocean using Ocean Model. Her research interests include numerical modelling, air–sea interaction, ocean dynamics and monsoon dynamics.

Dr. Dilip R. Kothawale is a senior scientist at the Short Term Climate Variability and Predication (STCVP), Indian Institute of Tropical Meteorology (IITM), Pune. He is currently working on the regional rainfall and temperature variability on different time scales. He obtained Ph.D. on the topic 'Surface and Upper air temperature Variability over India an its Influence on the Summer Monsoon rainfall from the University of Pune in 2005. He has published 30 research papers in international and national journals. All-India, homogeneous regions and 30 Met. Sub-divisions monthly and seasonal rainfall series for the period 1871–2013 and temperature series for 1901–2007 were prepared by him for the research work and made available on IITM website.

Dr. R. Krishnan is at present working as the Executive Director of the Centre for Climate Change Research, IITM Pune. He specializes in climate modelling on scientific issues relating to the *Dynamics and variability of the Asian monsoon.* His interests include monsoon dynamics and variability, phenomenon of monsoon-breaks and droughts; large-scale organization of monsoon convection; global climate change and impacts on the Asian monsoon and regional climate extremes. He is deeply involved in developing in-house capability in Earth System Modeling to address climate change and related scientific issues. He carried out Ph. D. research in Atmospheric Sciences at the Physical Research Laboratory, Ahmedabad and obtained Ph.D. degree from the University of Pune in 1994.

Dr. Anil V. Kulkarni is a distinguished Visiting Scientist at the Divecha Centre for Climate Change, Indian Institute of Science, Bangalore, India. Earlier he worked at the Indian Space Research Organization, Ahmedabad, India. He has developed numerous remote sensing-based methods and models to understand the Himalayan cryosphere. These methods are widely used to map seasonal snow cover, glaciers, moraine dammed lakes, glacier mass balance, snow albedo, stream runoff, and to understand influence of climate change on Himalayan cryosphere. He has published more than 65 papers in peer-reviewed journals and also published numerous atlases and reports. He has been conferred with the prestigious National Award in Polar Science and cryosphere in 2014 by Government of India; National Geosciences Award in 2009 by Government of India, as a recognition of his outstanding contribution in the field of Applied Geosciences. He has also received an award from Indian Space Research Organization for his contribution as a team leader of Snow and Glacier project.

Dr. Ashwini Kulkarni has a Ph.D. in Statistics (1992) from the Pune University. She has more than 25 years of research experience in the field of Atmospheric Science. Her main research interests include, Climate variations and tele-connections over South, East and Southeast Asia; Asian monsoon variability in coupled and regional climate model simulations and projections; Applications of Statistics in Climate Research. She has published 42 papers in reviewed scientific journals. She received the 10th IITM Silver Jubilee award for best research contribution in 1997; contributory author of the Chapter 14 of IPCC AR5 and "Adjunct Professor" of Department of Atmospheric and Space Science, Savitribai Phule Pune

University since 2000. She is a member of the editorial board of renowned scientific journals like Met Applications.

Prof. Shyam Lal is currently working as Honorary Scientist and as J. C. Bose National Fellow at the Physical Research Laboratory, Ahmedabad. His major contributions have been in atmospheric chemistry with particular emphasis on the distributions of ozone and trace gases over India and adjoin oceans. His pioneering studies on halons and methyl bromide and their temporal trends in the stratosphere have been recognized due to their importance in ozone depletion. He developed and led a national program for in-depth studies of tropospheric chemistry over different regions of India as a project of ISRO-GBP. He has supervised nine students for their Ph.D. dissertations and has about 115 peer-reviewed publications. Prof. Lal is a Fellow of all the three science academies of India and has been member of the scientific steering committees of International Global Atmospheric Chemistry Project (IGAC), International Commission on Atmospheric Chemistry and Global Pollution (iCACGP), International Ozone Commission (IOC) and a contributor to the first report of the Intergovernmental Panel on Climate Change (IPCC).

Dr. M. Mohapatra is serving as the Head of Weather Forecasting Services of India Meteorological Department (IMD). He is also the Head of the WMO recognized Regional Specialised Meteorological Centre-Tropical Cyclone, New Delhi. With his 23 years of experience in weather forecasting, he has made significant contribution in modernization and improvement of early warning services of IMD, especially the cyclone warning services over the south Asian region. He has authored 60 research papers in peer-reviewed journals and has edited 10 books and 3 journals. He has received a number of recognitions including 25th *Biennial Mausam Award* and the *Young Scientist Award* of the Ministry of Earth Sciences, Government of India and *Achievers Award* from IMD for his research and service contributions in the field of atmospheric sciences.

Dr. Milind Mujumdar is senior scientist at the Centre for Climate Research Centre (CCCR), Indian Institute of Tropical Meteorology (IITM), Pune. He is currently working on the Asian monsoons and its response to warming climate by using both diagnostic and modelling studies. He is also actively involved in the IITM Earth System Model (ESM) development programme. He coordinates the regional climate downscaling activity (CORDEX) over South Asian region. He obtained his Ph.D. on the topic "Studies on Initial and Boundary conditions in Climate Models using spectral techniques: Application to Asian Monsoon" from the University of Pune in 2002. Having published more than 25 research articles in national and international journals, he is also Adjunct Professor at the University of Pune.

Dr. P. Mukhopadhyay has been working for more than 20 years in the field of Numerical Modelling and problem of parameterization of clouds and convection. He did his Masters in Physics and M.Tech. in Atmospheric Science from Calcutta University and obtained his Ph.D. degree from University of Pune in the year 2005. He has guided four students for Ph.D. degree and six students for M.Sc./M.Tech.

degree. He has made significant contribution in the field of cloud and convection parameterization to improve monsoon simulation by CFSv2 model. He has published more than 30 papers in peer-reviewed journals

Ms. Smitha Anil Nair is working at the India Meteorological Department as a Junior Scientist. She has a postgraduate degree in Physics and pursuing for Ph.D. in Physics. She has carried out research work on heat and cold waves over the country.

Dr. Manish Naja is Senior Scientist at Aryabhatta Research Institutes of Observations Sciences (ARIES), Nainital. He is actively involved in surface based, balloon-borne and ship-borne observations of trace gases. His interest is also in the modelling of atmospheric chemistry and analysis of the data from various space-borne sensors. He has been involved in the Indian Ocean Experiment (INDOEX), Greenhouse Gases Observing SATellite (GOSAT) and Ganges Valley Aerosol Experiment (GVAX) projects. He is a principal investigator of ISRO-GBP projects at ARIES and leading a new project of DST sponsored Stratosphere-Troposphere (ST) Radar at ARIES. He did his postdoctoral research at Japan Marine Science and Technology (JAMSTEC), Japan and National Institute for Environmental Studies (NIES), Japan after completing his Ph.D. work at Physical Research Laboratory, Ahmedabad. He is a member of the scientific steering committees of International Commission on Atmospheric Chemistry and Global Pollution (iCACGP) and International Global Atmospheric Chemistry (IGAC). He has about 60 publications in the peer-reviewed journals.

Dr. B. Padmakumari is Senior Scientist at Indian Institute of Tropical Meteorology, Pune. She obtained her M.Tech. degree in Atmospheric sciences from Andhra University and Ph.D. in Physics from the Pune University. Her specialization includes (1) Aerosol-Cloud-Radiation and Climate interactions (2) Remote sensing and Aerosol-Radiation closure studies, Radiative Forcing and (3) Climate change. Her paper on solar dimming published in *GRL 2007* received G*lobal Publicity*. Received IITM Silver Jubilee Award for the best research paper in 2013 and best poster award (Gold Medal) during the INTROPMET 2014. Participated in various airborne missions/field operations and acted as a flight scientist as well as an instrument scientist. She also participated in ground observational campaigns conducted in India. She has published 30 research papers in various peer-reviewed journals.

Dr. D. Sivanand Pai is at present working at the India Meteorological Department, National Climate Centre. Dr. Pai has a Ph.D. in Physics from the University of Pune for his thesis on monsoon variability. He is specialized in long range forecasting and climate variability studies. He was involved in development of advanced statistical models for seasonal forecast models and also development of gridded rainfall products. He has more than 50 papers in his credit. His papers are mostly on monsoon variability and prediction. He has received the Young Scientist Award from the Ministry of Earth Sciences for his research contribution.

Dr. G. Pandithurai is senior scientist at Indian Institute of Tropical Meteorology, Pune. He is leading the projects of High-altitude cloud physics laboratory (HACPL) and Radar and Satellite meteorology. He received his M.Sc. (Physics) from Alagappa University and Ph.D. from the University of Pune. He worked as a postdoctoral fellow in Department of Meteorology, University of Maryland, USA during 1998–2000. He also received invitation fellowship from JSPS to work as a researcher in Centre for Environmental Remote Sensing, Chiba University, Japan in 2008–2009 on aerosol-cloud interaction studies. He has contributed to the understanding of aerosol radiative forcing and indirect aerosol effect on cloud microphysics using surface and aircraft observations. He has guided 4 Ph.D. students and published about 75 papers in peer-reviewed journals. He has received (i) MoES Certificate of Merit for his outstanding contribution in the field of Atmospheric Science and Technology, (ii) IITM Silver Jubilee awards for best papers.

Dr. Anant Parekh He received the Ph.D. degree in Physics from Gujarat University, Ahmadabad, India, in 2008. He carried out his Ph.D. work at Meteorology and Oceanography Group of Space Applications Centre (ISRO), Ahmadabad, India. Since 2005, he is Scientist with the Indian Institute of Tropical Meteorology (IITM), Pune, India. His research interests include ocean processes, ocean modelling, satellite oceanography, monsoon modelling, data assimilation and Tele-connections associated with the ENSO. He has published more than 30 peer-reviewed research papers, guided several students for M.Tech. as well as Ph.D. dissertation. He is also adjunct Assistant Professor with the Savitribai Phule Pune University, Pune, India.

Dr. S.K. Peshin obtained his Ph.D. degree in Physics from the University of Kashmir in 1986. Thereafter, he joined the India Meteorological Department (IMD) in 1986. Presently he is Scientist-F and Head, Environment Monitoring Research Centre and Polar Meteorology Research Division, India Meteorological Department, New Delhi. He has been associated with ozone monitoring and research for more than two decades and worked as a Rapporteur on Atmospheric Ozone for the region RA-II of WMO during 2000–2004 besides contributing to Ozone Reports of WMO-UNEP. He is actively involved with “System for Air Quality Forecasting & Research (SAFAR)” programme for the past five years. He has more than 50 publications to his credit in peer-reviewed journals and has also presented his work in national and international conferences.

Ms. S. Pratibha is Senior Research Fellow at Divecha Centre for Climate Change, Indian Institute of Science, Bangalore, India. She has a master’s degree in Remote Sensing and Geographic Information System. She studied glaciology at the University of Iceland, funded by Nehru-Grimsson Fellowship. Her topic of research is monitoring and modelling Himalayan glaciers using Remote sensing and GIS techniques. e-mail: 27pratibha.s@gmail.com

Mr. M.V.S. Ramarao is working as Project Scientist at the Centre for Climate Research Centre (CCCR), Indian Institute of Tropical Meteorology (IITM), Pune under IITM-IIHS collaborative project on "Climate Change Adaptation at Scale in Semi Arid Regions (ASSAR)". He is also pursuing his Ph.D. from Savitribai Phule Pune University. His Ph.D. research investigates the land surface hydrological response to changing South Asian monsoon in a warming climate using high resolution long-term climate model simulations. He is also actively involved in the data management and analysis of regionally downscaled climate simulations from CORDEX (Coordinated Regional Climate Downscaling Experiment) South Asia. He has published four research articles in national and international journals during his research tenure. He has obtained two postgraduation degrees with M.Tech. in Atmospheric Sciences and M.Sc. in Physics from Andhra University, Visakhapatnam.

Dr. R. Ramesh is currently an Outstanding Scientist in the Geoscience Division of the Physical Research Laboratory, Ahmedabad. He received Indian National Science Academy (INSA) Young Scientist Medal, 1987. SS Bhatnagar Award for Earth, Atmosphere, Ocean and Planetary Sciences, 1998. TWAS Prize for Earth Science 2006, Nobel Peace Prize citation 2007 and Ramanathan Gold Medal 2011. His present areas of interest include Stable isotope mass spectrometry, paleoclimate and paleoceanography, isotopic tracers in earth system, climate modelling, marine nitrogen cycle. He is a Fellow of the Indian National Science Academy, the Indian Academy of Sciences, the National Academy of Sciences, Third World academy of Science, and the Geological Society of India. Life member of the Indian Society for Mass spectrometry, Project Director, ISRO- GBP Paleoclimate Programme, Life member of the Gondwana Geological Society and the Indian Meteorological Society. He has published about 200 research papers and guided 20 students for Ph. D.

Dr. Jayashree Revadekar is a Scientist working at the Indian Institute of Tropical Meteorology (IITM). Her main research interests are Daily Weather Extremes, Monsoon Variability and Climate Change. She is an author of about 45 research articles. From the Indian side, she led the APN project on *Development and Application of Climate Extreme indices and indicators for Monitoring Trends in Climate Extremes and their Socio-economic Impacts in South Asian Countries which received* outstanding project award out of 80 projects reviewed. She received her Ph.D. from Pune University in Atmospheric and Space Science. She is an active member of Monsoon-On-Line Web of IITM on real-time analysis.

Dr. M.K. Roxy is Climate Scientist at the Indian Institute of Tropical Meteorology (IITM), and an Adjunct Associate Professor at the University of Pune. He holds a Ph.D. in Ocean and Atmospheric Dynamics from Hokkaido University in Japan. He previously worked as Research Scientist at the Euro-Mediterranean Centre for Climate Change in Italy, and as Visiting Scientist at the National Oceanic and

Atmospheric Administration (NOAA), Maryland. Dr. Roxy has been a leading member in charge of the development of the first Climate Model for South Asia—focusing on the climate variability and change over the South Asian domain and Indian Ocean, and to contribute to the future IPCC Assessment Reports. He is currently leading the research on Indian Ocean warming and its impact on the monsoon and the marine ecosystem. He is an Indian Ocean Panel member of the World Climate Research Program (WCRP). He was named the Kavli Fellow by the U.S. National Academy of Sciences.

Dr. J. Sanjay is Senior Scientist at the Centre for Climate Change Research (CCCR), Indian Institute of Tropical Meteorology (IITM), Pune, India. His area of specialization is Regional Climate Change with a focus on the generation of future climate change scenarios for the Indian monsoon region using dynamical downscaling techniques with high resolution regional climate models (RCMs). He is leading the CCCR team for co-ordinating the data archiving, management and dissemination activities of the South Asia component of the international co-ordinated Regional climate Downscaling Experiment (CORDEX) initiative by the World Climate Research Program (WCRP) of WMO. He is a scientific knowledge partner on climate science for a segment of the International CARIAA research program on climate change Adaptation at Scale in Semi-Arid Regions (ASSAR) of India. He has 40 publications, including 21 papers in peer-reviewed journals. Prior to joining IITM in 1988, he completed M.Sc. in Meteorology at the Cochin University of Science and Technology.

Prof. S.K. Satheesh is Professor at the Centre for Atmospheric and Oceanic Sciences and Divecha Centre for Climate Change, Indian Institute of Science, Bangalore. He obtained his Ph.D. in 1998 from the University of Kerala and carried out postdoctoral research at the Scripps Institution of oceanography, USA. He also held visiting scientist positions at NASA and University of Bern, Switzerland. He has published more than 130 journal articles. He has received a number of awards including the TWAS Prize 2011 in Earth Sciences and S.S. Bhatnagar Prize in 2009. Others include START (USA), INSA (India), WMO (Geneva), Asian Aerosol Research Assembly (Japan), SCOPUS (UK) awards, Rajib Goyal Prize 2010 in Physical Sciences, PRL award 2007, Swarna Jayanti fellowship award 2006, J.C. Bose fellowship and among several others. He is an elected fellow of all the three science academies in India and also The World Academy of Sciences (TWAS). He was Associate Editor of Journal of Geophysical Research (USA) and member of Editorial Advisory Board, Atmospheric Environment (Elsevier). He is Associate Editor of Current Science (India). He is Lead Author of Chapter-7 (Clouds and Aerosols) of IPCC WG-1 Assessment Report-5 (2013) and was member of Scientific Committee (SC) of International Geosphere Biosphere program (IGBP). His research areas include Atmospheric Aerosols, Clouds; Radiation; Remote Sensing, Climate.

Dr. Vijay Kumar Soni did M.Sc. in Physics from IIT Roorkee and received Ph.D. in Atmospheric and Space Science from the University of Pune. He joined the India Meteorological Department in 1999 as Meteorologist. He worked in Air Pollution Division, IMD, Pune till March 2011. Since April 2011, he is working as Scientist, Environmental Monitoring and Research Centre, IMD, New Delhi. He is also leading Polar Meteorological Research Division, IMD, New Delhi. He contributed significantly to establish aerosol monitoring network in India called Skynet-India and Black Carbon Monitoring Network of IMD. His research areas include aerosols, precipitation chemistry, solar radiation and air quality. He has published more than 26 research papers in peer-reviewed journals, four meteorological monographs on various subjects and two chapters in books.

Dr. A.K. Srivastava is serving as a scientist in the Research Division of the India Meteorological Department (IMD), Pune. He joined IMD in 1992 and has actively contributed towards operational weather forecasting and research. His area of research is in climate change and climate forcings affecting the monsoon performance with special emphasis on the extra-tropical circulation and monsoon relationship. Since the inception of the *National Climate Centre* of IMD, he was involved in different climate monitoring/research activities. He has served as the Head of the Climate Centre for a few years. He was instrumental in bringing out the IMD daily gridded temperature data and other operational gridded products. He worked as one of the selected members of the WMO panel on Tropical Cyclones and Climate change. He has published more than thirty research papers in peer-reviewed high impact international journals.

Mr. S. Venkataramani has been working at Physical Research Laboratory, Ahmedabad since 1986 after completing his M.Sc. degree in physics. He is presently Scientist-F in the Space and Atmospheric Sciences Division. He is a hard core experimentalist. He started his career with the measurements of vertical distributions of various CFCs by collecting in-situ air samples using an indigenously built cryogenic air sampler flown on high-altitude balloons from Hyderabad. He has been involved in the measurements of ozone and related trace gases as well as greenhouse gases using ground based, ship, aircraft, balloon and rocket borne sensors. He has participated in many national and international programs (ISRO-GBP, JGOFS, INDOEX, etc). He is currently looking after "Atmospheric Trace gases-Chemistry, Transport and Modelling (ATCTM)" project of ISRO-GBP. He has about 45 publications in peer-reviewed journals.

Dr. M.G. Yadava is Scientist (SF) and in charge of the Radiocarbon Dating Laboratory in Geosciences Division at Physical Research Laboratory, Ahmedabad. His major research focus has been to reconstruct past climate through stable isotopes of oxygen, hydrogen and carbon (δ^{18}O, δD and δ^{13}C respectively) and trace elements supported by radiocarbon and uranium–thorium-based chronology. He has also used sedimentary deposits to find evidences of past climate change and soils to understand dynamics of organic carbon. Presently, he is involved in setting up a 1MeV Accelerator Mass Spectrometry (AMS) system for measuring ^{14}C, ^{10}Be and ^{26}Al in variety of natural samples.

Executive Summary

The summary on the variability and change of the regional climate system based on the 16 chapters in this book is as follows.

Rainfall

The analysis of rainfall over India is based on the quality-controlled district rainfall data for the period 1901–2010.

1. The all-India southwest monsoon rainfall does not show any long-term trend, but it exhibits significant multi-decadal variability.
2. However, there are significant regional trends in southwest monsoon rainfall. Monsoon rainfall in the meteorological subdivisions of Jharkhand, Chhattisgarh and Kerala have shown significant decreasing trends. However, monsoon rainfall over Gangetic West Bengal, West Uttar Pradesh, Jammu and Kashmir, Konkan and Goa, Madhya Maharashtra, Rayalaseema, Coastal Andhra Pradesh and North Interior Karnataka showed increasing trends.
3. Monsoon rainfall in the month of July has shown decreasing trends over most parts of central India. However, June and August rainfall has shown increasing trend over the central and southwestern parts of the country.
4. Northeast monsoon (October–December) over southeast peninsula does not show any significant trend, but shows multi-decadal variability.

Temperature

The analysis on surface air temperatures is made using gridded monthly temperature data for the period 1901–2010.

1. Annual mean, maximum and minimum temperatures averaged over the country as a whole showed significant warming trend of 0.60, 1.0 and 0.18 °C per 100

years respectively. The rate of warming trend in the annual mean temperatures since 1980s is much sharper, 0.2 °C per decade.
2. On the seasonal scale, the highest increasing trend is observed in the post-monsoon and winter seasons. Monsoon season shows the lowest increasing trend.
3. The warming is mostly confined to the northern, central and eastern/northeastern parts of the country. Peninsular India experienced the least warming.
4. The warming trend is observed in upper air temperatures also with significant warming trends at the lower tropospheric levels, viz. 850 hPa (1.5 km) and 700 hPa (3.1 km) levels.

Extreme Rainfall and Rainstorms

The analysis on extreme rainfall is made using daily gridded rainfall data for the period 1901–2010 and rainstorms based on daily gridded data for the period 1951–2015.

1. Frequency of very light rain and light to moderate rain events during the monsoon season has decreased over most of the country.
2. However, frequency of very heavy and extreme rainfall events over northern parts of the country has increased significantly. They also show multi-decadal variability, possibly associated with the tropical oceans, especially the equatorial Indian Ocean.
3. During the period, 1901–2010, heavy rainfall events (rainfall exceeding 15 cm in 24 hours) over northern parts of the country show an increasing trend of about 6 % per decade.
4. Frequency of rainstorms (weather systems with potential of causing large scale floods) has shown an increasing trend of 4 rainstorms in 65 years (1951–2015). Duration of rainstorms has shown a substantial increase of about 15 days during the same period.

Heat and Cold Waves

The analysis on heat and cold waves has been made based on daily temperature data of 103 stations for the period 1961–2010.

1. During the hot weather season (April–May–June), heat waves are experienced over the north, northwest, central, east India and northeast peninsula. Similarly, during the cold weather season (December–January–February), northern parts of the country including Jammu and Kashmir experience cold waves.
2. Significant decadal variation is observed in the frequency, spatial coverage and area of maximum frequency both in heat wave and cold wave.

3. Noticeable increase (decrease) in frequency of heat wave days is observed during the El Nino (La Nina) events. Exactly opposite association was observed in case of CW days.
4. Significant increasing (decreasing) trends in heat waves (cold waves) are observed during the hot (cold) weather season over most parts of the country.

Meteorological Droughts

The analysis on meteorological droughts is made using monthly rainfall data of the period 1901–2010 based on the Standardized Precipitation Index and Standard Precipitation Evapotranspiration Index (SPEI).

1. The all-India droughts are observed almost once in five years. More intense droughts are mainly observed over north and northwest India.
2. There is a significant increasing trend in the intensity and areal coverage of moderate droughts of various accumulated periods over India during the recent years, since 1950s. This increase could be also linked to warming of the equatorial Indian Ocean.

Tropical Cyclones and Monsoon Depressions

The analysis on tropical cyclones and monsoon depressions is made using the data of 1901–2010 period. A separate analysis during the satellite era (1961–2010) also has been made.

1. Over the north Indian Ocean, significant decreasing trends are observed in the frequencies of cyclonic disturbances and tropical cyclones during the southwest monsoon season.
2. Significant increasing trend in severe tropical cyclones is observed during the post-monsoon season (October–December).
3. During the satellite era (1961–2010), cyclonic disturbances, tropical cyclones and severe tropical cyclones over the north Indian ocean and Bay of Bengal show significant decreasing trends during the monsoon and post-monsoon seasons.
4. Monsoon low pressure systems forming over the Bay of Bengal during the monsoon season contribute significantly to the seasonal rainfall over the central parts of the country.

Total Cloud Cover

The analysis on cloud cover is made using 195 surface stations data for the period 1951–2010.

1. There is a general decrease in mean total cloud cover over most of India, but an increase in the Indo-Gangetic plains and northeast India.
2. The annual mean total cloud cover shows significant decreasing trend of 0.44 % per decade, mainly contributed by the monsoon season, where the declining rate is 0.93 % per decade.
3. Out of the total number of stations showing decreasing trends, 59, 38, 48, 58 and 33 % of the stations show significant decrease in total cloud cover for annual, winter, summer, monsoon and post monsoon seasons respectively.
4. There is a negative relationship between total cloud cover and diurnal temperature range suggesting decrease in cloud cover may be responsible for increase in diurnal temperature range.

Moisture Content

The analysis of surface moisture and soil moisture data has been made using the data of 1969–2012.

1. The all-India averaged specific humidity shows significant increasing trend during all the seasons, which is consistent with the surface warming trends.
2. The magnitude of seasonal trends in specific humidity is the highest in summer where it is increasing at the rate +0.32 g/Kg per decade and the lowest in monsoon season where the rate of increase is +0.20 g/Kg per decade.
3. The regional re-analysis data sets suggest an increase in atmospheric moisture content over the region during the same period.
4. Out of 27 stations considered for soil moisture analysis, 15 stations showed increasing trends in soil moisture.

Radiative Fluxes

The analysis on surface and top of the atmosphere radiative fluxes over the region is made using the data of the period 1981–2010.

1. The analysis of global irradiance data from India, from 1981 to 2006 showed a significant decreasing trend of 0.89 Wm^{-2} yr^{-1} under all sky conditions, called solar/global dimming.
2. The analysis of long-term global irradiance data from 1971 to 2010 showed that the declining trend of all-sky global irradiance over India as a whole was 0.6 $Wm^{-2}yr^{-1}$ during 1971–2000 and 0.2 $Wm^{-2}yr^{-1}$ during 2001–2010.
3. There is an indication of solar brightening after 2001, consistent with the decrease in total cloud cover.

Sea Surface Temperature, Ocean Circulation

The analysis on SST, Ocean Circulation and Ocean heat content has been made using the data of 1958–2015.

1. The tropical Indian Ocean is warming at a faster rate compared to other tropical oceans. The warming trend show a basin scale warming with peak warming in the central equatorial Indian Ocean (0.2 °C per decade).
2. There is a strengthening of mean westerly winds over the equatorial Indian Ocean in the recent years. There is also an evidence of strengthening of the mean eastward currents along the equator and westward currents in the equator.
3. The upper north Indian Ocean is gaining heat during the recent 60 years or so at an alarming rate, which is supported by the ocean circulation and dynamics.

Sea Level

The analysis on sea level has been made using data of varying lengths.

1. The sea level rise over the Arabian Sea and Bay of Bengal from the TOPEX/Poseidon altimeter observations depict that the rate over the Arabian Sea is about 0.5 mm/year to 3 mm/year and over the Bay of Bengal is 0.75 mm/year to 6 mm/year. The rise is of steric origin and are also driven by short-term climate variability.
2. The sea level rise in the Indian ocean over the last 60 years amounts to 1.5 mm/year, which is slightly less than the global average.
3. The AR5 projected sea level rise for all the scenarios with the highest emission scenario (RCP8.5) projecting sea-level rise in the range of 0.45–0.82 m for the late twenty-first century (average over 2081–2100) for the Indian Ocean

Glaciers and Snow Cover

The analysis of Glaciers and Snow cover has been made using different data sets of varying length.

1. The areal extent of glaciers in the Himalayas is estimated as 24,697 ± 3,260 Km^2.
2. Glacier retreat is estimated by numerous investigations and the mean loss for 83 glaciers for past four decades is estimated at 550 ± 419 m.
3. The retreat estimation has high uncertainty due to large variability in individual retreat.
4. An overall loss of area of 12.6 ± 7.5 % for 40 years from 1960 onwards.

5. The mass balance data is available for few glaciers and cumulative mass loss for the past 44 years is estimated as 20 ± 6 m.
6. The average snow cover in the Himalaya from 2000 to 2011 varied between $\sim$0.3 and 0.03 million km^2. The average volume of snow is estimated at $\sim$54.5, $\sim$9.3 and $\sim$14.5 billion cubic metres in the Indus, Ganga and Brahmaputra basins, respectively

Atmospheric Aerosol

The analysis of atmospheric aerosols is made using different data sets of varying lengths.

1. There is a phenomenal increase in aerosol loading over India. There is a statistically significant increasing trend in aerosol optical depth over India with a significant seasonal variability.
2. Seasonally, the rate of increase is consistently high during the dry months (from December to March). The trends are weak during the pre-monsoon and monsoon seasons.
3. There is a clear evidence of effect of anthropogenic activities in increase of aerosol loading.
4. Black carbon observations at a remote coastal location shows a decreasing trend, which is perceptible after 2004.

Ozone and Related Trace Gases

The analysis on atmospheric ozone has been done using different data sets of varying lengths.

1. There are indications of increasing total ozone at six sites over India except at Varanasi, where a decreasing trend was observed.
2. Ozonesonde data suggests at Pune, there are increasing trends of ozone at 9.7 % per year in the planetary boundary layer.
3. Surface ozone observations also show an increase of 1.45 % per year.

Paleoclimatic Records of Past Monsoons

The analysis on past monsoons is based on proxy data from different sources (tree rings and speleothems).

1. Monsoon fluctuations have been documented from teak trees in Kerala for the past 553 years.
2. The proxy data suggests that monsoon rainfall has been steadily increasing during the Holocene (the past 10000 years).

Regional Climate Change Scenarios

Future regional climate change scenarios are created for the period 1950–2100 by downscaling the simulations of four coupled climate models.

1. The projections indicate significant temperature increases (more than 1.5 °C) over the central and northern parts of India in the mid-term (2031–2060) period. The annual warming range over South Asia land areas is 1.8–3.0 °C by 2060.
2. However, the summer monsoon season precipitation change over India is uncertain not just in magnitude but also in sign.

Chapter 1
Observed Variability and Long-Term Trends of Rainfall Over India

Pulak Guhathakurta and Jayashree Revadekar

1 Introduction

Asian monsoon circulation dominates the climate of south Asia. South-west (SW) monsoon season (June–September) over India is of paramount importance as it contributes about 75 % of the annual rainfall. SW monsoon rainfall over India shows variations in all time scales from diurnal to multi-decadal. It exhibits considerable year to year variability with years of excess rainfall and years of droughts. SW monsoon rainfall plays a crucial role for the sectors like agriculture, disaster management, water resources and power management. Though there is a growth in the service sectors, the Indian economy is still dependent on agriculture and thus the SW monsoon. Instances of droughts and famines associated with weak or deficient monsoons become very critical to the country.

There is a need to review the variability and observed changes of the Indian rainfall and to re-examine the analysis based on longer period of data, 1901–2010. The present analysis deals rainfall variability and trends during the south-west and north-east monsoon (October–December) seasons. Attempts are also made to identify the observed abrupt changes in monsoon rainfall at different spatial scales that may be the result of some abrupt changes in climate.

P. Guhathakurta (✉)
India Meteorological Department, Earth System Science Organization, Pune, India
e-mail: hspune05@gmail.com

J. Revadekar
Indian Institute of Tropical Meteorology, Earth System Science Organization, Pune, India

M.N. Rajeevan and S. Nayak (eds.), *Observed Climate Variability and Change Over the Indian Region*, Springer Geology, DOI 10.1007/978-981-10-2531-0_1

2 Brief Review of the Past Work

The interannual variability of monsoon rainfall has been rigorously examined by several researchers in the past (Pramanik and Jagannathan 1954; Parthasarathy and Dhar 1978; Parthasarathy 1984; Mooley and Parthasarathy 1984; Parthasarathy et al. 1993; Rupa Kumar et al. 1992; Pant and Rupa Kumar 1997; Kripalani and Kulkarni 2001; Guhathakurta and Rajeevan 2008 and others). The existence of 30-year multi-decadal epoch in the Indian monsoon rainfall is documented in many studies. The presence of the multi-decadal epochal variability in all India south-west monsoon rainfall has been documented by Guhathakurta and Rajeevan (2008) and later updated by Guhathakurta et al. (2015). However, these 30 years epochs (1901–30, 1931–60, etc.) are not in the same phases for the all the four broad homogeneous regions of India. Weakening of the monsoon circulation over India is documented in the recent studies (Naidu et al. 2011; Bawiskar 2009; Chung and Ramanathan 2006; Ramesh and Goswami 2007; Joseph and Simon 2005). Most of these studies used the data set for the period 1951–2004/2006. Decreasing trend in rainfall has been documented for the period 1951–2004 due to the wet decades in the period 1931–1960 followed by the dry decades during the period 1961–90. Zhang and Zhou (2011) also assessed long-term changes in global monsoon precipitation. They used three sets of rain gauge precipitation data for the period 1901–2001 and suggested a decrease in monsoon rainfall over different regions of the world.

3 Data Used and Methodology

India Meteorological Department (IMD) has recently generated new monthly rainfall data for more than 640 districts (Guhathakurta et al. 2011a, b) from the station rain gauge data. This rainfall time series was prepared using 5894 rain gauge stations spread over the country. However, there is a variation of total number of stations on every year. On an average, more than 3100 stations are available during the period of analysis. The average number of stations during the period 1951–2010 is around 3800. In addition, the IITM all India rainfall time series (http://www.tropmet.res.in) for the same period are also used in the present analysis. This data set is based on fixed network of 306 stations spread over the country (Parthasarathy et al. 1993). In this time series, the hilly regions of the country are not considered for the analysis.

Existence of monotonic trend in the rainfall time series has been examined using the least square linear fit, and its significance is tested using the Student's t-distribution. In addition, we have also used the well-known nonparametric Mann–Kendal trend test to detect the significance trend in the rainfall time series. The statistics used here for change point detection is a nonparametric test, developed by Pettitt (1979). This test is very useful for evaluating the presence of abrupt or sudden changes in climate records (Smadi and Zghoul 2006; Sneyers 1990; Tarhule and Woo 1998; Wijngaard et al. 2003) than the other tests used by Guhathakurta and Srivastava (2004), Leith (1978), Yamamoto et al. (1985).

4 Observed Changes in Rainfall Pattern

4.1 Trend and Variability in All India Rainfall

Descriptive statistics along with linear trend (in mm/decade) of all India area weighted rainfall during the SW monsoon season (June–September) for the period 1901–2010 are given in Table 1. All India mean south-west monsoon rainfall is 887 mm which is 75 % of the annual rainfall. Around 11 and 10 % of the annual rainfall is received during the pre-monsoon and post-monsoon seasons, respectively. Coefficient of variability is about 10 % for the south-west monsoon whereas maximum variability is about 34 % during the winter season. No significant trend in all India rainfall is observed in any of the four seasons as well as for the annual rainfall. However, increasing trend in the all India south-west monsoon (June–September) rainfall is seen during the period 1901–50 and insignificant decreasing trend during the period 1951–2011 as shown by Guhathakurta et al. (2015).

Figure 1 shows the 36 meteorological subdivisions of India along with the four homogeneous regions. Figure 2 shows the all India south-west monsoon rainfall during the period 1901–2010 and also the all India south-west monsoon rainfall of IITM based on the fixed network of 306 stations. The variations in the IMD and IITM time series are almost the same. The 31-year moving average of the all India south-west monsoon rainfall is shown in Fig. 3. The all India monsoon seasonal rainfall shows significant multi-decadal variation with epochs of above normal and below normal rainfall. Out of the four homogeneous regions, only north-east India

Table 1 Descriptive statistics of all India rainfall

	Winter	Pre-monsoon	Monsoon	Post-monsoon	Annual
Months	JF	MAM	JJAS	OND	
Mean (mm)	42.2	128.5	887.0	121.9	1179.6
% of annual rainfall	3.6	10.9	75.2	10.3	100
Standard dev (mm)	14.2	21.9	87.1	32.2	107.1
Coeff. var (%)	33.7	17.1	9.8	26.4	9.1
Maximum (mm)	89.2 **(2005)**	210.7 **(2005)**	1084.3 **(1917)**	206.1 **(1956)**	1463.9 **(1917)**
Minimum (mm)	11.6 **(1902)**	83.5 **(1910)**	674.3 **(1972)**	52.7 **(1908)**	933.8 **(2002)**

Figures in bracket indicate year of occurrence for maximum in green colour and minimum rainfall value in red colour.

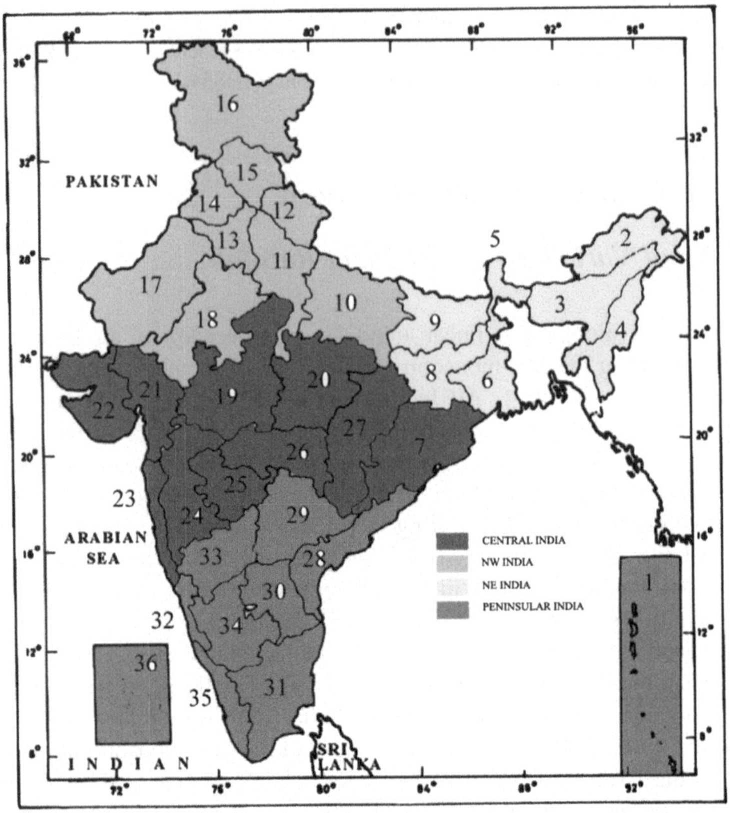

1. Andaman & Nicobar Island
2. Arunachal Pradesh
3. Assam & Meghalaya
4. N M M T
5. S H W B & Sikkim
6. Gangetic W. Bengal
7. Orissa
8. Jharkhand
9. Bihar
10. East U P
11. West U P
12. Uttaranchal
13. Har Delhi Chandigarh
14. Punjab
15. Himachal Pradesh
16. Jammu & Kashmir
17. West Rajasthan
18. East Rajasthan
19. West M P
20. East M P
21. Gujarat Region
22. Saurashtra & Kutch
23. Konkan & Goa
24. Madhya Maharashtra
25. Marathwada
26. Vidarbha
27. Chhatisgarh
28. Coastal A P
29. Telengana
30. Rayalseema
31. Tamil Nadu
32. Coastal Karnataka
33. North Interior Karnataka
34. South Interior Karnataka
35. Kerala
36. Lakshdweep

Fig. 1 36 meteorological subdivisions of India and the four homogeneous regions, viz. north-west India, central India, north-east India and peninsular India considered by the India Meteorological Department (IMD)

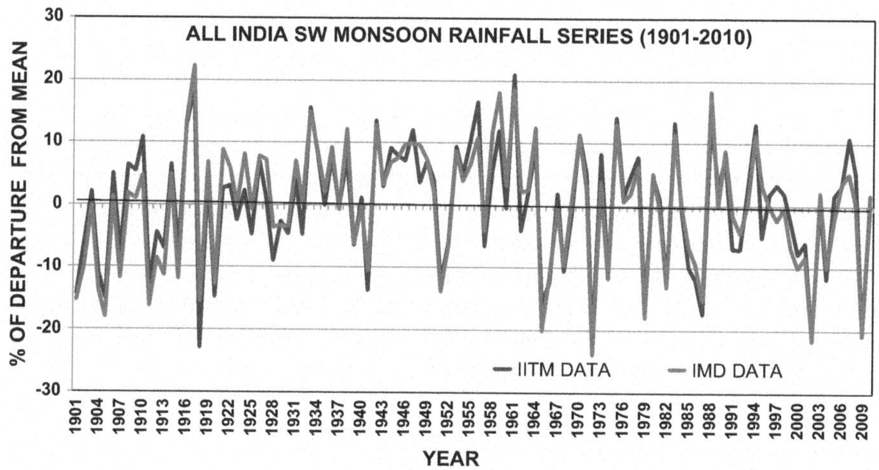

Fig. 2 All India south-west monsoon rainfall (as % departure from mean) time series for the period, 1901–2010 with IMD data (*red* in colour) and IITM data (*blue* in colour) (color figure online)

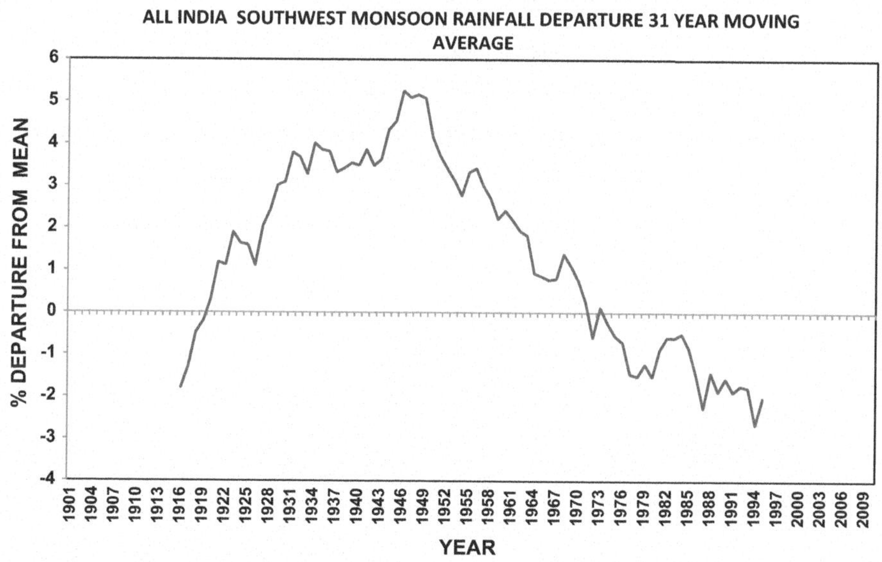

Fig. 3 31-year moving average of all India south-west monsoon rainfall as per cent departure from the long-term mean. The year shown is the central year of the 31-year period

(NEI) has shown significant decreasing trends in SW monsoon rainfall during the period, 1901–2010 as well as during the period, 1951–2010 (not shown).

Epochal average rainfall for 30 years and trends in each of the 30 years epoch of the monsoon rainfall averaged over the country as a whole are shown in Table 2. The epoch 1901–1930 was found to be drier than normal. The decadal means of all India and the four homogenous regions of India are shown in Fig. 4. All India SW monsoon rainfall was in dry epoch during the first two decades (1901–10, 1911–20). However, the next five decades (1921–30, 1931–40, 1941–50, 1951–60 and 1961–70) were in epochs with above normal rainfall. The present dry epoch started from the period 1971–1980. The presence of multi-decadal variation was also seen in the monsoon rainfall of the four homogeneous regions. Rainfall deficiency was the largest during the decade 2001–2010 for NE India.

Table 2 All India SW monsoon rainfall during 30 years epoch

All India south-west monsoon rainfall in mm				
30-year period	1901–30	1931–60	1961–90	1991–2010
Epochal average in mm	868.4	929.6	881.0	861.9
Trends (slope in mm per year) in 30 years epoch	4.03*	0.23	−0.32	−2.21

The mark * denotes statistically significant trends

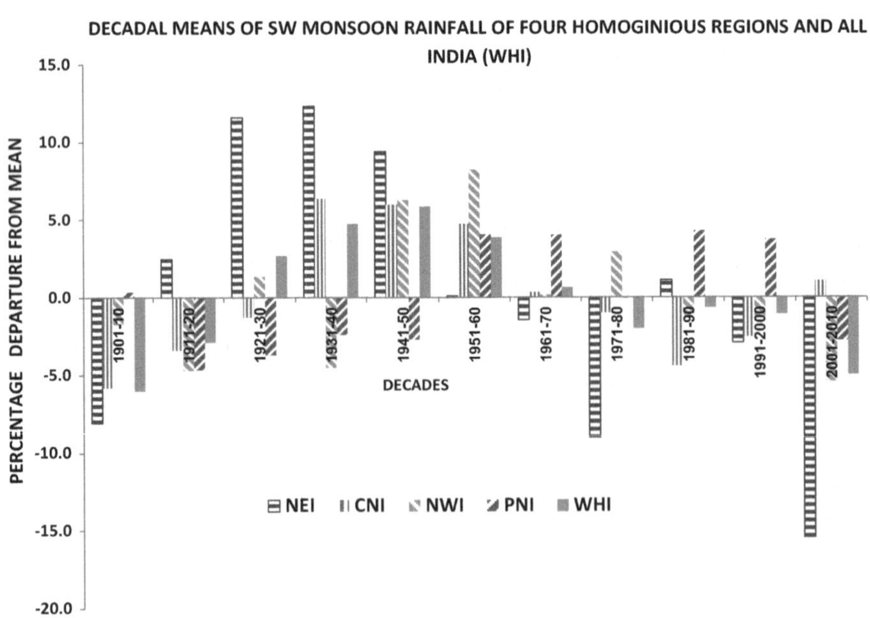

Fig. 4 Decadal means of south-west monsoon rainfall (percent departure from mean) for all India and the four homogeneous regions of India

4.2 Trends and Variability on Regional Scale

Due to the presence high spatial variability of rainfall, significant trends have been observed in annual as well as the seasonal rainfall on subdivision scale. This has been observed in the data for the period 1901–2003 by Guhathakurta and Rajeevan (2008). The analysis of district rainfall data for the period 1901–2011 by Guhathakurta and Saji (2013) revealed variability and trends in monthly and seasonal rainfall of further smaller regions. In order to study the secular variations of regional rainfall, a trend analysis was carried out for the monthly rainfall series of June, July, August and September and also for the four seasons and annual rainfall for all the 36 subdivisions (Figs. 5 and 6).

Figure 5 shows the spatial variation of trends in rainfall for the four monsoon months for the period 1901–2010. In June, eight subdivisions all along the western parts show increasing trends in rainfall while the subdivisions, viz. Kerala, Uttarakhand, Sub-Himalayan West Bengal and Sikkim and all three subdivisions of north-east India showed decreasing trends. In July, six subdivisions, viz. Himachal Pradesh, east Madhya Pradesh, Jharkhand, Arunachal Pradesh, NMMT and Kerala have shown decreasing trends. Major changes are observed in August rainfall where eleven subdivisions have reported significant trends of increasing rainfall. Rainfall over eight subdivisions mostly from eastern, north-eastern and northern parts of the India shows significant trends of decreasing rainfall. Significant decrease was observed over the two subdivisions from north-eastern region, viz. Arunachal Pradesh, Nagaland, Manipur, Mizoram and Tripura in all the four months and the season.

Spatial variations of trend in rainfall for the four seasons and annual time scales are shown in Fig. 6. Winter rainfall decreased significantly in 21 subdivisions while increasing trend has been noticed only over Jammu and Kashmir and sub-Himalayan West Bengal and Sikkim. In the pre-monsoon season, when rainfall is mainly dominated by convective activities, five subdivisions showed decreasing trends while two subdivisions showed increasing trends. Rainfall increased significantly in eight subdivisions, viz. Madhya Maharashtra, Saurashtra and Kutch, south interior Karnataka, coastal Karnataka, Konkan and Goa, coastal Andhra Pradesh, Lakshadweep and Gangetic West Bengal during southwest monsoon. Ten subdivisions viz. Chhattisgarh, Jharkhand, Uttarakhand, Himachal Pradesh, Arunachal Pradesh, NMMT (Nagaland, Manipur, Mizoram and Tripura), sub-Himalayan West Bengal, Kerala, east Uttar Pradesh and east Madhya Pradesh have shown significant decreasing trends in monsoon rainfall. No significant change in post-monsoon rainfall has been observed over any subdivisions except over Arunachal Pradesh. The annual pattern is almost the same as the monsoon pattern where eight subdivisions (out of which five in peninsular India) reported increasing trend and also eight (mostly from central and eastern parts) subdivisions reported decreasing trends.

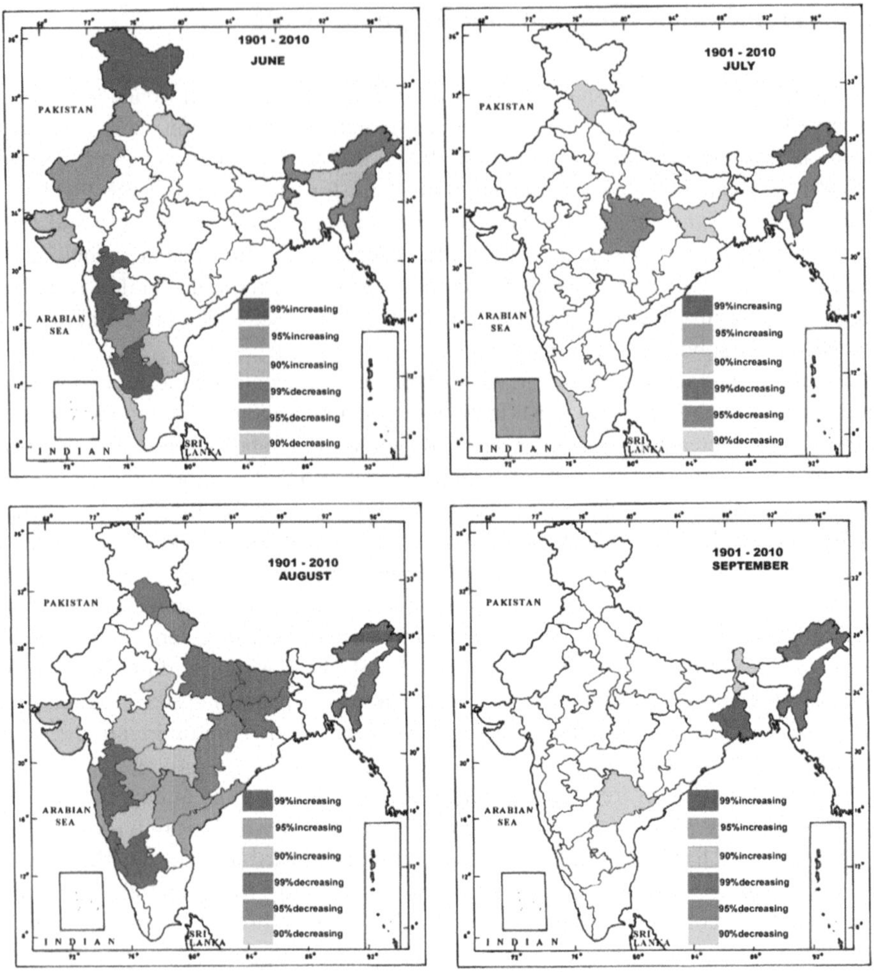

Fig. 5 Trends in the monthly rainfall (June, July, August and September) for the 36 meteorological subdivisions of India for the period 1901–2010. Trends shown here are qualitative in nature

5 Trends in North-east Monsoon Rainfall

South-eastern parts of the country including Tamil Nadu receive significant amount of rainfall during the north-east monsoon season (October–December). This is the major rainfall period for the south peninsular India, particularly over the five meteorological subdivisions, viz. Coastal Andhra Pradesh, Rayalseema, Tamil Nadu and Puducherry, south interior Karnataka and Kerala accounting for about 32, 31, 48, 21 and 16 %, respectively, of the annual rainfall. Many coastal districts of Tamil Nadu receive nearly 60 % of the annual rainfall and the interior districts

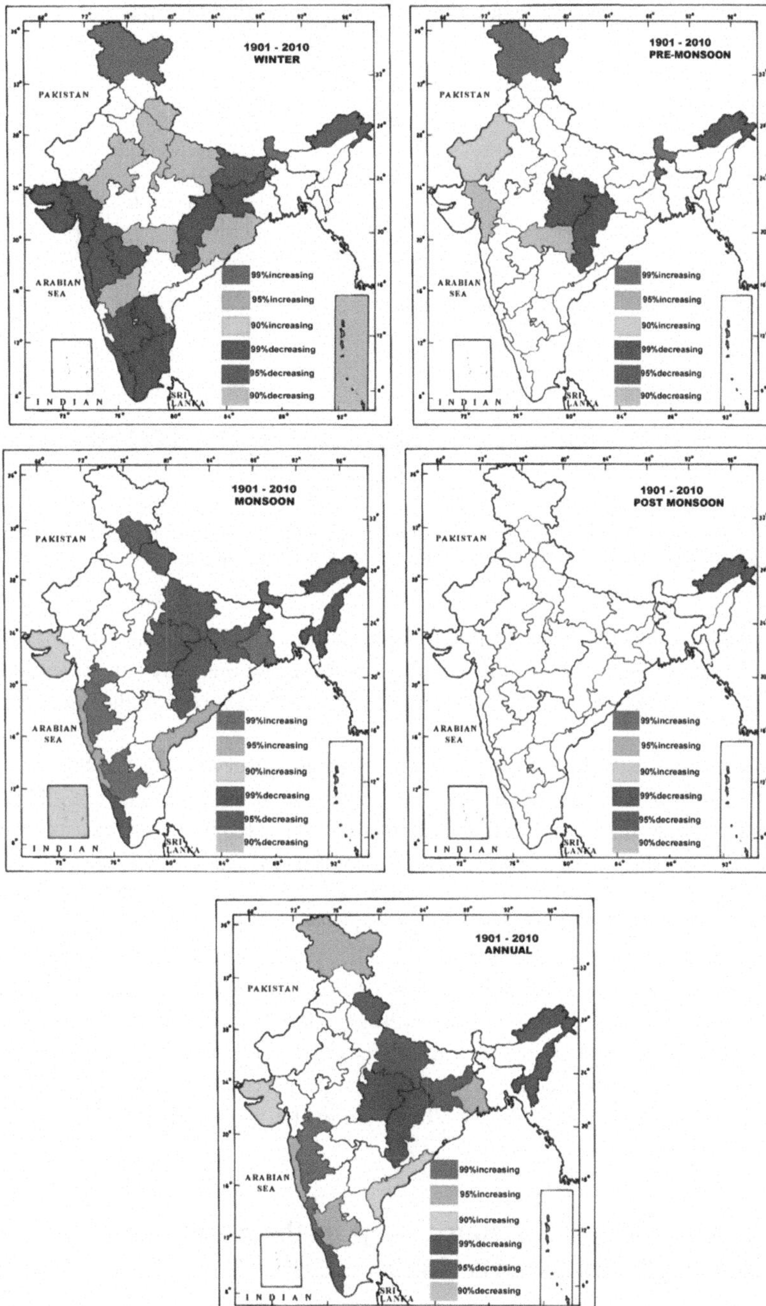

Fig. 6 Trends in the seasonal and annual rainfall for the 36 meteorological subdivisions of India for the period 1901–2010. Trends shown here are qualitative in nature

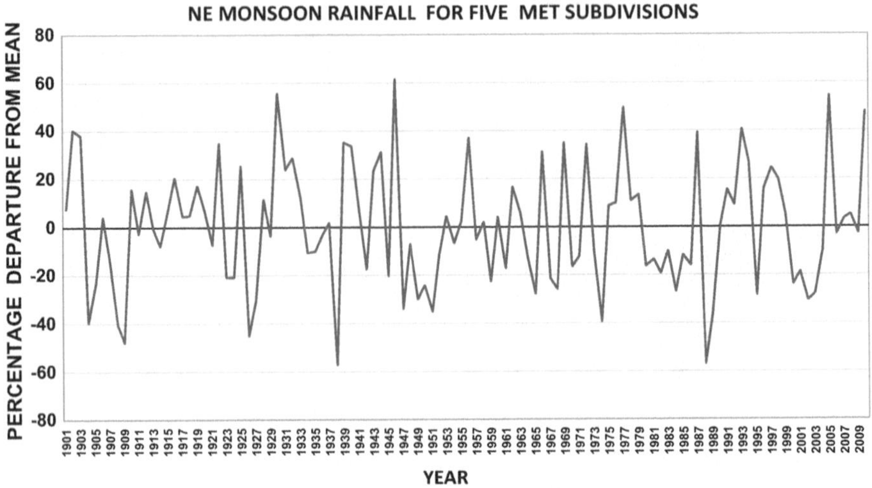

Fig. 7 Area weighted north-east monsoon (October–December) rainfall (as percentage departure from mean) averaged over the five met-subdivisions of south peninsula

receive about 40–50 % of the annual rainfall during this season. Thus, the variability of rainfall in NE monsoon season over these regions plays an important role in agriculture, water resources and many other sectors controlling the economy and social life.

Figure 6 suggests that no significant trend is observed in the NE monsoon seasonal rainfall in any of these five meteorological subdivisions. The area weighted rainfall and the fitted trend line for the NE monsoon season rainfall averaged over the five meteorological subdivisions is shown in Fig. 7. The long-period average of NE monsoon rainfall is 335.3 mm with standard deviation 85.9 mm which is 25.6 % of the long-period average. There is no significant trend observed during the period 1901–2010. The year 1946 was the wettest as the percentage departure was more than 61 % while in the years 1938 and 1988 rainfall was less than −57 % of long-period average. However, the existence of decadal variability in north-east monsoon rainfall is clearly noticed (Fig. 8). The decade 1981–90 was the driest decade (decadal average percentage departure of −15 %) with immediate next decade (1991–2000) being the wettest decade.

6 Change Point or Climate Jump in the South-west Monsoon Rainfall Over India

Detection of change point in time series data helps to identify the time and also the zone from which there was sharp climate jump has occurred. Guhathakurta et al. (2015) have shown the existence of climate jump in the all India SW monsoon

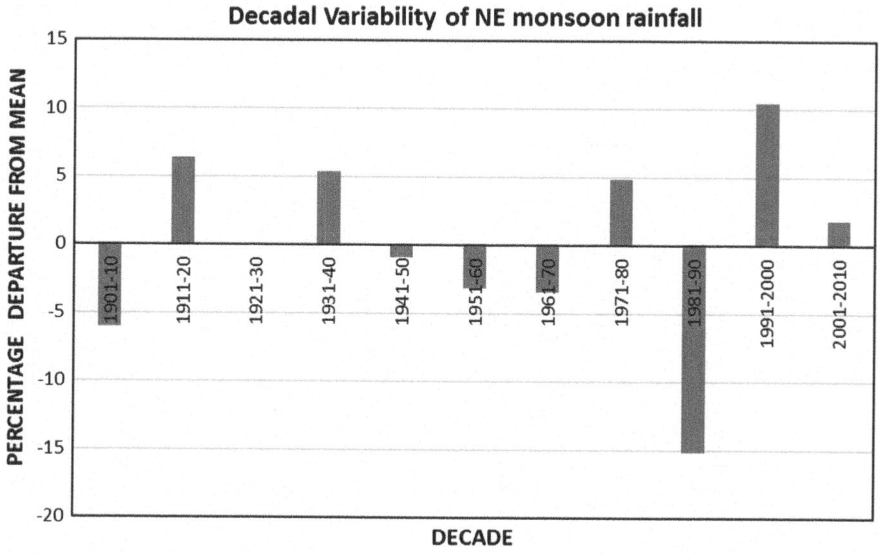

Fig. 8 Decadal variability of north-east monsoon rainfall (percentage departure from mean) averaged over five meteorological subdivisions, viz. Coastal Andhra Pradesh, Rayalseema, Tamil Nadu and Puducherry, south interior Karnataka and Kerala

rainfall. The All India south-west monsoon rainfall had experienced significant downward shift in the year 1965. They have also carried out the Pettit test to detect the existence of significant change point in the SW monsoon rainfall for all the four homogeneous regions. Significant upward shift in SW monsoon rainfall is detected over north-west India in the year 1942 and peninsular India in the year 1946. Central India had also experienced upward shift but not significant in the year 1926. Significant downward shift in the SW monsoon rainfall was observed over NE India in the year 1961.

We have also used the Pettit test to detect the change point in SW monsoon rainfall over 36 meteorological subdivisions of India. Figure 9 shows the change point (year) for each of these 36 meteorological subdivisions. Red colour indicates statistically significant change point (at the 95 % confidence level). Black number indicates changes which are not statistically significant. Figure 9 indicates an upward climate shift over the Peninsular India around 1945–1955. Downward shift was observed around 1960 over Chhattisgarh, east Madhya Pradesh, Vidarbha, Orissa, Jharkhand, Bihar, sub-Himalayan west Bengal and most of the north-eastern regions. As a result, as shown in Fig. 10, the all India SW monsoon rainfall has shown significant downward shift in the year 1965. It is important to understand why this climate shift has happened and the underlying physical mechanisms. Probably, the variability of the tropical oceans may be playing an important role.

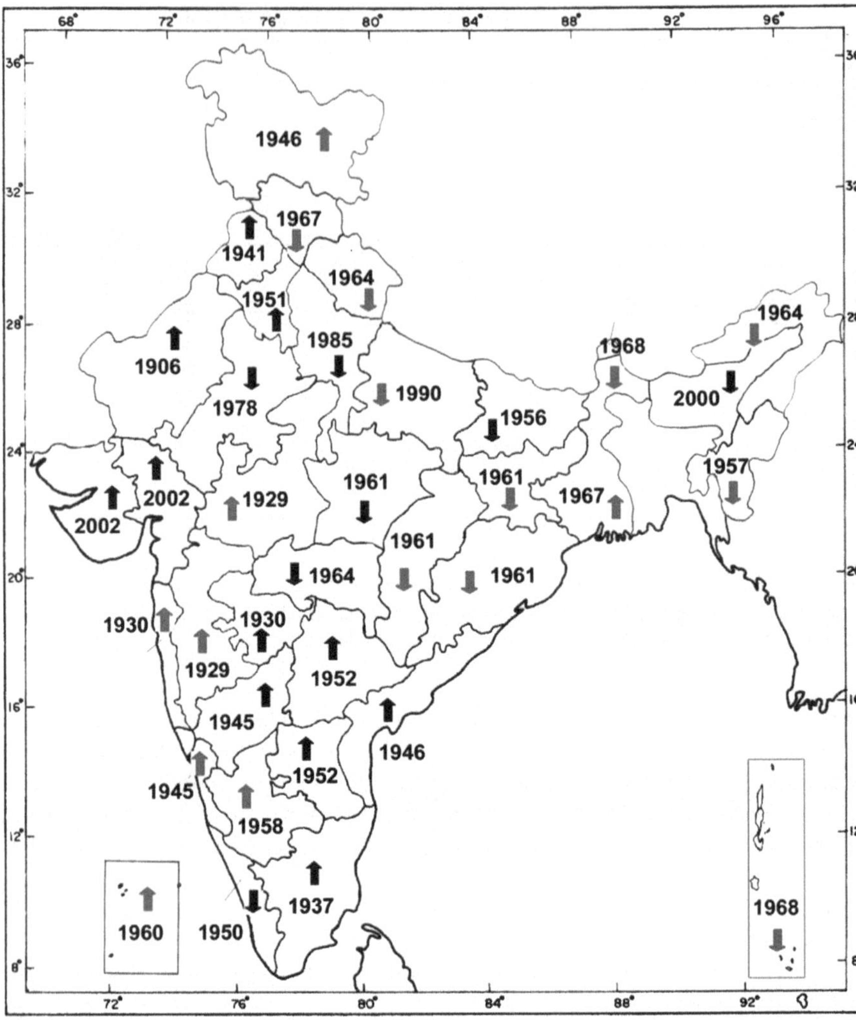

Fig. 9 Climate jump or change point (year) in south-west monsoon rainfall over the 36 meteorological subdivisions in India. *Red colour* indicates the change is significant (95 % confidence level). Upward and *downward arrow* represent upward and downward shift, respectively (color figure online)

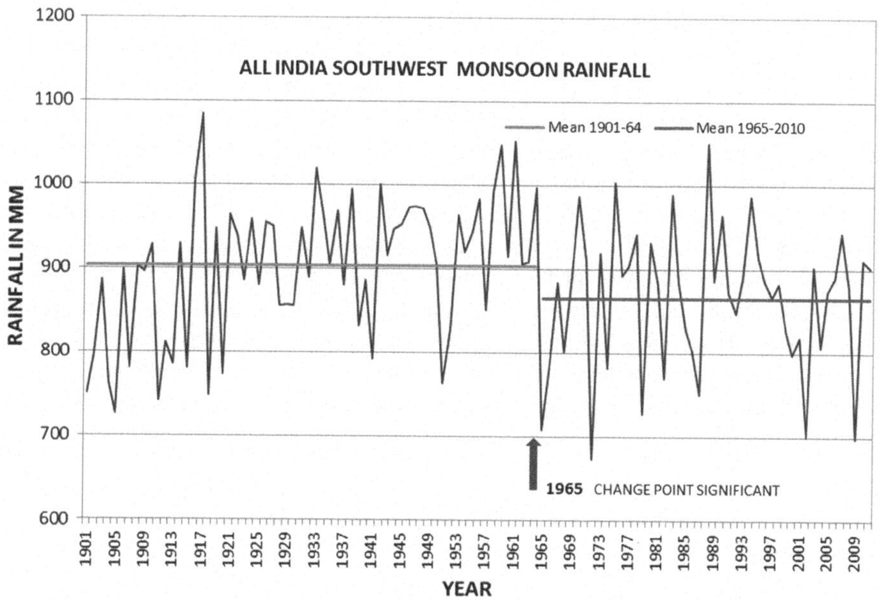

Fig. 10 Climate jump or change point (year) in the south-west (June–September) monsoon rainfall averaged over the country

7 Conclusions

The present analysis brings the variability and trends in the rainfall pattern over the country based on the monthly and seasonal data. Multi-decadal epochs having frequent droughts and flood years with alternate sequence have been observed in the all India monsoon rainfall data. The first two decades were drier than normal. The decades 1921–1930 to 1961–1970 were wet periods. The decades 1971–1980 onwards were drier than normal with the recent decade 2001–2010 being the driest. Decreasing trend in rainfall during the month of July is observed over most parts of the central India. However, increasing trends are observed in rainfall during June and August over the central and south-western parts of the country. Significant decreasing trend in rainfall during the south-west monsoon season is observed over ten subdivisions viz. Chhattisgarh, Jharkhand, Uttarakhand, Himachal Pradesh, Arunachal Pradesh, NMMT (Nagaland, Manipur, Mizoram and Tripura), sub-Himalayan West Bengal, Kerala, east Uttar Pradesh and east Madhya Pradesh whereas eight subdivisions viz. Madhya Maharashtra, Saurashtra and Kutch, south interior Karnataka, coastal Karnataka, Konkan and Goa, coastal Andhra Pradesh, Lakshadweep and Gangetic West Bengal showed significant increasing trends.

The contribution of June rainfall towards the annual rainfall has shown an increasing trend in 20 subdivisions of the country. Significant increasing trend in the annual rainfall is also observed over the subdivisions Konkan and Goa, Madhya

Maharashtra, north interior Karnataka, Rayalseema, coastal Andhra Pradesh, Gangetic West Bengal, Assam and Meghalaya and Jammu and Kashmir.

The analysis of north-east monsoon rainfall over the five met-subdivisions of the Peninsular India reveals no significant long-term trend. However, the presence of decadal variability has been clearly observed. The shift in rainfall pattern has been detected by performing change point detection analysis. Most of the subdivisions in northern, central, eastern and north-eastern India have shown downward shift during the south-west monsoon rainfall around the 1960s while all the subdivisions in the peninsular India (except Kerala) have shown upward shift around the 1930s.

Climate jump analysis identified an upward climate shift over the Peninsular India around 1945–1955 while downward shift was observed around 1960 over Chhattisgarh, east Madhya Pradesh, Vidarbha, Orissa, Jharkhand, Bihar, sub-Himalayan west Bengal and most of the north-eastern regions in southwest monsoon rainfall.

References

Bawiskar SM. 2009, Weakening of lower tropospheric temperature gradient between Indian landmass and neighboring oceans and its impact on Indian monsoon. *Journal of Earth System Science* **118**(4), 273–280.

Chung CE, Ramanathan V. 2006, Weakening of North Indian SST Gradients and the Monsoon Rainfall in India and the Sahel. *Journal of Climate,* **19,** 2036-2045.

Guhathakurta P, Srivastava AK. 2004, Evidence of climate jump in annual temperature in India. *Mausam* **55**(3), 419-428.

Guhathakurta P, Rajeevan M. 2008, Trends in rainfall pattern over India. *International Journal of Climatology* **28,** 1453–1469.

Guhathakurta P, Koppar AL, Krishan U, Menon P. 2011a., New rainfall series for the districts, meteorological sub-divisions and country as whole of India. *National Climate Centre Research Report No*: 2/2011, India Meteorological Department.

Guhathakurta P, Sreejith OP, Menon P 2011b., Impact of climate changes on extreme rainfall events and flood risk in India. *Journal of Earth System Science* **120,** 359-373.

Guhathakurta P, Saji E 2013. Detecting changes in rainfall pattern and seasonality index vis-à-vis increasing water scarcity in Maharashtra, *Journal of Earth System Science* **122,** 3, 639–649.

Guhathakurta P, Rajeevan M, Sikka DR, Tyagi A 2015, Observed changes in southwest monsoon rainfall over India during 1901–2011", Int. J. Climatol., 35, 1881–1898

Joseph PV, Simon A. 2005, Weakening trend of southwest monsoon current through Peninsular India from1950 to present. *Current. Science* **89** (4), 687–694.

Kripalani RH, Kulkarni A. 2001, Monsoon rainfall variations and teleconnections over South and East Asia. *International Journal of Climatology,* 21, 603–616.

Leith GE. 1978. Predictability of climate, *Nature* 276 (5686), 352-355.

Naidu CV, MuniKrishna K, Rao SR, Bhanu Kumar OSRU, Durgalakshmi K, Ramakrishna SSVS, 2011, Variations of Indian summer monsoon rainfall induce the weakening of easterly jet stream in the warming environment. *Global and Planetary Change,* **75,** 21–30.

Pant GB, Rupa Kumar K. 1997. *Climates of South Asia.* John Wiley & Sons, Chichester, 320 pp.

Parthasarathy B. 1984, Inter-annual and long term variability of Indian summer monsoon rainfall. *Proceedings of the Indian Academy of Sciences (Earth Planetary Sciences),* **93,** 371–385.

Parthasarathy B, Dhar ON. 1978, Climate fluctuations over Indian region–rainfall: a review. *Research Report no. RR-025.* Indian Institute of Tropical Meteorology: Pune.

Parthasarathy B, Rupa Kumar K, Munot, A. 1993,. Homogeneous Indian monsoon rainfall: variability and prediction. *Proceedings of the Indian Academy of Sciences (Earth Planetary Sciences)* **102** : 121–155.

Pettitt AN. 1979, A non-parametric approach to the change point problem. *J. Appl. Statist.***28** (2) 126-135.

Pramanik SK, Jagannathan P. 1954, Climate change in India–I rainfall. *Indian Journal of Meteorology Geophysics* **4**, 291–309.

Mooley DA., Parthasarathy B. 1984, Fluctuations in all-India summer monsoon rainfall during 1871–1978. *Climatic Change* **6,** 287–301.

Ramesh KV, Goswami P. 2007, Reduction in Spatial and Temporal Extent of Monsoon rainfall, *Geophy Res Lett* **34**, L 23704 doi:10.1029/2007GL031613.

Rupa Kumar K, Pant GB, Parthasarathy B, Sontakke NA. 1992, Spatial and sub-seasonal patterns of the long-term trends of Indian summer monsoon rainfall. *International Journal of Climatology* **12,** 257–268.

Smadi MM, Zghoul A 2006, A Sudden Change in Rainfall characteristics in Amman, Jordan during the Mid 1950's. *Am. J. Environ. Sci.* **2,** 84-91.

Sneyers S 1990. On the statistical analysis of series of observations; Technical note no. 143, WMO No. 725 415, Secretariat of the World Meteorological Organization, Geneva, 192p.

Tarhule A, Woo M. 1998.Changes in rainfall characteristics in northern Nigeria; *International Journal of Climatology* **18**, 1261-1271.

Wijngaard JB, Klein Tank M, Konnen GP. 2003. Homogeneity of 20th century European daily temperature and precipitation series. *International Journal of Climatology* **23**, 679-692.

Yamamoto R, Iwashima T, Kadi SN, Hoshiai M. 1985. Climate jump: A hypothesis in climate diagnostic. *J. Met. Soc. of Japan* 63, 1157-1160.

Zhang L, Zhou T. 2011. An assessment of monsoon precipitation changes during 1901-2001, *Climate Dynamics* **37**, 279-296.

Chapter 2
Variability and Long-Term Changes in Surface Air Temperatures Over the Indian Subcontinent

A.K. Srivastava, D.R. Kothawale and M.N. Rajeevan

1 Introduction

Surface air temperature is one of the most important meteorological parameters of the climate system. It is widely used to detect the first signal of climate change. The term, "global warming," frequently discussed today, is a consequence of the substantial rise in surface temperatures across the globe. With the steep rise in surface temperatures over the past few decades, global warming is being viewed as a triggering source of the observed and projected rise in the frequency and intensity of many extreme weather events. Owing to this observed and projected temperature rise, climate models indicate larger changes in the climate system, on regional and planetary scales. Probable changes in the climate systems affecting the Indian subcontinent, due to rise in regional surface temperatures, are of great concern. Changes in seasonal temperatures over the region, especially trends in temperature gradients, may induce a significant change in monsoon performance and its effect on crop production. Hence, it is desirable that a true assessment of the rise in the surface temperatures over the Indian subcontinent is carried out. In this chapter, we document the past observed changes in surface air temperatures over the region as a whole and on different parts of the region. The analysis has been done for (a) for the whole data period 1901–2010 and (b) recent 30 years 1981–2010.

A.K. Srivastava (✉)
India Meteorological Department, Earth System Science Organization (ESSO), Pune 411005, India
e-mail: aks_ncc2004@yahoo.co.in

D.R. Kothawale
Indian Institute of Tropical Meteorology, Earth System Science Organization (ESSO), Pune 411008, India

M.N. Rajeevan
Earth System Science Organization (ESSO), Ministry of Earth Sciences, New Delhi 110003, India

M.N. Rajeevan and S. Nayak (eds.), *Observed Climate Variability and Change Over the Indian Region*, Springer Geology, DOI 10.1007/978-981-10-2531-0_2

2 Review of Past Work Done

Past studies reported that the global as well as regional mean surface air temperatures have increased significantly in the last century (Jones and Moberg 2003). The rise has been steep and more pronounced in the recent decades (IPCC-AR4 2007; IPCC-AR5 2014, and the references therein). With the availability of sufficiently long instrumental temperature records, attempts to investigate temporal variations of temperatures over India were made in as early as in the 1950s. Pramanik and Jagannathan (1954) examined the trends of maximum and minimum temperatures of 30 Indian stations for the period 1880–1950 to conclude that generally there was no significant trend in maximum and minimum temperatures. In one of the earliest studies in the context of contemporary global warming, Hingane et al. (1985) reported that the mean annual temperature has increased by about 0.4 °C/100 years in India during the period 1901–1982. Srivastava et al. (1992) studied decadal trends in the climatic parameters over India for the 1901–1986 period. The study highlighted that there was in general, a widespread cooling over northern India and warming over southern India. Rupa Kumar et al. (1994) pointed out that the increase in annual mean temperatures over India during 1901–87 was mainly contributed by the increase in maximum temperatures with the minimum temperatures remaining practically trendless. After filtering out the global effects of greenhouse gases and natural variability, Krishnan and Ramanathan (2002) found that all India surface air temperature during the drier part of the year (January–May) cooled by as much as 0.3 °C since 1971. Subsequent analysis by Kothawale and Rupa Kumar (2005) brought out that both the all India annual maximum and minimum temperatures rose significantly during the 1971–2003 period. Kothawale et al. (2010) examined the temperature data of the period 1901–2007 to conclude that there was significant rise in all India mean, maximum, and minimum temperatures, and pace of the warming was more in recent years. Recently, Kothawale et al. (2012) examined the spatial and temporal asymmetry of temperature trends over India and the possible role of aerosols. Thus, there are several studies which examined and highlighted temperature trends over India.

However, most of the above studies used temperature data of limited number of stations and did not use temperature data of all the stations for which temperature records are available in the archives of the India Meteorological Department (IMD). Moreover, above studies examined the trend in temperature over different time spans. Further, spatial variation of the temperatures was also not examined by most of the studies. Therefore, there is not only a need to assess actual rise in the temperature data over India using long series of temperature data of all the stations but also to examine the spatial variability. The present analysis aims to do the same by utilizing data of all the stations for which data are recorded, quality controlled, and archived.

2.1 Data and Methodology

For the present study, monthly maximum and minimum temperature data of 395 available stations for the period 1901–2010 were collected from the IMD archives. Figure 1 shows the spatial distribution of stations whose data were used in this study. The data were subjected to basic quality checks such as rejecting values, greater than exceeding known extreme values, minimum temperature greater than the maximum temperature, and same temperature values for many consecutive days. Unusual high values were flagged by putting a filter suggested by Sellers and Liu (1988). The flagged values showing spatial continuity were accepted and only the isolated values were rejected. Using station data, 1.0° × 1.0° gridded data for the Indian region were prepared. The development of data set was based on the methodology of Jones and Moberg (2003) and Srivastava et al. (2009) by interpolating anomaly series (based on 1961–1990 period). Monthly temperature series for the country (India) were computed by area-weighted average of the grid point data series.

Annual/seasonal, winter (January and February), premonsoon (March–May), monsoon (June–September), and postmonsoon (October–December) temperature

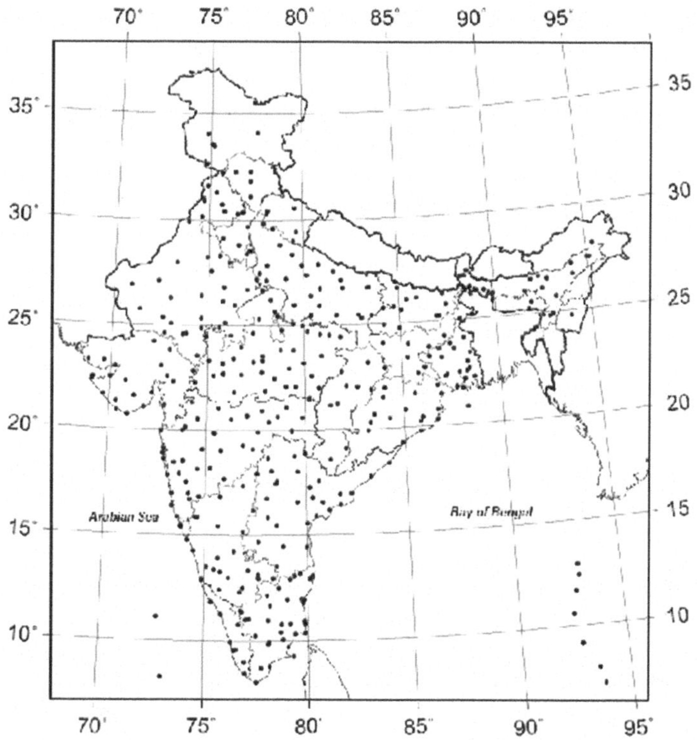

Fig. 1 Network of 395 stations selected for developing the gridded data set

series for the period 1901–2010 were constructed by taking arithmetic average of the months of the year/season. Temperature series were examined for long-term variations and trends for the 1901–2010 and 1981–2010 periods. The trend was estimated by fitting a simple linear regression to the time series. A trend in the annual temperature (maximum and minimum) series over different moving 30-year periods such as 1901–1930 and 1911–1940 was also examined. The statistical significance of a trend was assessed by testing significance of regression coefficients of the fitted lines in the different annual and seasonal temperature time series. In addition to the trend analysis, confidence intervals of the trends are also calculated.

3 Results and Discussions

3.1 Annual Trend of All India Temperatures

The all India annual mean, maximum, and minimum temperature series are shown in Fig. 2. A significant increasing trend of 0.6 °C per 100 years is observed in the all India annual mean temperatures for the period 1901–2010. It may be seen that the annual maximum temperature rose consistently over the data period and rate of increase is 1.0 °C per 100 years for the period 1901–2010. Similarly, in the annual minimum temperature series, a significant increasing trend of 0.18 °C per 100 years for the period 1901–2010 is noticed. We may observe that the minimum temperatures did not witness any significant rise over most of the data period except during the recent few decades. Therefore, the rise in annual mean temperatures is primarily contributed by the increase in maximum temperatures. These trends are statistically significant at the 95 % confidence level.

From Fig. 2, we may also observe that the rate of warming as 0.18 °C per decade. Thus, although maximum temperatures increased gradually and consistently over the entire period, the rate of rise during the past 30 years was almost double. It may also be seen that rise in minimum temperatures was very spectacular for the period 1981–2010 as annual minimum temperatures (which was practically trendless till 1980) increased at the rate of 0.17 °C per decade.

3.2 Seasonal Temperatures

Seasonal mean, maximum, and minimum temperature anomaly series and their trends are shown in Fig. 3a–d. During the period 1901–2010, there was a significant rise in the mean temperatures in all the seasons. Increase in the mean temperatures was the highest in the postmonsoon (0.79 °C per hundred years) and winter (0.70 °C per hundred years) seasons, followed by the premonsoon (0.55 °C per hundred years) season. The mean temperature during the monsoon season

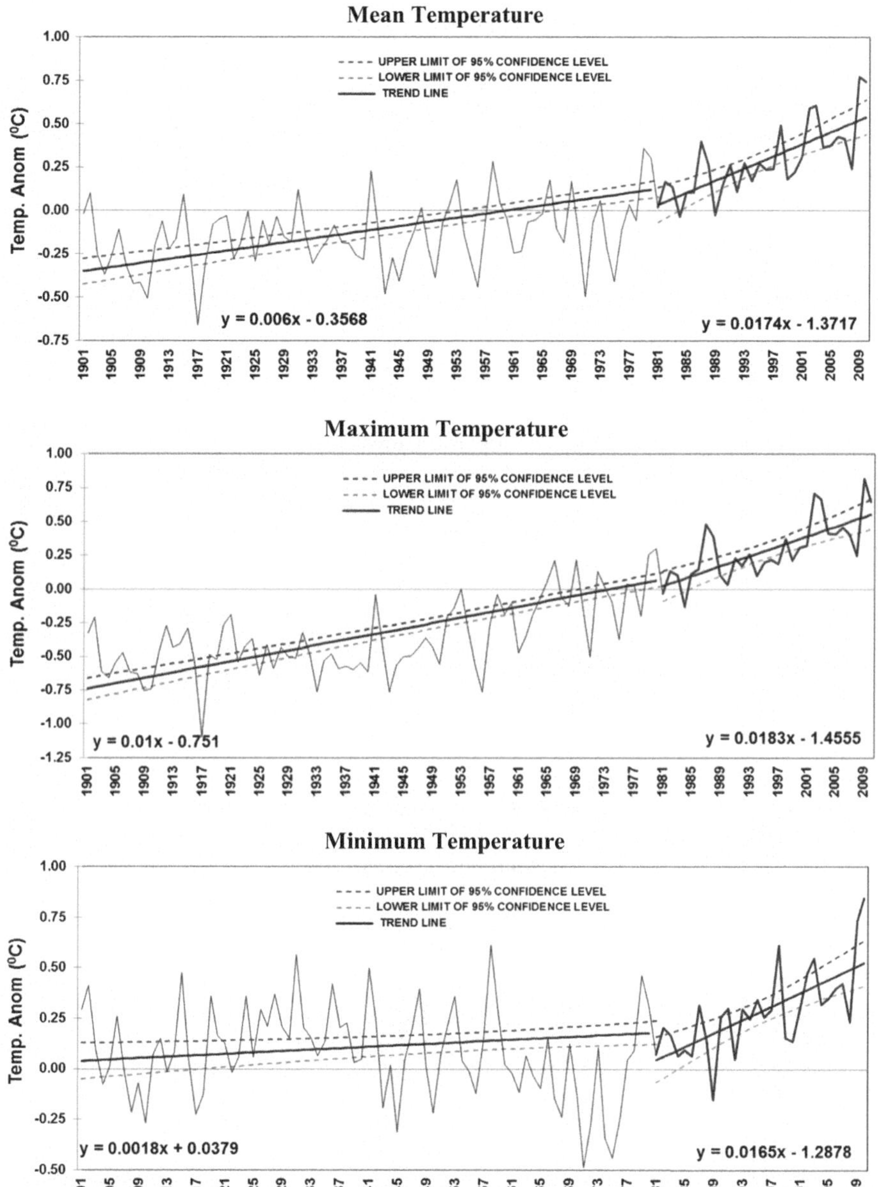

Fig. 2 All India annual mean, maximum, and minimum temperature anomaly series and the linear trends

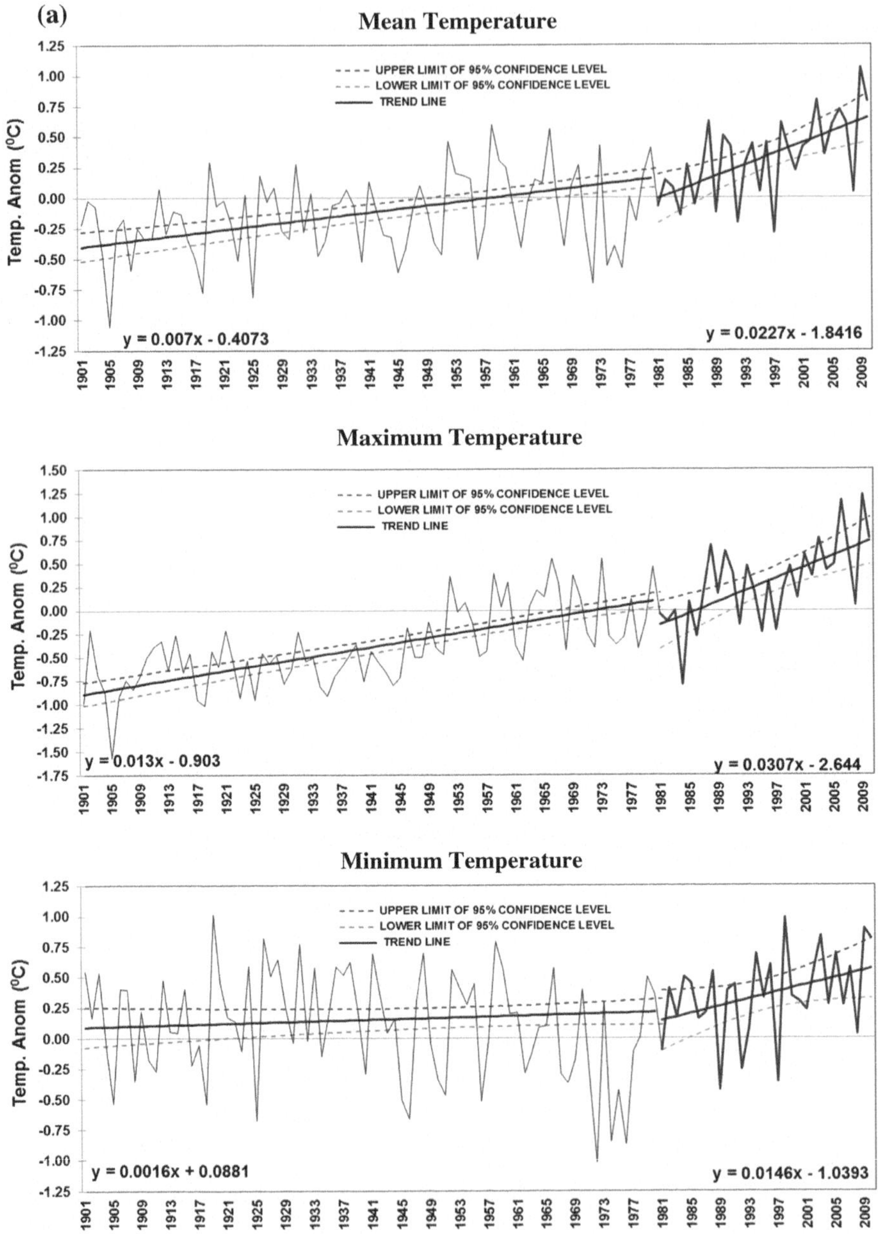

Fig. 3 **a** All India mean, maximum, and minimum temperature anomaly series and the linear trend for the winter season. **b** All India mean, maximum, and minimum temperature anomaly series and the linear trend for the premonsoon season. **c** All India mean, maximum, and minimum temperature anomaly series and the linear trend for the monsoon season. **d** All India mean, maximum, and minimum temperature anomaly series and the linear trend for the postmonsoon season

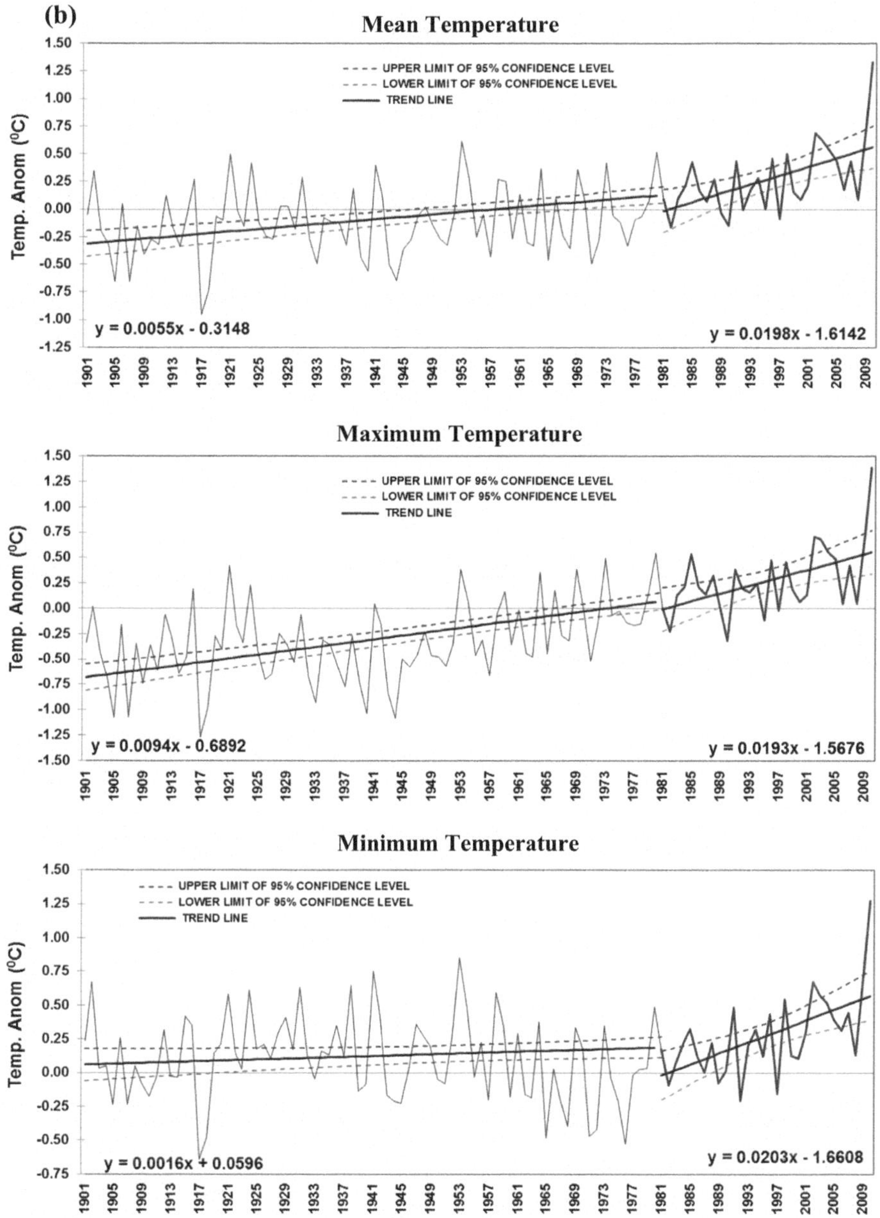

Fig. 3 (continued)

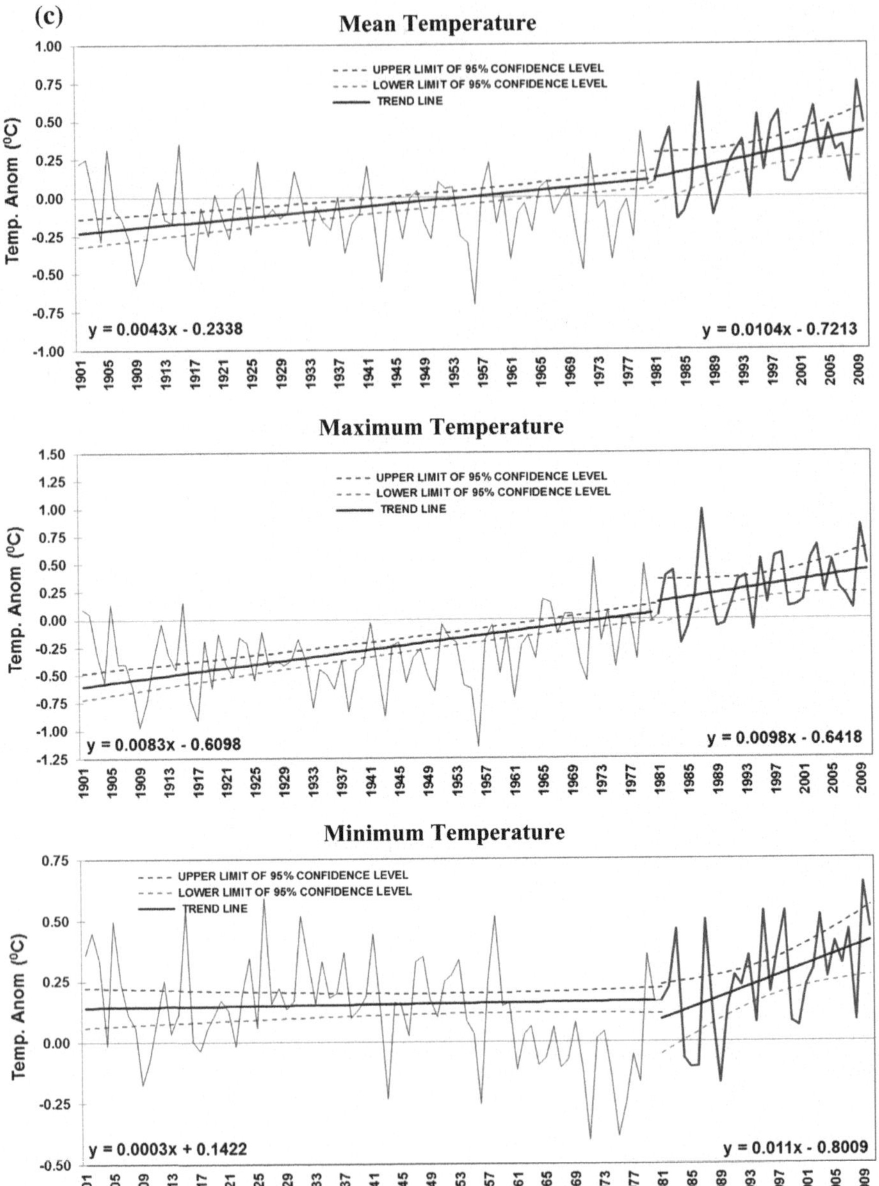

Fig. 3 (continued)

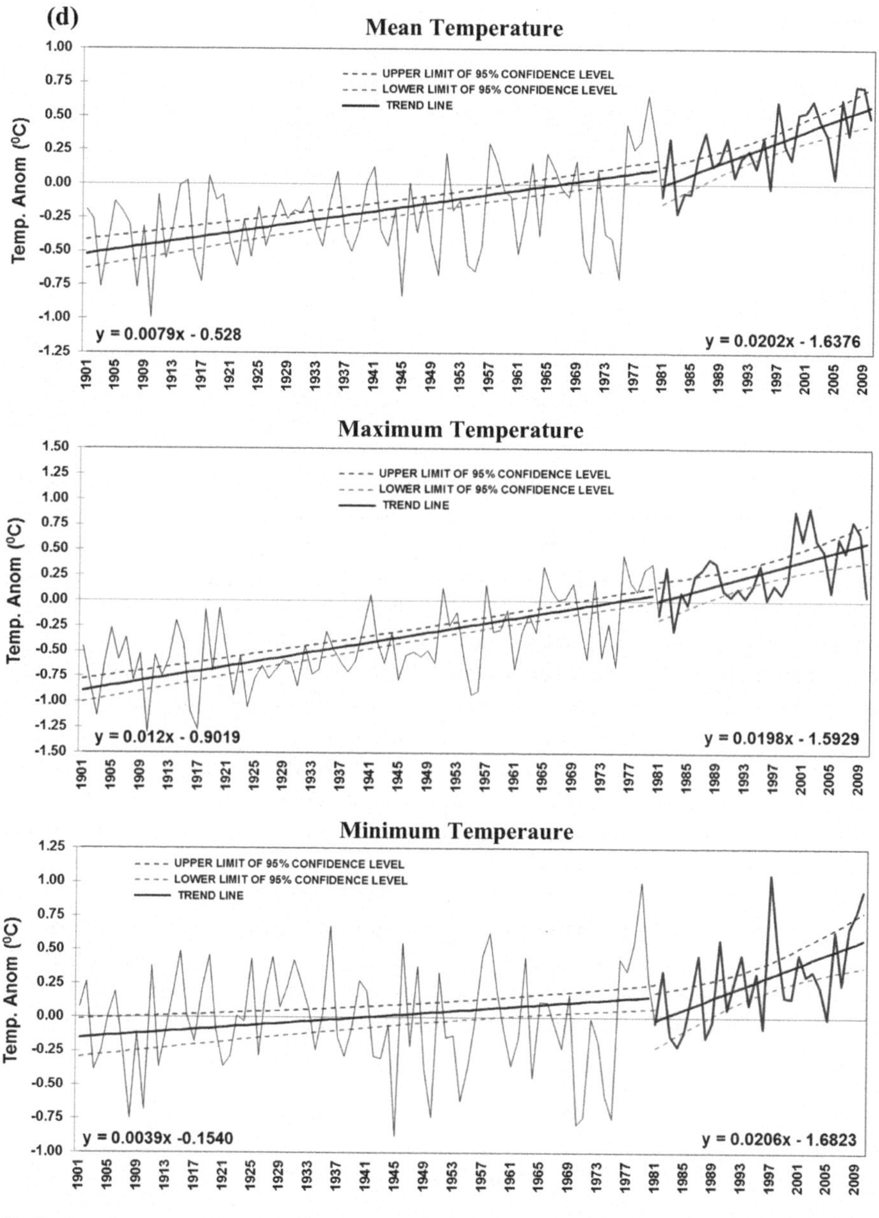

Fig. 3 (continued)

Table 1 Surface air temperature trends for the country as a whole

Season	1901–2010			1981–2010		
	Mean temperature (°C) per 100 years	Maximum temperature (°C) per 100 years	Minimum temperature (°C) per 100 years	Mean temperature (°C) per decade	Maximum temperature (°C) per decade	Minimum temperature (°C) per decade
Annual	0.60	1.00	0.18	0.17	0.18	0.17
Winter (Jan–Feb)	0.70	1.30	0.16	0.22	0.31	0.15
Premonsoon (Mar–May)	0.55	0.94	0.16	0.20	0.19	0.20
Monsoon (Jun–Sep)	0.43	0.83	0.03	0.10	0.10	0.10
PostMonsoon (Oct–Dec)	0.79	1.20	0.39	0.20	0.20	0.20

showed the least rise of 0.43 °C per hundred years. Similarly, there is a significant rise in the maximum temperature in all the seasons and the rate of increase is also almost constant around 1.0 °C per 100 years, except during the monsoon season in which the rise was less than 1.0 °C per 100 years. The minimum temperatures were practically stationary (trendless till 1980) in all the seasons except the postmonsoon season during which significant rise was observed (0.39 °C per 100 years).

Further, during the recent three decades (1981–2010), the mean temperature rose significantly in all the seasons. The rate of rise in the mean temperature is about 0.2 °C per decade in all the seasons except the monsoon season during which an increase of 0.1 °C per decade was observed. Similarly, during the period, maximum temperature rose at about 0.3 °C per decade for the winter season and at 0.2 °C per decade for the pre- and postmonsoon seasons. The rise was the least at 0.1 °C per decade during the monsoon season. During the period, minimum temperatures also increased significantly in all the seasons and the rate of increase is the highest during the pre- and postmonsoon seasons (around 0.2 °C per decade). The rise in minimum temperature in the winter season during the period is around at 0.15 °C per decade, while the increase in the monsoon season during the period is the least (0.1 °C per decade). Trend values in the annual and seasonal temperatures are given in the Table 1.

3.3 *Spatial Patterns of Temperature Trends*

The spatial pattern of temperature trends in the annual/seasonal, mean, maximum, and minimum temperature series for the periods 1901–2010 and 1981–2010 was examined and its statistical significance was tested. Spatial patterns of the trend in the annual and all the four seasons in respect of mean, maximum, and minimum temperatures during the periods 1901–2010 and 1981–2010 are shown in Figs. 4 and 5a–d.

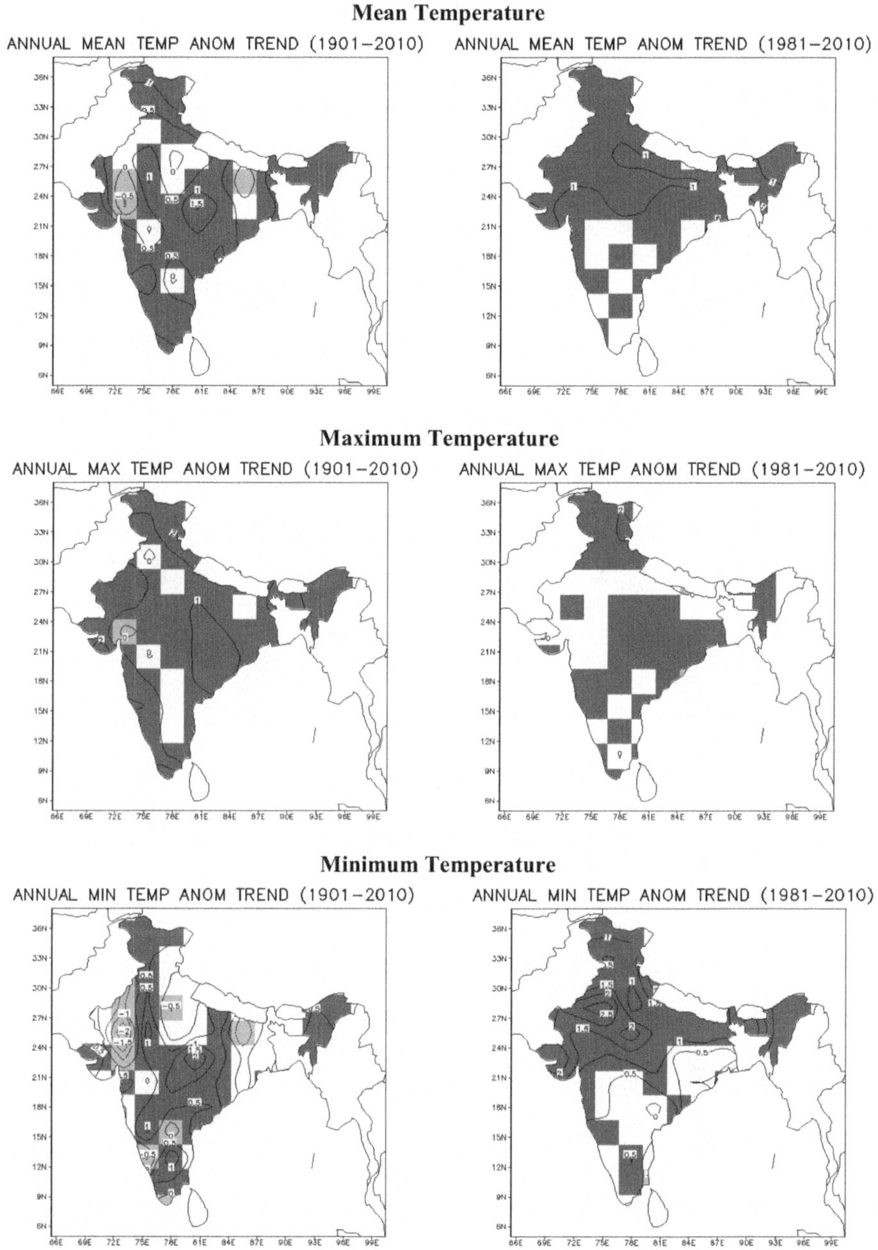

Fig. 4 Spatial pattern of the trend in annual mean, maximum, and minimum temperatures for the 1901–2010 and 1981–2010 periods. *Shaded regions* represent the regions where trends are significant at the 95 % confidence level. *Red* and *blue* regions show significant increase and decrease, respectively, and magnitude of trend during the periods is depicted by contour lines (color figure online)

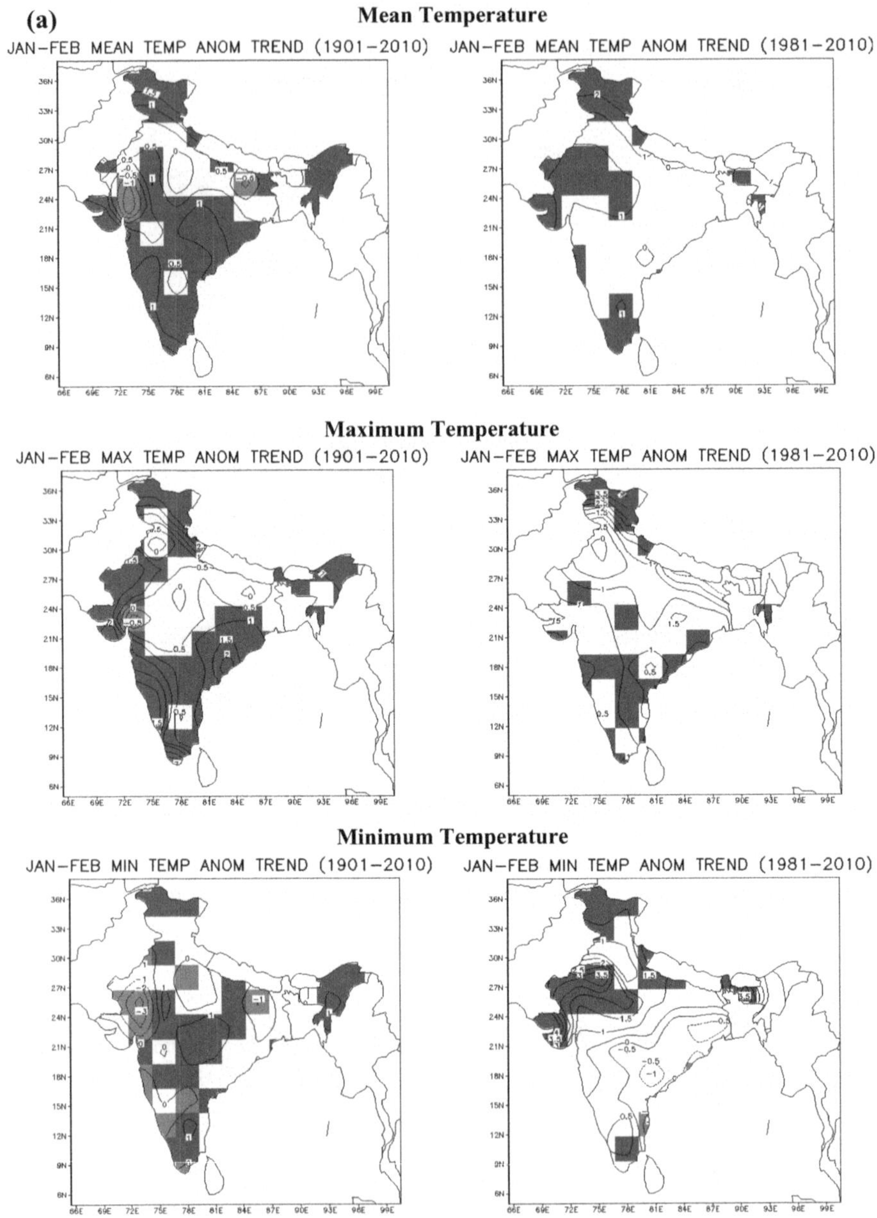
(a)
Mean Temperature
JAN-FEB MEAN TEMP ANOM TREND (1901-2010)
JAN-FEB MEAN TEMP ANOM TREND (1981-2010)
Maximum Temperature
JAN-FEB MAX TEMP ANOM TREND (1901-2010)
JAN-FEB MAX TEMP ANOM TREND (1981-2010)
Minimum Temperature
JAN-FEB MIN TEMP ANOM TREND (1901-2010)
JAN-FEB MIN TEMP ANOM TREND (1981-2010)

◀ **Fig. 5** **a** Spatial pattern of the trend in mean, maximum, and minimum temperatures during the postmonsoon season for the periods 1901–2010 (*left*) and 1981–2010 (*right*). *Shaded regions* represent the regions where trends are significant at the 95 % confidence level. Magnitude of total trends during the periods is depicted by *contour lines*. **b** Spatial pattern of the trend in mean, maximum, and minimum temperature during the winter season for the periods 1901–2010 (*left*) and 1981–2010 (*right*). *Shaded regions* represent the regions where trends are significant at the 95 % confidence level. Magnitude of total trends during the periods is depicted by *contour lines*. **c** Spatial pattern of the trend in mean, maximum, and minimum temperature during the premonsoon season for the periods 1901–2010 (*left*) and 1981–2010 (*right*). *Shaded regions* represent the regions where trends are significant at the 95 % confidence level. Magnitude of total trends during the periods is depicted by *contour lines*. **d** Spatial pattern of the trend in mean, maximum, and minimum temperatures during the monsoon season for the periods 1901–2010 (*left*) and 1981–2010 (*right*). *Shaded regions* represent the regions where trends are significant at the 95 % confidence level. Magnitude of total trends during the periods is depicted by *contour lines*

Annual Temperatures

During the 1901–2010 period, significant increasing trends in the annual mean temperature are observed over a large part of the country. However, significant negative trend is observed over northwestern parts of Central India (Fig. 4) and some isolated parts in the country. For the period, significant increasing trends in the annual maximum temperature are observed over almost the entire country. Similarly, the spatial pattern of trend in annual minimum temperatures data for the whole period exhibits a significant increasing trend over most parts of the Peninsular India and some northern, south-central regions of the country. Significant negative trends are also observed over parts of Gujarat, western Rajasthan, and some parts of the peninsula.

For the period 1981–2010, the spatial pattern of trend in the annual mean temperature suggests significant warming over most parts of the country except the Peninsular India. Similarly, in the annual maximum temperatures, a significant rising trend over parts of northern, east-central, and Peninsular India was also observed. During the period, significant rise in annual minimum temperatures was observed over most parts of the country except the peninsula.

Spatial Trends in the Seasonal Temperatures

For the period 1901–2010, significant increasing trend was observed in the spatial pattern of mean temperatures during the winter and postmonsoon seasons. However, warming was more widespread during the postmonsoon season. During the premonsoon and monsoon seasons, significant rise was limited to some parts of the Peninsular and Central India and the eastern region. Similarly, in the maximum temperatures, significant increasing trend was observed over most areas of the country during the postmonsoon season. In the winter, premonsoon, and monsoon seasons, significant rise was observed over the peninsular and northern parts, and

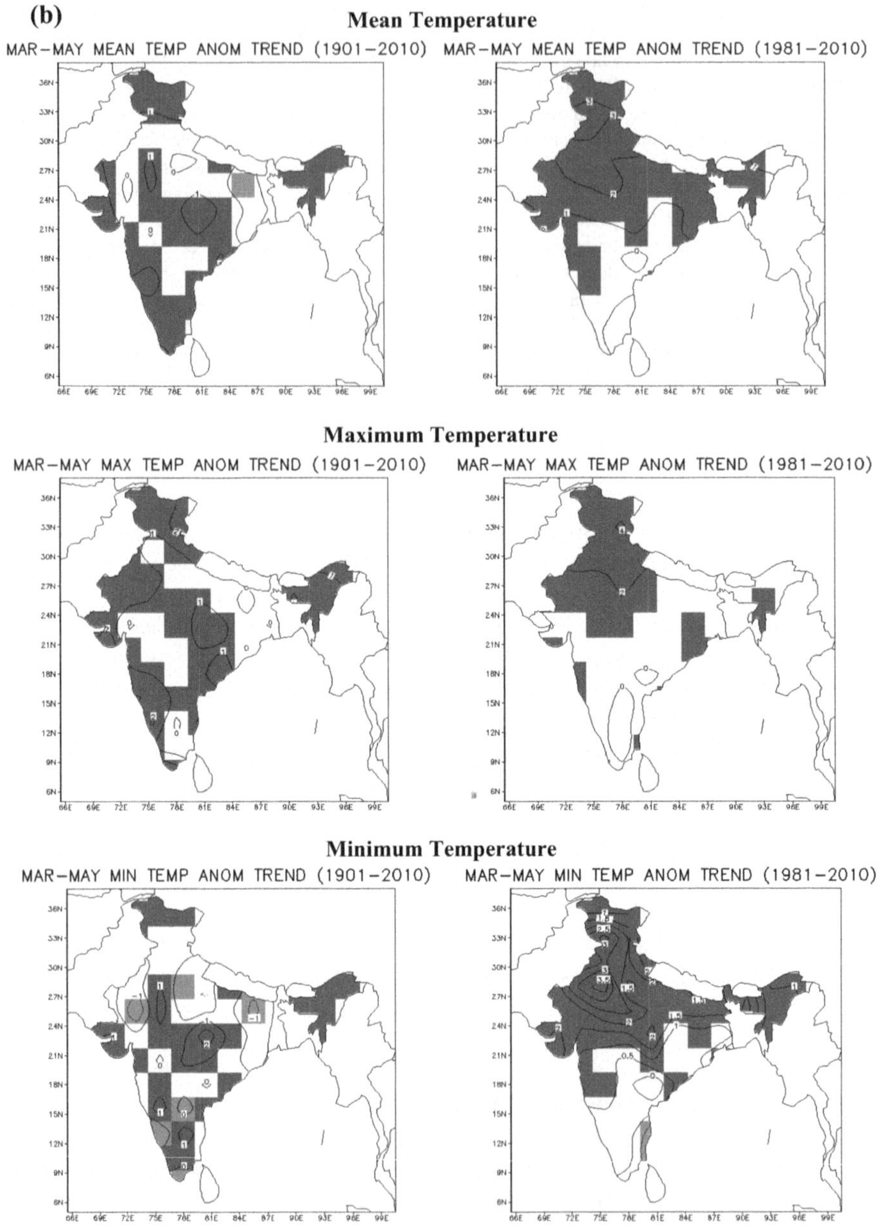

Fig. 5 (continued)

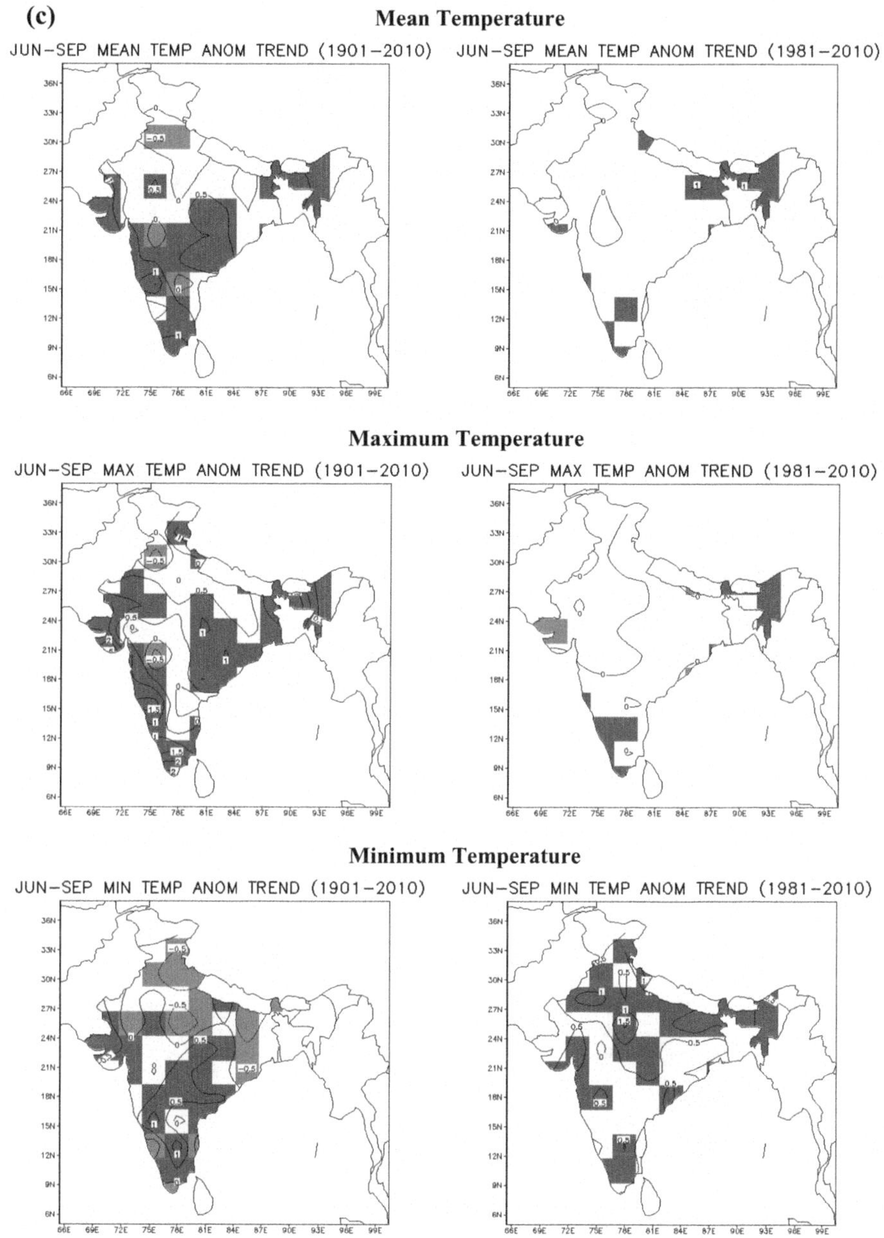

Fig. 5 (continued)

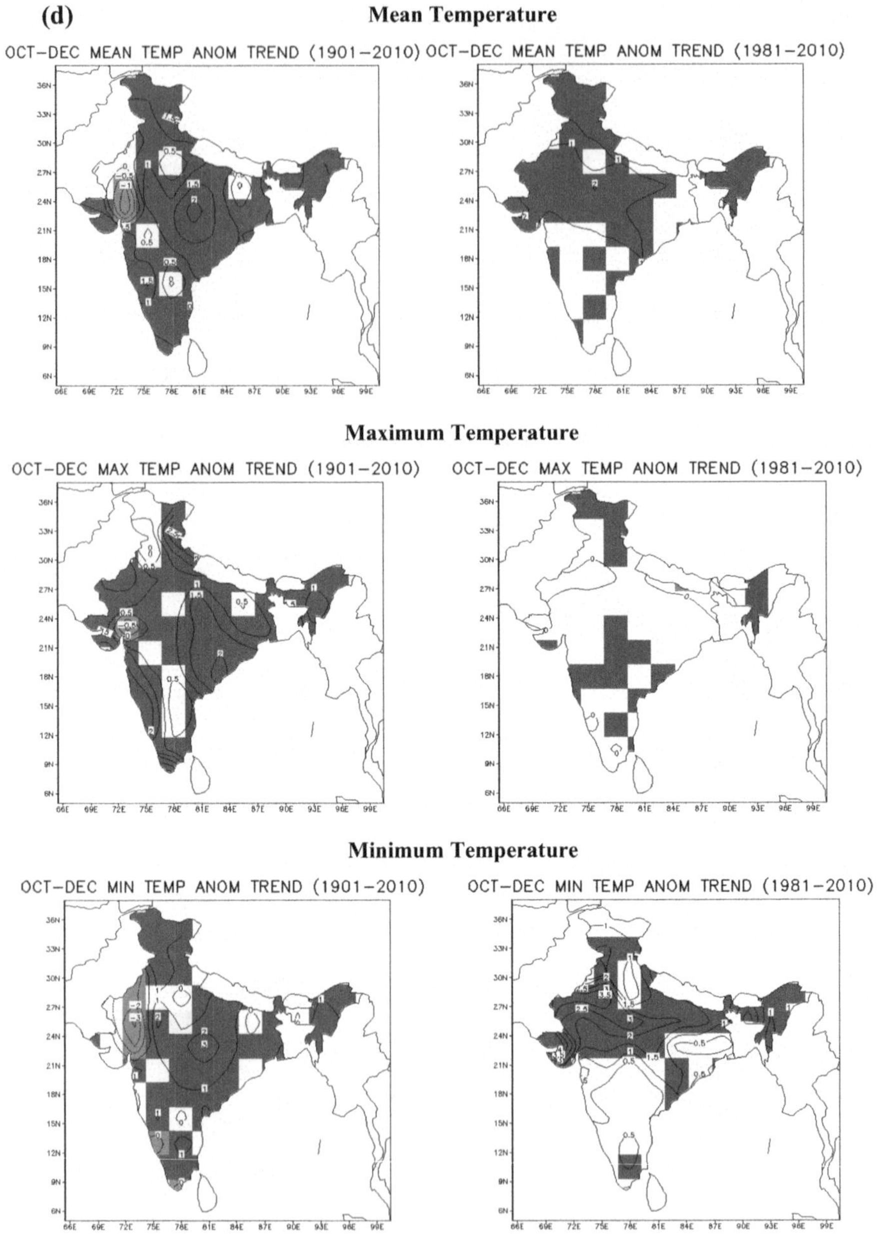

Fig. 5 (continued)

no significant trend was noticed over the central and adjoining parts. Spatial pattern of trend of the minimum temperature data for the whole period suggests a significant increasing trend over larger parts of the country (except some central and northwestern regions) in the postmonsoon season. In the other remaining seasons, rising trend was widespread over the peninsular parts and over some isolated regions.

For the period 1981–2010, spatial trend in mean temperatures showed a significant warming over most parts of the country (except Peninsular India) in the pre- and postmonsoon seasons. In the winter season, significant rise was limited to the northwestern, extreme northern, and peninsular parts, whereas during the monsoon season, no region witnessed any significant trend. Similarly, spatial pattern of trend in the maximum temperatures for the period 1981–2010 shows a significant rise over the extreme northern parts and Peninsular India. In the pre- and postmonsoon seasons, maximum temperatures witnessed significant increasing trend over most parts of northern India. Spatial pattern of trend in minimum temperatures for the period 1981–2010 exhibits significant rise over the northern, central, and northeastern parts during the pre- and postmonsoon seasons. Winter season witnessed significant increasing trend over parts of northwestern India and some isolated region, whereas significant rise was observed over some parts of the northern, central, and eastern regions of the country during the monsoon season.

4 Trend in Upper-Air Temperatures

In order to assess changes in the upper-air temperatures, the all India annual/seasonal tropospheric temperature series were computed by simple averaging of upper-air temperature data (for the year/season), recorded at RS/RW stations located in the country. Temperature data series of upper air, viz. 850, 700, 500, 200, and 150 hPa levels, for the period 1971–2007 were examined for assessing the trends, if any. The statistical significance of the trend was assessed by applying the Mann–Kendall rank test. The analysis suggests that the annual upper-air temperature averaged over the country as a whole showed a significant increasing trend at 850 and 700 hPa levels (Fig. 6). The annual temperature at the 500 hPa level also showed an increasing trend but the same was not statistically significant. Temperature data of other upper-air levels did not show any significant trend. On the seasonal scale, only for the winter season, significant increasing trend was observed for all the above three levels. During the pre- and postmonsoon seasons, an increasing trend (not significant) at the three levels, viz., 850, 700, and 500 hPa levels, was observed. There was no trend in the upper-air temperatures during the monsoon season.

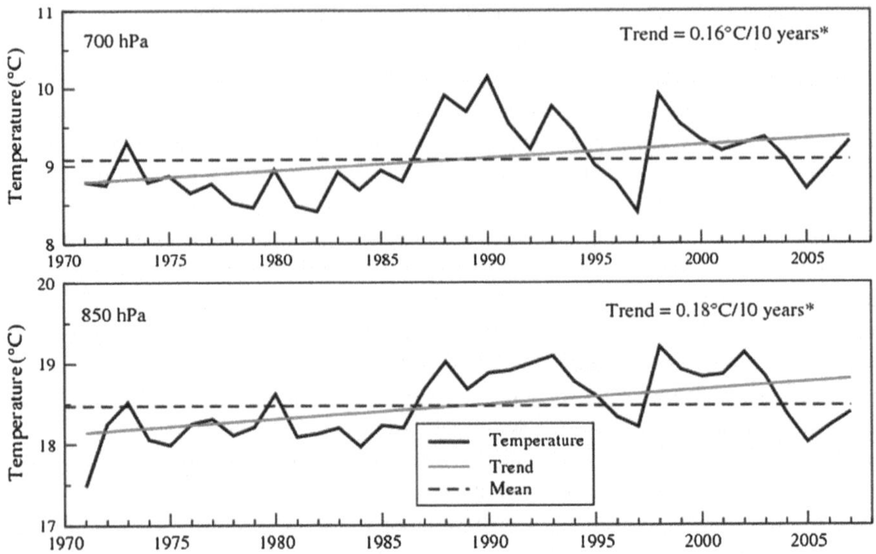

Fig. 6 All India annual upper-air temperature series and its linear trend at the 850 hPa (1.5 km) and 700 hPa (3.1 km) levels. The trends are significant at the 95 % confidence level

5 Summary

Annual mean and maximum temperatures for the country as a whole have increased significantly by around 0.6 and 1.0 °C per hundred years, respectively, during the period 1901–2010. Similarly, the annual minimum temperatures for the country as a whole increased by 0.18 °C per hundred years. On the seasonal scale also, mean and maximum temperatures for the country as a whole increased significantly in all the seasons, while only the postmonsoon season witnessed significant rise in respect of minimum temperatures.

Maximum temperature series exhibited an accelerated warming during the recent 30-year period (1981–2010). Minimum temperature series also witnessed a very rapid rise during the last thirty years, similar to that observed in maximum temperatures. Therefore, the rate of rise in the mean temperature for the country as a whole is almost three times as that of the same for the whole period (1901–2010). Most of the rise in maximum and minimum temperatures during the recent 30 years is more prominent over the northern, central, and eastern/northeastern parts of the country. Peninsular India witnessed the least warming during the recent 30-year period. Annual upper-air temperatures averaged over the country as a whole for the period 1971–2007 also showed a significant increasing trend at the 850 and 700 hPa levels. This suggests that the warming is not only confined on the surface, but also extends up to 3 km in the troposphere.

References

Jones PD, Moberg, A.,, 2003, hemispheric and large-scale surface air temperature variations: an extensive revision and update to 2001. J Climate, 16, 206–223.

Hingane LS, Rupa Kumar K, Ramana Murthy Bh.V, 1985, Long-term trends of surface air temperature in India. Int. J. of Climatol., 5, 521-528.

IPCC-AR4, 2007, Climate Change 2007, The Scientific Basis, Contribution of Working Group-I to the Fourth Assessment Report of Intergovernmental Panel on Climate Change (IPCC). Cambridge University Press, Cambridge.

IPCC-AR5, 2014, Climate Change 2014, The Scientific Basis, Contribution of Working Group-I to the Fourth Assessment Report of Intergovernmental Panel on Climate Change (IPCC). Cambridge University Press, Cambridge.

Kothawale DR. Rupa Kumar K., 2005, On the recent changes in surface temperature trends over India. Geophysical Research Letters 32, L18714, doi:10.1029/2005GL023528.

Kothawale DR. Munot AA, Krishna Kumar K, 2010, Surface air temperature variability over India during 1901-2007and its association with ENSO. Clim Res.42.89-104 doi:10.3354/cr00857

Kothawale DR. Krishna Kumar K, Srinivasan G., 2012, Spatial asymmetry of temperature trends over India and possible role of aerosols. Theor.Appl.Climatol. 110:263-280, doi 10.1007/s00704-012-0628-8

Krishnan R, Ramanathan V, 2002, Evidence of surface cooling from absorbing aerosols. Geophys. Res. Lett., 29, No. PP 54-1 54-4.

Pramanik SK, Jagannathan, P., 1954, Climatic Change in India (II) - Temperature. Ind. J. Met. Geophys., 5, 1-19.

Rupa Kumar K, Krishna Kumar K, Pant GB, 1994, Diurnal asymmetry of surface temperature trends over India.,Geophys. Res. Let., 21, 677-680.

Sellers WD and Liu Wen., 1988, Temperature patterns and trends in theupper troposphere and lower troposphere. *Journal of Climate,* 573–581.

Srivastava HN, Dewan BN, Dikshit SK, Prakash Rao GS, Singh SS. and Rao KR., 1992, Decadal trends in climate over India, Mausam, 43, 7-20.

Srivastava AK, Rajeevan M and Kshirsagar SR, 2009, Development of a high resolution daily gridded temperature data set (1969–2005) for the Indian region, *Atmos. Sci. Let.*, DOI:10.1002/asl.232.

Chapter 3
Variability and Trends of Extreme Rainfall and Rainstorms

Pulak Guhathakurta, D.S. Pai and M.N. Rajeevan

1 Introduction

During the southwest monsoon season (June–September), low-pressure systems like monsoon depressions form over the Bay of Bengal and move across northern parts of India, causing widespread rainfall activity over the central region and the west coast. At some occasions, sustained heavy rainfall activity in 2–3 days leads to large-scale floods over different river basins. Generally, the worst floods occur in the Indo-Gangetic Plains. The other equally worst flood-hit region is the north-eastern parts of the country. Floods occur in the Central India region by rivers such as the Mahanadi, the Naramada, the Godavari, and the Krishna. Floods cause several deaths and economic losses every year. Damages from floods take several forms, including destruction of footbridges that often provide the only link between remote villages, demolition of irrigation diversions, and damages to flood plain agricultural land by erosion and sedimentation. Warmer climates, owing to increased water vapor, lead to more intense precipitation events and therefore increase risks of floods.

Given the population density and economic conditions, occurrence of extreme rainfall events and associated floods over India are viewed very seriously. The heavy rainfall event over Mumbai on July 26, 2005 (Chang et al. 2009) and the recent floods over the hilly regions of northern Indian states of Uttarakhand (Joseph et al. 2014) and Jammu and Kashmir in September 2014 are just few examples of extreme hydrometeorological events which caused devastating damages. Dash et al. (2009)

P. Guhathakurta · D.S. Pai
India Meteorological Department, Earth System Science Organization (ESSO), Shivajinagar, Pune 411005, India

M.N. Rajeevan (✉)
Earth System Science Organization (ESSO), Ministry of Earth Sciences, New Delhi 110003, India
e-mail: rajeevan61@yahoo.co.in

M.N. Rajeevan and S. Nayak (eds.), *Observed Climate Variability and Change Over the Indian Region*, Springer Geology, DOI 10.1007/978-981-10-2531-0_3

examined possible changes in the frequency of rain events in India for the period 1951–2004. The study suggested that long spells show a significant decreasing trend over India as a whole while short and dry spells indicate an increasing tendency. They attributed these changes to the weakening of the summer monsoon circulation over India. Dash et al. (2011) examined the changes in short and long spells of rainfall during the summer monsoon season (June to September) over India. The trend analysis suggested that short spell (less than 4 days) rain events with heavy intensity have increased over India as a whole. Recently, Guhathakurta et al. (2014) using the station data for the period 1901–2011 have shown significant decreasing trends in the frequency of moderate rain (24-h rainfall between 5 and 10 cm), no significant change in the frequency of heavy rainfall events (24-h rainfall greater than 10 cm), and significant increasing trend in the frequency of very heavy rainfall events (24-h rainfall greater than 15 cm) over the monsoon core region of India.

Revadekar and Preethi (2012) analyzed extreme precipitation events over India using the $1° \times 1°$ gridded rainfall data over India (Rajeevan et al. 2006) and their relationship with agricultural output. They found that extreme precipitation events have an adverse effect on agricultural output over India. There are many other studies (Rakhecha and Pisharoty 1996; Rakhecha and Soman 1994; Stephenson et al. 1999; Sen Roy and Balling 2004; Rakhecha and Soman 1994; Guhathakurta et al. 2011) addressing heavy rainfall events during the southwest monsoon season. Sen Roy and Balling (2004) have considered parameters such as total annual precipitation, 5-day total precipitation, and 30-day total precipitations in their extreme rainfall analysis. In this chapter, we discuss the observed variability and long-term changes in heavy rainfall events and rainstorms over India.

2 Data and Methodology

In the present analysis, extreme rainfall events and flood risk are analyzed using data from well-distributed rain gauge stations during the 110-year period (1901–2010). To study the changes in the frequency of rainfall events of different intensities, the following categories of rainfall intensity based on daily rainfall are followed:

1. Wet days: 24-h rainfall $\geq$ 0.1 mm. Alternately dry days if the day is not wet or rainfall = 0.0 mm.
2. Very light rain: 0.1 mm $\leq$ 24-h rainfall $\leq$ 2.4 mm.
3. Light-to-moderate rain: 2.5 mm $\leq$ 24-h rainfall $\leq$ 64.4 mm.
4. Heavy rainfall: 64.5 mm $\leq$ 24-h rainfall $\leq$ 124.4 mm.
5. Very heavy rainfall: 124.5 mm $\leq$ 24-h rainfall $\leq$ 244.4 mm.
6. Extremely heavy rainfall: 24-h rainfall $\geq$ 244.5 mm.

Daily rainfall data of all the available stations during the period 1901–2010 are collected from the National Data Centre of the India Meteorological Department. From the list of all available 9294 stations, we have computed month-wise frequency of all the above six events for each of the 30 states of the country. From the monthly data, seasonal [four seasons, viz. winter (January–February), pre-monsoon (March–May), SW monsoon (June–September), and post-monsoon (October–December)] and annual time series of the frequency of all the six rainfall events for all the states are constructed. Linear trend analysis was carried out on the percentage frequency time series for each state of the six events to see the changes in the percentage frequencies of the events. In this way, we have considered all the occurrences of rainfall events throughout the country without rejecting/losing any events including significant ones.

In addition to station rainfall data, we have also used high-resolution $0.25° \times 0.25°$ gridded daily rainfall data set (Pai et al. 2014) and the $1° \times 1°$ gridded rainfall data of Rajeevan et al. (2008) to see the significant trends if any in the frequency of extreme rainfall events. With the gridded data, a threshold of 15 cm was considered as the extreme rainfall.

3 Result and Analysis

3.1 Changes in the Frequency of Rainfall Events of Different Intensities in States

Monthly/seasonal/annual frequency of events in six categories, viz. wet days (24-h rainfall $\geq$ 0.1 mm), light-to-moderate rain (2.5 mm $\leq$ 24-h rainfall $\leq$ 64.4 mm), heavy rain (64.5 mm $\leq$ 24-h rainfall $\leq$ 124.4 mm), very heavy rain (124.5 mm $\leq$ 24-h rainfall $\leq$ 244.4 mm), and extremely heavy (24-h rainfall $\geq$ 244.5 mm), for each of the states and each year for the period 1901–2010 are computed. To remove the effect of missing data or variable number of stations in each state, the frequencies in each category are calculated in percentage of total frequency of all the events in the state.

Figure 1 gives the trends in the percentage of frequencies of all the six indices for the states during the winter season (January–February). The trends have been calculated for the period 1901–2010. Frequencies of wet days, very light, and light-to-moderate rain have decreased over most of the states except a few in South India and Jammu and Kashmir. Frequencies of very heavy rainfall have increased over Jammu and Kashmir and Uttaranchal during the winter season. Frequencies of wet days have decreased in many states over Central and Western India (Fig. 2) during the pre-monsoon season. Almost the similar pattern is observed in the frequencies of very light rain and light-to-moderate rain events. Increase in heavy/very heavy/extremely heavy rainfall events is observed over Jammu and Kashmir, West Bengal, Bihar, Tamil Nadu, Uttarakhand, Assam, and Manipur.

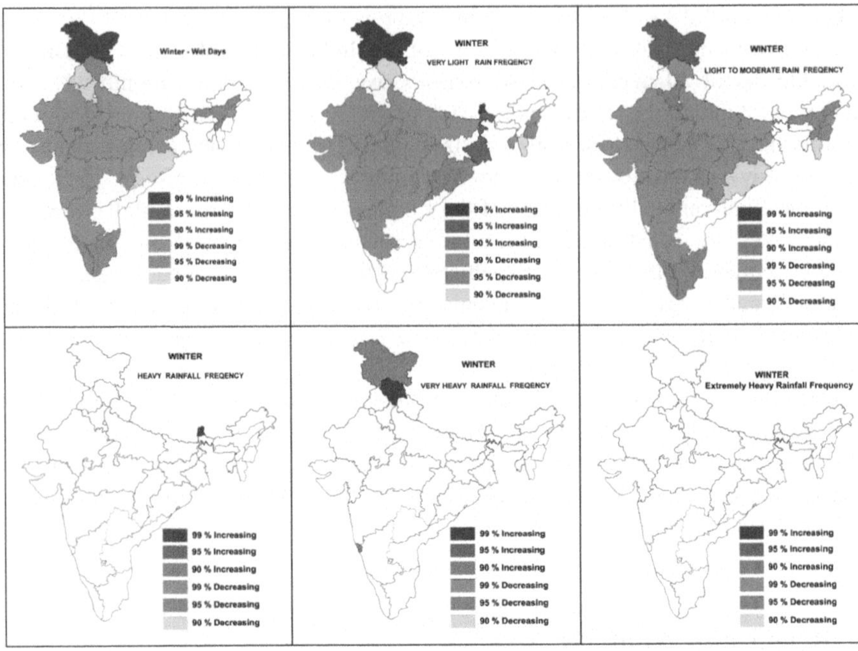

Fig. 1 Trends in the frequencies of different rainfall events over the states during the winter season (January–February)

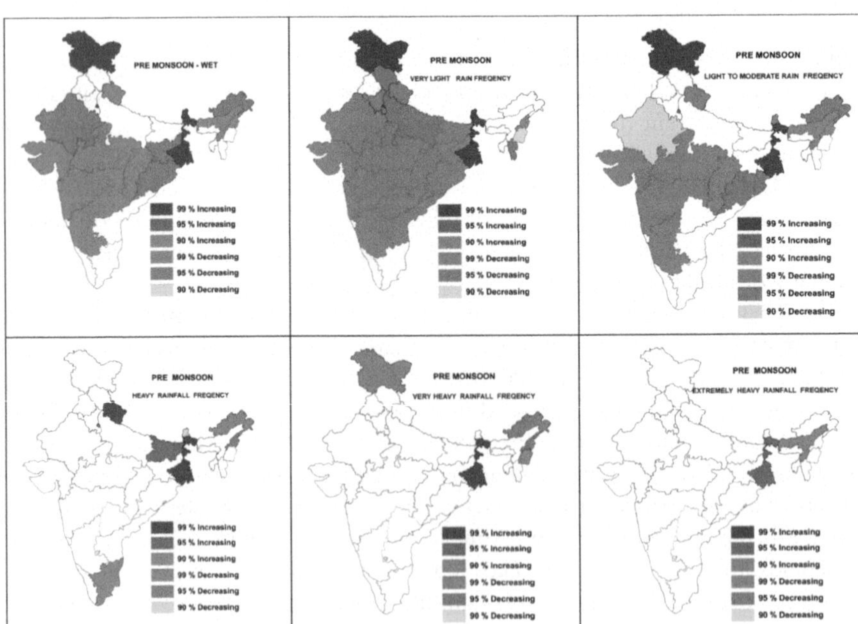

Fig. 2 Trends in the frequencies of different rainfall events over the states during the pre-monsoon season (March–May)

Figure 3 gives the trends in the percentage of the frequencies of all the six events during the southwest monsoon season (June–September). The general inference is the decrease in frequencies of lower intensities and increase in higher intensities (like heavy to very heavy rainfall) during the southwest monsoon season. Frequencies of very light-to-moderate rain have decreased in most of the states in Central India, Peninsular and northeast India. Significant increase in heavy rainfall event is observed over the nine states (West Bengal, Tripura, Haryana, Orissa, Andhra Pradesh, Telangana, Karnataka, Goa, and Tamil Nadu). Frequency of very heavy rainfall events has increased over West Bengal, Assam, Tripura, Sikkim, Punjab, Jammu and Kashmir, Gujarat, Maharashtra, Orissa, Andhra Pradesh, Telangana, Karnataka, Goa, and Tamil Nadu and extreme heavy rainfall events over West Bengal, Assam, Tripura, Uttar Pradesh, Punjab, Jammu and Kashmir, Maharashtra, Orissa, Andhra Pradesh, Karnataka, Goa, and Tamil Nadu.

During the post-monsoon season, no much changes are observed in the states except Jammu and Kashmir (increasing), West Bengal (increasing), and Karnataka and Arunachal Pradesh (decreasing) (Fig. 4). However, significant decreasing trend is noticed in the states such as Karnataka (both very light and light-to-moderate rain), Telengana (very light rain), and Kerala and Tamil Nadu (light-to-moderate rain) which are affected by the northeast monsoon during this season.

The annual pattern (Fig. 5) is almost similar to that of the southwest monsoon season. Frequencies of very light, light-to-moderate and wet days have decreased in most of the states. Frequencies of dry days have also increased significantly during

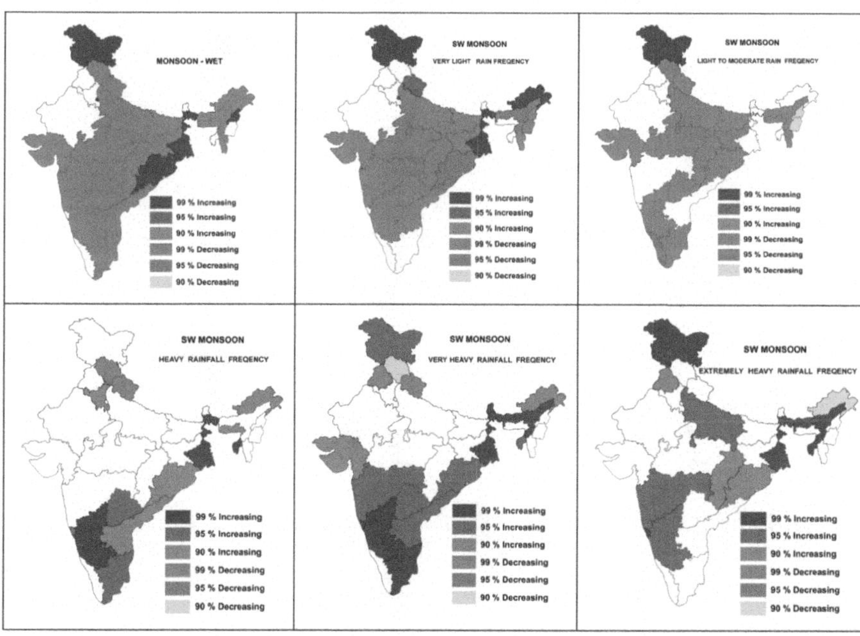

Fig. 3 Trends in the frequencies of different rainfall events over the states during the southwest monsoon season (June–September)

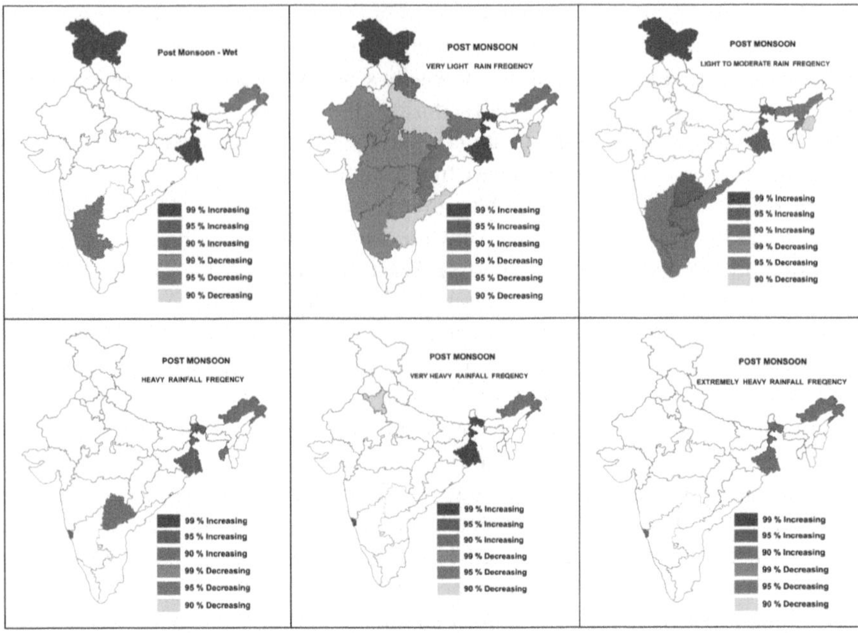

Fig. 4 Trends in the frequencies of different rainfall events over the states during the post-monsoon season (October–December)

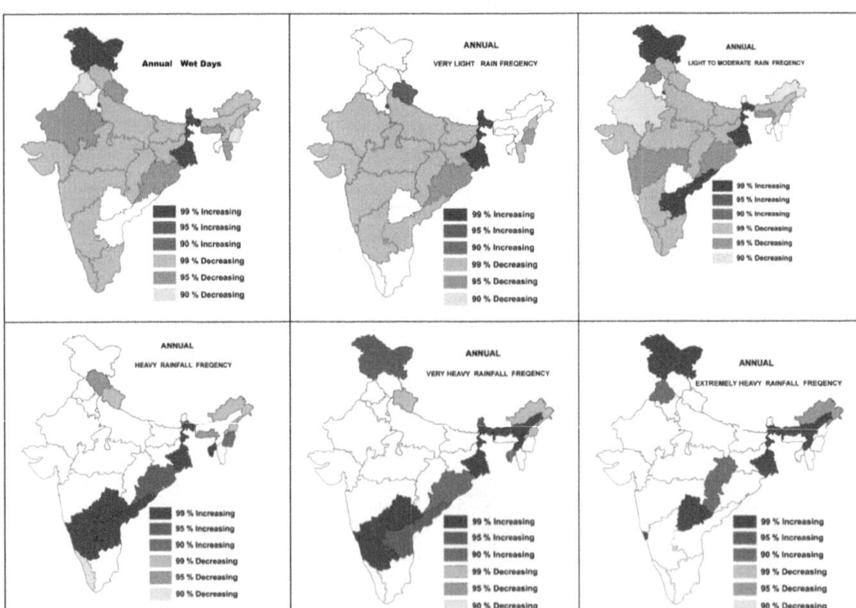

Fig. 5 Trends in the annual frequencies of different rainfall events over the states (January–December)

the period 1910–2010. Significant increase in heavy rainfall event is observed over the eight states (West Bengal, Tripura, Manipur, Andhra Pradesh, Telengana, Karnataka, Goa, and Orissa. Frequency of very heavy events has increased over West Bengal, Tripura, Manipur, Sikkim, Andhra Pradesh, Telengana, Karnataka, Goa, Jammu and Kashmir, and Orissa and extreme rainfall events over West Bengal, Assam, Punjab, Jammu and Kashmir, Chhattisgarh, Goa, and Telengana.

Figure 6 gives the decadal variability of annual frequencies of rainfall events of different intensities. Trend analysis shows significant increase in the frequency of dry days during the period 1901–2010 while significant decrease in the frequencies of very light and light-to-moderate rainfall events. No significant changes in the frequency of heavy rainfall events are observed. However, significant increase in the annual frequency of very heavy and extremely heavy rainfall events is observed over the country as a whole. Decadal variability shows the decade 1981–1990 has the highest percentage of extremely heavy rainfall event, while the decade 1991–2000 has the highest percentage frequencies for both heavy and very heavy rainfall events among all the decades.

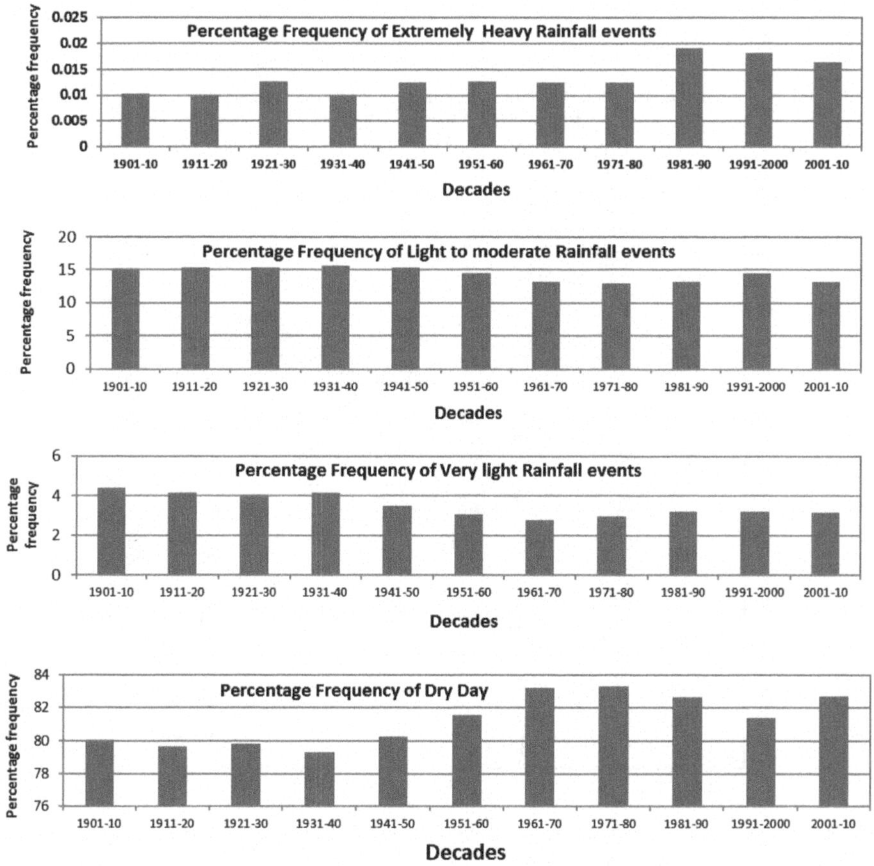

Fig. 6 Decadal variability of annual frequencies of different rainfall events over India

3.2 Trends of Extreme Rainfall Events Based on Gridded Data

Goswami et al. (2006) using a rainfall data set for the period 1951–2000 showed that there are significant rising trends of extreme rainfall trends over the northern parts of India. Using a longer time series of daily gridded data set of 1901–2004 period, Rajeevan et al. (2008) documented an increasing trend of heavy precipitation exceeding 15 cm per day over Central India. But the magnitude of the increasing trend was much smaller than the trend documented by Goswami et al. (2006) using 50 years (1951–2000) of rainfall data. Rajeevan et al. (2008) further documented that there are significant multi-decadal variation in heavy precipitation events over Central India, which may be associated with the variations in sea surface temperature anomalies over the tropical Indian Ocean.

Figure 7 shows the spatial variation in very heavy rainfall (VHR) events (exceeding 15 cm in 24 h) during the monsoon season for the period 1901–2009. The time series of average frequency of VHR events over the Central India for the period 1901–2009 is shown in Fig. 8. From Fig. 8, it can be seen that VHR events

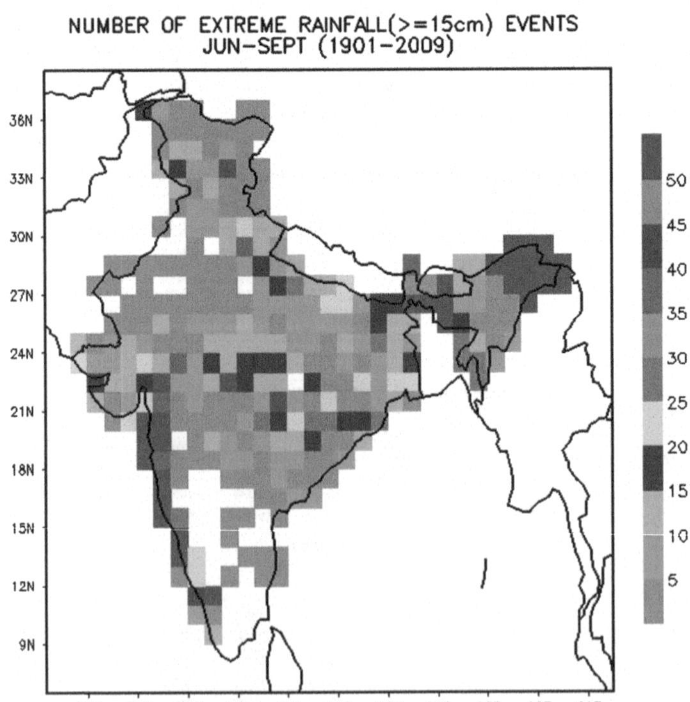

Fig. 7 Spatial variation in frequency of extreme rainfall events (rainfall exceeding 15 cm per 24 h) during the monsoon season (June–September), period: 1901–2009

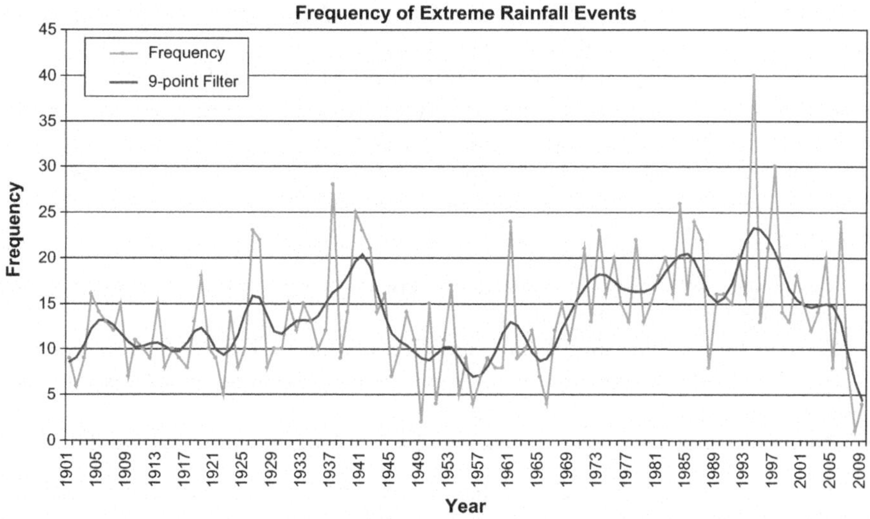

Fig. 8 Time series of extreme rainfall events (Exceeding 15 cm in 24 h) averaged over Central India (monsoon core zone) for the period 1901–2009. The smoothed line with 9 point filter is also shown in *red color* (color figure online)

show a multi-decadal variation. They were more frequent in 1920s and 1930s and then in 1980s and 1990s. However, VHR events were below average in 1940s and 1950s. Over the period 1901–2009, there is an increasing trend of about 0.8 events per decade or 6 % per decade. The trend obtained by Goswami et al. (2006) for the period 1951–2000 is 10 % per decade. The trend for the recent period 1951–2004 is 2.2 events per decade or about 14.5 %. These trends are significant at the 99 % confidence level. These variations could be related to global SST variations (Rajeevan et al. 2008).

Pai et al. (2014) using the $0.25° \times 0.25°$ gridded data examined long-term trends of the extreme rainfall over three regions over the country. The study found that during the recent decades, there has been significant decrease in moderate rainfall events, while heavy and very heavy rains have increased in frequency. These results are consistent with the results of Goswami et al. (2006) and Rajeevan et al. (2008). Therefore, the results on extreme rainfall are independent of the resolution of gridded data used for the analysis.

3.3 *Rainstorms*

Using station rainfall data, Dhar and Nandargi (1995) examined different aspects of rainstorms and the characteristics of major floods over India. The spells of heavy rain due to synoptic weather systems which cover large areas are usually referred to as “rainstorms.” Using objective criteria, Dhar and Nandargi (1995) identified 231 rainstorms during the period 1880–1990. During the monsoon season, July month contributes to maximum rainstorms, followed by August and September.

In this section, we discuss the long-term trends of rainstorms over India. Rainstorms are basically synoptic systems which can potentially cause large-scale floods over the country. Using the gridded daily rainfall data, we have used the following criteria to define a rainstorm during the monsoon season (June–September):

1. A rainstorm with a closed isohyetal pattern around its heavy rainfall center, the grid point with maximum rainfall.
2. The rainstorm center should have received rainfall of 125 mm/day or more.
3. The closed isohyetal pattern should have an area extending over 50,000 km^2 or more with rainfall of 25 mm/day or more.
4. The above three conditions are satisfied for at least two consecutive days.

The criterion 3 above on areal extent is considered to include large rainstorms, which cause large-scale floods. The criteria are based on the study by Karuna Sagar (2016). The analysis has been made using the gridded rainfall data of 1951–2015.

Figure 9 shows the spatial distribution of origin of rainstorms (location of maximum rainfall on day 1 of the storm) over land during the period 1951–2015. It suggests that the majority of rainstorms form over the northern parts of the country along the seasonal monsoon trough (continental tropical convergence zone). These rainstorms include synoptic disturbances like low-pressure systems which form over the northern Bay of Bengal and move along the monsoon trough. Another pocket of rainstorms is found along the foothills of the Himalayas. On some occasions, the monsoon trough gets shifted toward the foothills of the Himalayas, and heavy precipitation may occur along the foothills of the Himalayas.

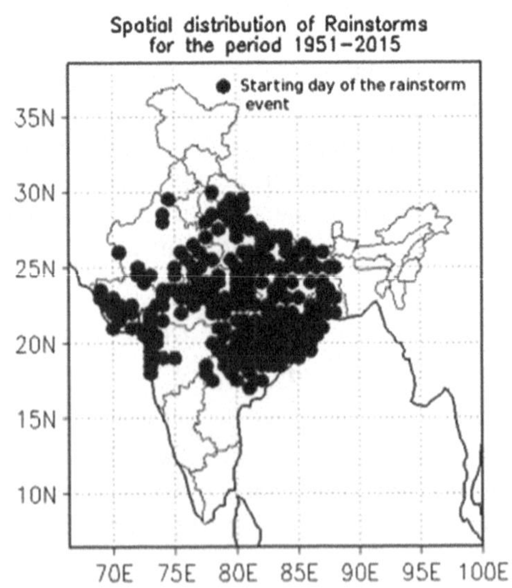

Fig. 9 Spatial variation in origin of rainstorms over India during the southwest monsoon season (June–September), period: 1951–2015

Figure 10 shows the time series (1951–2015) of frequency of rainstorms averaged over the northern parts of the country (18°N–27°N and 75°E–85°E) where the majority of rainstorms generally occur. The time series (1951–2015) of rainstorm days (total number of days with the presence of rainstorms) is also shown in Fig. 10. Both the time series (frequency and rainstorm duration) suggest statistically significant increasing trends. Frequency of rainstorms has shown an increasing trend of 4 rainstorms in 65 years (1951–2015). Similarly, duration of rainstorms has shown an increase of about 15 days during the period 1951–2015, which is also statistically significant. These increasing trends are statistically significant and suggest the possible impact of causing large-scale floods. During the 65-year period (1951–2015), the maximum number of rainstorms (16) occurred in the year 2013. On the other hand, there were 9 rainstorms, but total duration of rainstorms was 36 in 2006. This suggests that these 9 rainstorms that formed in 2006 were long-lived.

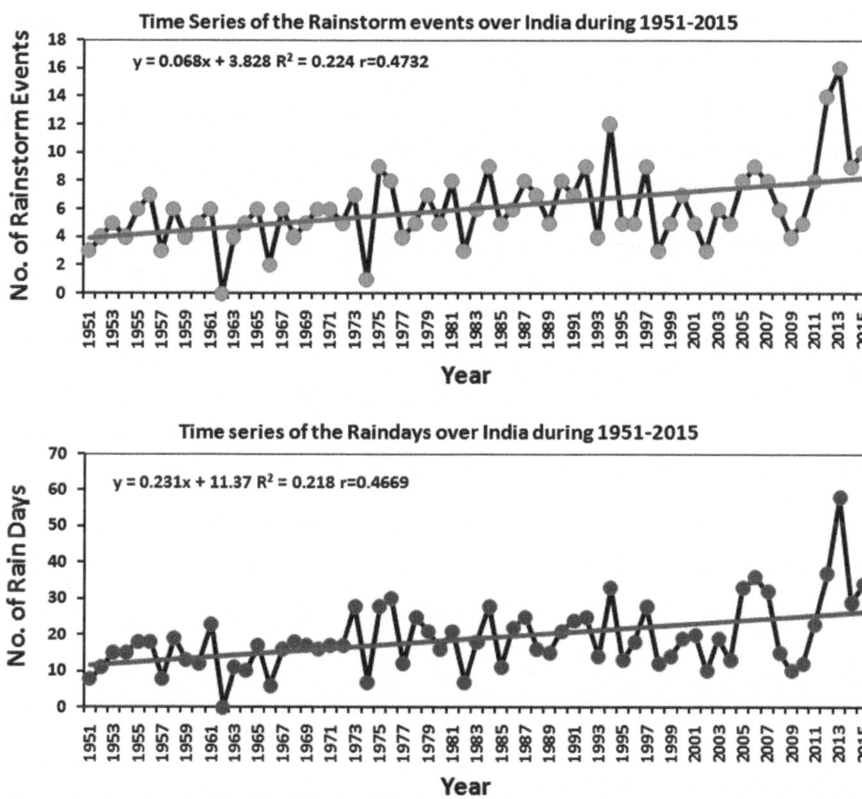

Fig. 10 Time series of frequency of rainstorm events over India during 1951–2015 averaged over North India (*above*) and rainstorm days (*below*). Period: 1951–2015

4 Conclusions

The present study documents the changes in the extreme rainfall events that occurred during the period 1901–2010. The trend analyses revealed increasing trends in the frequency of dry days in most parts of the country during the winter, pre-monsoon, and southwest monsoon seasons. Frequencies of very light rain and light-to-moderate rain events have decreased significantly over most of the states. Both the station and gridded data have shown significant increasing trends of very heavy to extremely heavy rainfall events over most parts of the country. Over Central India, extreme rainfall events show a significant decadal variation which could be related to variations in sea surface temperatures over the tropical oceans. There is also an increase in extreme rainfall events over the region during the period 1901–2009. Over the period 1901–2009, the frequency time series, however, shows an increasing trend of about 0.8 events per decade or 6 % per decade. Further analysis is made on rainstorms over North India where majority of rainstorms cause heavy rainfall. These rainstorms are responsible for causing large-scale floods over the country. The majority of rainstorms form along the monsoon trough region. These rainstorms include synoptic disturbances like low-pressure systems which form over the northern Bay of Bengal and move along the monsoon trough. Frequency of rainstorms has shown an increasing trend of 4 rainstorms in 65 years (1951–2015). Similarly, the duration of rainstorms has shown an increase of about 15 days during the period 1951–2015, which is also significant. Both the increases are statistically significant at the 99 % confidence level. Ajayamohan and Rao (2008) analyzed the role of east equatorial Indian Ocean on heavy rainfall events over Central India. The ongoing warming trend of the Indian Ocean (Chap. 10) coupled with the increase in number of positive IOD events in recent decades could be the primary causes for the increases in rainstorms over the Indian Ocean. Increase in atmospheric moisture content (Chap. 8) in response to surface warming over the region (Chap. 2) also could enhance atmospheric instability and severity of rainstorms.

References

Ajayamohan RS, and Suryachandra A RAO, (2008), Indian Ocean dipole modulates the number of extreme rainfall events over India in a warming environment, J. Meteor. Soc. Japan, 86, 245–252.

Chang HI, Kumar A, Niyogi D, Mohanty UC, Chen F, and Dudhia J, (2009), The role of land surface processes on the mesoscale simulation of the July 26, 2005 heavy rain event over Mumbai, India. Global and Planetary Change, 67, 87-103.

Dash SK, Kulkarni MA, Mohanty UC and Prasad K, (2009), Changes in the characteristics of rain events in India, J. Geophys. Res. 114, DOI:10.1029/2008JD010572.

Dash S. K., Archana A. Nair, Makarand A. Kulkarni and U. C. Mohanty, (2011), Characteristic changes in the long and short spells of different rain intensities in India, Theor. Appl. Climatol., 105:563-570, DOI:10.1007/s00704-011-0416-x.

Dhar ON, and Nandargi S, (1995), On some characteristics of severe rainstorms of India, Theor. Appl. Climatol., 50, 205-212.

Goswami BN, Venugopal V, Sengupta D, Madhusoodanan MS, Xavier PK. (2006). Increasing Trend of Extreme Rain Events over India in a Warming Environment. *Science* **314**, 1442-1445.

Guhathakurta P, Sreejith OP, Menon P. (2011). Impact of climate changes on extreme rainfall events and flood risk in India. *Journal of Earth System Science* **120** (3), 359-373.

Guhathakurta P, Rajeevan M, Sikka D R and Tyagi, A. (2014). Observed changes in southwest monsoon rainfall over India during 1901–2011, *Int. J. Climatol.* DOI:10.1002/joc.4095

Joseph Susmitha, Sahai AK, Sharmila S, Abhilash S, Borah N, Chattopadhyay R, Pillai PA, Rajeevan M, and Arun Kumar, (2014), North Indian heavy rainfall event during June 2013: diagnostics and extended range prediction, Clim.Dyn., 44, 2049-2065.

Karuna Sagar, S., M. Rajeevan, and S.V.Bhaskara Rao, 2016, On increasing monsoon rainstorms over India, Natural Hazards, In Press.

Pai D.S., Sridhar Latha, Rajeevan M, Sreejith O.P., Satbhai N.S., Mukhopadhyay B. (2014) Development of a new high spatial resolution ($0.25° \times 0.25°$) long period (1901–2010) daily gridded rainfall data set over India and its comparison with existing data sets over the region. Mausam 65(1):1–18.

Rajeevan M, Jyoti Bhate and Kale JD, (2006), High resolution daily gridded rainfall data for the Indian region: Analysis of break and active monsoon spells, Curr. Sci., 91, 296-306.

Rajeevan M, Bhate J, Jaswal AK. (2008). Analysis of variability and trends of extreme rainfall events over India using 104 years of gridded daily rainfall data, *Geophy Res Lett* **35**, L18707 doi:10.1029/2008GL035143.

Rakhecha, PR,. Pisharoty PR (1996). Heavy rainfall during monsoon season: Point and spatial distribution. Current Science, 71, 177-186.

Rakhecha PR, Soman MK (1994). Trends in the annual extreme rainfall events of 1 to 3 days duration over India, Theoretical and Applied Climatology, 48, 4.

Revadekar JV and Preethi B, (2012), Statistical analysis of the relationship between summer monsoon precipitation extremes and food-grain yield over India. Int. J. Climatol., 32: 419–429,doi: 10.1002/joc.2282.

Sen Roy S, Balling RC. (2004). Trends in extreme daily precipitation indices in India, Int. J. Climatol. 24, 457–466.

Stephenson DB, Rupa Kumar K, Doblas-Reyes FJ, Royer JF, Chauvin F, Pezzulli S. (1999) Extreme Daily Rainfall Events and their Impact on Ensemble Forecasts of the Indian Monsoon, Monthly Weather Rev., 127, 1954-1966.

Chapter 4
Heat and Cold Waves Over India

D.S. Pai, A.K. Srivastava and Smitha Anil Nair

1 Introduction

During the period 1901–2012, almost the entire globe has warmed with a rise of about 0.84 °C in the annual surface air temperature averaged over the globe (land + ocean) leading to significant changes in the frequency and intensity of extreme weather and climate events, such as heat waves (HWs) and cold waves (CWs), droughts and floods, hurricanes, tornadoes and thunderstorms, over various parts of the world (IPCC 2013). Changes in the extreme weather and climate events pose significant threats on our environment as well as on the life, health and well-being of the human society (Easterling et al. 2000; Meehl and Tebaldi 2004; Balbus and Malina 2009; Coumou and Rahmstorf 2012). The HWs and CWs are one of the less-known forms of extreme weather as they are not visible as other forms of severe weather. However, these abnormal temperature events can impose severe physiological stress on the human body as the body operates best within a fairly normal temperature range. There is a marked relationship between human mortality and thermal stress. During unusually cold/hot episodes, deaths from different causes can rise significantly with the elderly at greater risk than others (WMO 1996). In USA, 3829 of the 8015 heat-related deaths recorded during the period 1979–1999 were attributed to weather conditions (Donoghue et al. 2003). In 2003, Europe witnessed the hottest summer on record since 1540, which had huge adverse impacts (García-Herrera et al. 2010).

In India, manifold increase in the human deaths was observed during various heat waves (HWs) of 1995, 1998 and 1999 (De 2001). Similarly, an intense cold wave over North India during the first to third week of January 2003 caused deaths of more than 900 people of which 813 were from Uttar Pradesh (De et al. 2005). Figure 1 shows the deaths caused by HWs and CWs during the period 1971–2010. In Fig. 1, time series of deaths due to both HWs and CWs does not show any

D.S. Pai (✉) · A.K. Srivastava · S.A. Nair
India Meteorological Department, Pune, India
e-mail: sivapai@hotmail.com

M.N. Rajeevan and S. Nayak (eds.), *Observed Climate Variability and Change Over the Indian Region*, Springer Geology, DOI 10.1007/978-981-10-2531-0_4

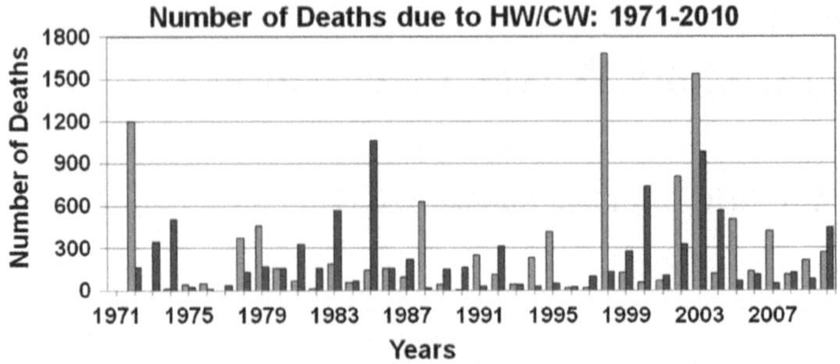

Fig. 1 Number of deaths in India annually due to HW/CWs for the period 1971–2010 obtained from media reports and IMD's annual disaster weather reports. The *red* (*blue*) *bars* corresponds to deaths due to HW/SHWs (CW/SCWs) (color figure online)

noticeable trends. However, number of deaths due to heat wave events was the highest during the recent decade (2001–2010) compared to the previous three decades.

The heat (cold) waves represent rise (fall) of maximum (minimum) temperature at a given place by a certain amount from its climatological value. In India, the HWs are generally observed during March to July, and CWs are observed during November to March with each of these extreme events mostly experienced during the middle 3-month period of the respective seasons; hot weather season of April–June (AMJ) and cold weather season of December to February (DJF), respectively. Due to significant impact of HWs and CWs on the human health and observed changes in their frequency, intensity and persistency over various parts of the globe, there have been many studies on these extreme temperature events over India and their impact on human mortality (Raghavan 1966, 1967; Rai Sircar and Datar 1963; Natarajan 1964; Bedi and Parthasarathy 1967; Bedekar et al. 1974; Subbaramayya and Surya Rao 1976; Chaudhury et al. 2000; De 2001; De et al. 2005; Pai et al. 2004, 2013). However, each of these studies used different threshold temperature values to describe the HWs and CWs.

Here, we describe various characteristics such as frequency, persistency and spatial coverage of HWs and CWs in the country during the seasons of AMJ and DJF, respectively, and the associated decadal variations. In addition, trends in the frequencies of these extreme temperature events and changes associated with the two ENSO phases (El Nino and La Nina) have been also highlighted.

2 Definition of HW and CW

There is no universal definition for HW or CW. Different definitions of HW and CW are used in different countries. However, a HW (CW) over a region represents an interval of hotter (colder) than normal weather over the region. The criteria used

Table 1 Criteria used in this study for defining heat and cold waves

Criteria for declaring heat wave based on maximum temperature (T_{max})
Heat wave over a station is declared only when the actual T_{max} of the station is ≥ 40 °C for plains and ≥ 30 °C for Hilly regions. However, when the T_{max} is ≥ 40 °C for coastal stations and ≥ 45 °C for other stations, conditions are declared as heat wave
The following criteria are used for defining severity of the heat wave When normal $T_{max} \leq 40$ °C and i. if (actual T_{max} − normal T_{max}) is 5–6 °C: heat wave ii. if (actual T_{max} − normal T_{max}) is ≥ 7 °C: severe heat wave When normal $T_{max} > 40$ °C and i. if (actual T_{max} − normal T_{max}) is 4–5 °C: heat wave ii. if (actual T_{max} − normal T_{max}) is ≥ 6 °C: severe heat wave
Criteria for declaring cold wave based on minimum temperature (T_{min})
When the T_{min} is ≤ 10 °C for coastal stations, conditions are declared as cold wave
The following criteria are used for defining severity of the cold wave over all the stations When normal $T_{min} \geq 10$ °C and i. if (actual T_{min} − normal T_{min}) is −5 to −6 °C: cold wave ii. if (actual T_{min} − normal T_{min}) is ≥ -7 °C: severe cold wave When normal $T_{min} < 10$ °C and i. if (actual T_{min} − normal T_{min}) is −4 to −5 °C: cold wave ii. if (actual T_{min} − normal T_{min}) is ≥ -6 °C: severe cold wave

for defining the HW and CW are given in the Table 1, which signify increase (decrease) in the daily maximum (minimum) temperature at a station by a certain threshold compared to a climatological value. In this study, we have used the base period of 1971–2000 for computing the climatological values. HW (CW) of relatively higher intensity is classified as severe HW (severe CW) or SHW (SCW).

3 Data and Methodology

The HW/SHW information of 103 stations uniformly distributed over the country for the AMJ season was derived from the daily maximum temperature data for the period 1961–2010. Data for at least 90 % of days were available in 95 out of 103 stations. For deriving CW/SCW information, daily minimum temperature data of 86 stations uniformly distributed over the country for DJF season for the period 1971–2010 were used. Data for at least 90 % of days were available in 90 out of 86 stations. Both the above data sets were obtained from the archive of the Climate Services Division, IMD, Pune.

Climate Prediction Center (CPC), USA, uses the Oceanic Nino Index (ONI) [3-month running mean of ERSST.v3b Sea surface temperature (SST) anomalies in the Nino 3.4 region (5°N–5°S, 120°W–170°W)] a threshold of ±0.5 °C for defining El Nino/La Nina events. The events are declared when the threshold is met for a minimum of 5 consecutive overlapping 3-month seasons. In this study, for HW (CW) case, a reference year was classified as an El Nino/La Nina year, when any of the three 3-month seasons of March to May, April to June and May to July (November of previous year to January of reference year, December of previous year to February of reference year and January to March of reference year) is part of the 5 consecutive overlapping seasons. As per this definition, for HWs during the period 1961–2010, 14 years (1963, 1965, 1966, 1969, 1972, 1982, 1983, 1987, 1991, 1992, 1997, 1998, 2002 and 2010) satisfied the El Nino criteria and 12 years (1964, 1971, 1973, 1974, 1975, 1976, 1985, 1988, 1989, 1999, 2000 and 2008) satisfied the La Nina criteria. Similarly for the CWs case, during 1971–2010, the 13 years (1973, 1977, 1978, 1983, 1987, 1988, 1992, 1995, 1998, 2003, 2005, 2007 and 2010) satisfied El Nino criteria and 15 years (1971, 1972, 1974, 1975, 1976, 1984, 1985, 1989, 1996, 1999, 2000, 2001, 2006, 2008 and 2009) satisfied the La Nina criteria. Differences in the spatial distribution of HWs/CWs associated with the El Nino and La Nina events were examined by preparing composite spatial maps of frequency of HWs/CWs for the respective El Nino/La Nina years.

The nonparametric Mann–Kendall test (Gilbert 1987) was employed to test the monotonic increasing or decreasing trends in the frequency and spatial coverage of the HWs/CWs. The Mann–Kendall test assumes the trends to be monotonic and thus no seasonal or other cycle is present in the data.

4 Heat Wave (HW)/Severe Heat Wave (SHW)

4.1 Climatology and Decadal Variation

Frequency of the HW/SHW Days

Figure 2a, b, respectively, depicts the spatial variation of mean number (frequency) of HW and SHW days over the country expressed as days per season for the AMJ season. Most areas of the country except north-east India and large parts of Peninsula have experienced on an average ≥ 2 HW days (Fig. 2a). This region hereafter called the core HW zone (CHZ) covers states of Punjab, Himachal Pradesh, Uttarakhand, Delhi, Haryana, Rajasthan, Uttar Pradesh, Gujarat, Madhya Pradesh, Chhattisgarh, Bihar, Jharkhand, West Bengal, Orissa and Telengana and met subdivisions of Marathwada, Vidharbha, Madhya Maharashtra and coastal Andhra Pradesh. Many areas of west Rajasthan, Punjab, Haryana, Vidharbha, coastal Andhra Pradesh, northern parts of East Rajasthan, Madhya Pradesh, Orissa, Telengana and southern parts of Uttar Pradesh and Chhattisgarh have experienced on an average ≥ 8 HW days. As shown in Fig. 2b, mean SHW days of 1–3 days were mainly experienced over isolated areas of north and eastern parts of the CHZ.

The decade wise spatial distribution of mean HW days for the AMJ season shows (Fig. 3a–e) that in all the 5 decades, on an average, ≥ 8 HW days/season were experienced by many areas from eastern parts of Peninsula, north and north-west India, and some areas of Central India. The overall decadal mean

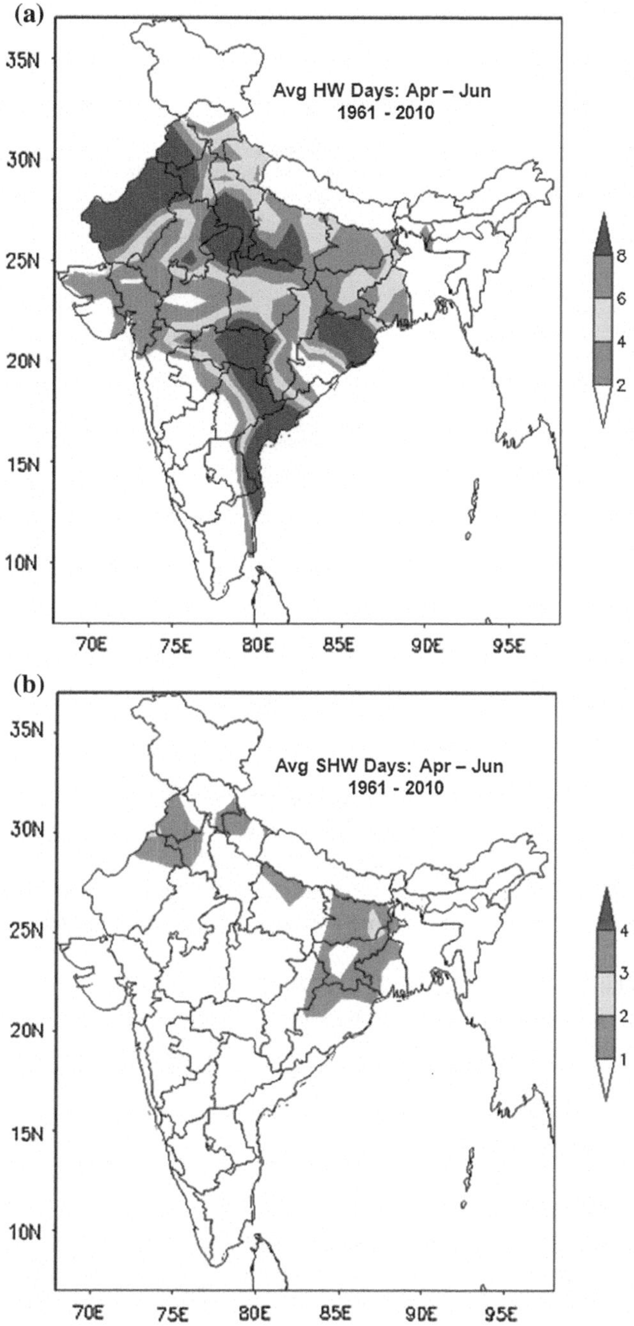

Fig. 2 **a** Average number of HW days over India during the hot weather season (AMJ) (1961–2010). **b** Same as **a** but for the SHW days

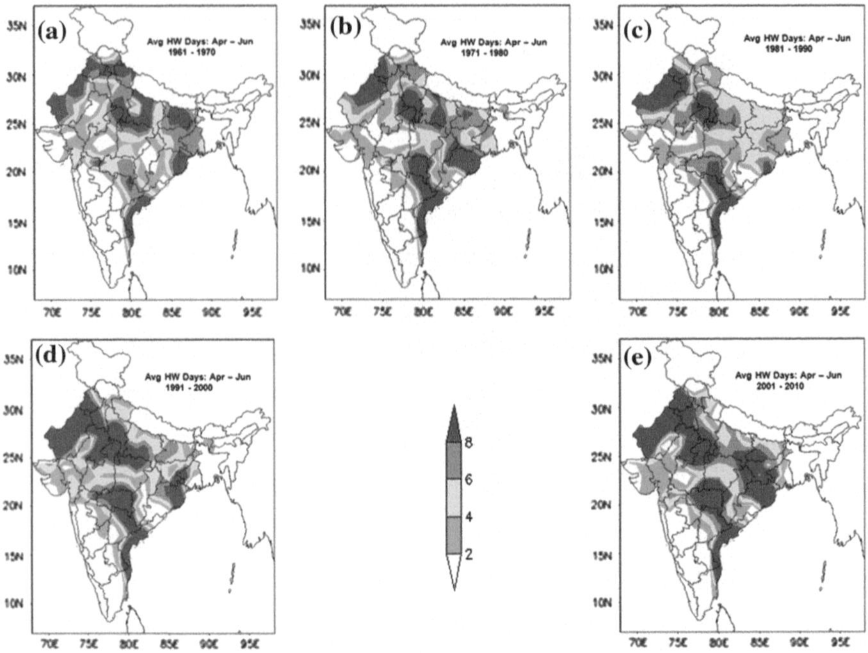

Fig. 3 Mean number of HW days over India during the hot weather season (AMJ) for the 5 decades viz. **a** 1961–70. **b** 1971–80. **c** 1981–90. **d** 1991–2000 and **e** 2001–2010

patterns of HW days during the last two decades were similar to that of middle decade (1981–90), but with areas of ≥ 8 HW days spread southwards over North India and northwards over the Peninsular. Among the 5 decades, the areas of ≥ 8 HW days over the north and north-west India during the recent two decades (1991–2000 and 2001–10) were slightly more and relatively south than that during the first two decades (1961–70 and 1971–80).

The decadal variation of the SHW days (figure not presented) showed increase in the frequency and areal coverage in the average SHW over north and eastern parts of CHZ during the recent two decades compared to previous three decades.

Persistency of HW/SHW Conditions

Duration of HW/SHW event is as important as its intensity as far as thermal stress on the human body. Therefore, it is interesting to know how many days these conditions are likely to persist or prevail once the conditions are set in. It was found that during all the decades and for most of the stations, most frequent HW/SHW spells were of about 1–2 days duration. However, few HW/SHW spells persisted for very long duration ($\geq 10/\geq 5$ days) in some stations. Fig. 4a, b depicts length of duration of the longest HW (SHW) spell in each station with duration ≥ 10

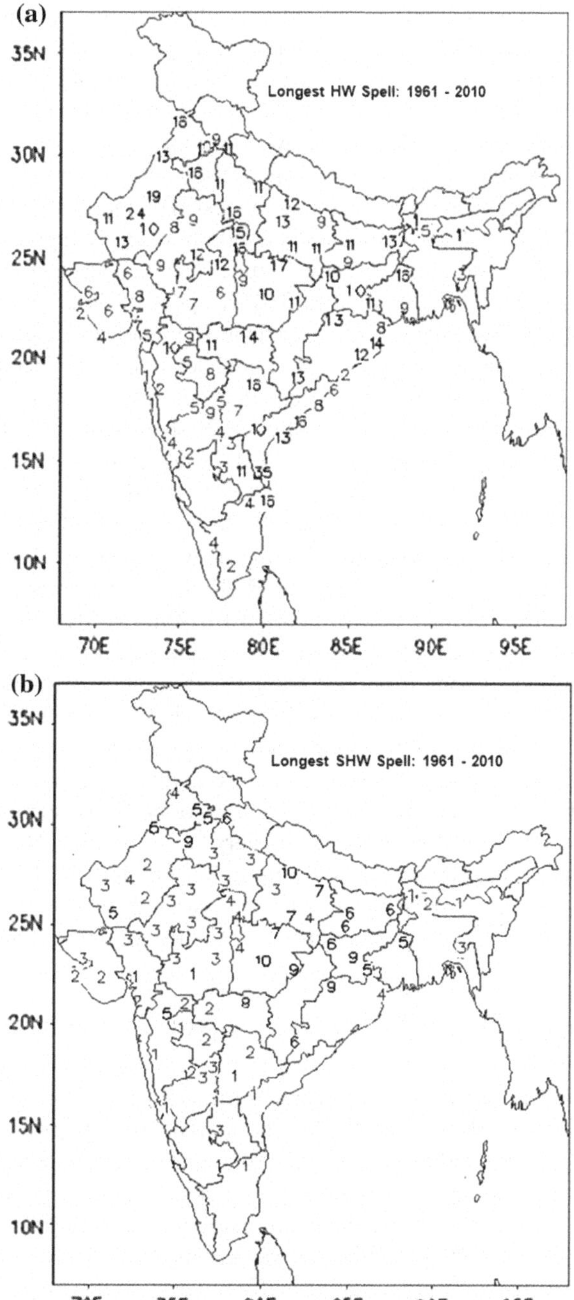

Fig. 4 **a** Station-wise duration of the longest HW spells during 1961–2010. The HW spells of ≥ 10-day duration are shown using *red colour*. **b** Same as **a**, but for the SHW spells. The duration of SHW spells ≥ 5 days are shown using *red colour* (color figure online)

(≥ 5) days shown in red. Most of the stations north of 23°N and east of 78°E have experienced longest HW spells of duration ≥ 10 days (Fig. 4a) **with** 8 stations of central and the north-west India having longest HW spells of duration ≥ 15 days. Phalodi (West Rajasthan) among these 8 stations experienced the longest HW spell of duration of 24 days (17 May–9 June, 1994). Three stations from coastal regions of Andhra Pradesh and Tamil Nadu and 1 station each from West Bengal and Telengana have experienced longest HW spells of duration ≥ 15 days with Nellore (NLR) station from coastal Andhra Pradesh reporting the longest HW spell of 35 days duration (6 May–9 June, 1996). During the data period, all the stations together have experienced 38 HW spells of duration of ≥ 15 days. However, 20 of these HW spells were experienced in the last 2 decades compared to 18 spells experienced during the first 3 decades indicating increase in the persistency of HW conditions during the recent decades. Further, during only 23 years of the data period, duration of the longest HW spell was ≥ 15 days. Nellore alone has experienced 18 HW spells of ≥ 15 days duration. The longest HW spell during 4 of the five calendar decades, 1961–70 (30 days), 1981–90 (22 days), 1991–2000 (35 days) and 2001–2010 (22 days), was also experienced by Nellore. However, Bikaner had experienced longest HW spell of 19 days in the decade 1971–80. Thus, among all the stations, Nellore is most prone to longer HW spells.

The stations which reported the longest SHW spell of ≥ 5 days were mostly belonged to north-west, east India and neighbouring Central India (Fig. 4b). Most of the stations from east Uttar Pradesh and east Madhya Pradesh reported longest SHW spells of ≥ 7 days. Bhariach (BRC) in 2005 (11th–20th June) and Jabalpur (JBP) in 2009 (19th June–28th June) experienced the longest SHW spell of 10-day duration each. However, it was noticed that no stations from coastal Andhra Pradesh (including Nellore) had reported SHW spell of ≥ 7 days. This shows that whereas Nellore is prone to very long duration HWs, it rarely experiences long duration SHWs. During the entire data period, 10 SHW spells of duration ≥ 7 days of were reported by various stations with 6 of these SHW spells (all of duration ≥ 9 days) being experienced in the recent decade 2001–2010.

Spatial Coverage of the HW/SHW

As the hot season progresses, the areal coverage of HW/SHW shows noticeable increase (Raghavan 1966). The spatial coverage of a HW/SHW varies from event to event as well as year to year. Whenever a station reports HW/SHW conditions, area surrounding the station also experiences similar conditions. Therefore, number of stations affected by the event can be used as the proxy for the spatial coverage of the event. Here, spatial coverage of the HW (SHW) per season was computed as the number of stations affected by at least one spell of the HW (SHW) during that season. During the data period, the spatial coverage of HW varied from a few stations to many stations covering large part/whole subdivision to a group of subdivisions. During the total period (1961–2010), on an average around 55 (14) out of 103 stations used in the study were affected by HW (SHW). Highest (71 stations), and lowest (27 stations) spatial coverage of the HWs was in 1972 (1990). Highest (41 stations) and lowest (0) spatial coverage of SHWs was in 2010 (1990). However, no trends were observed in both the time series.

4.2 *HW/SHW Days and ENSO*

Composite seasonal spatial maps of mean HW days over India for the 14 El Nino years as well as that of 12 La Nina years for the April–May–June season during the data period were prepared. Noticeably, higher frequency and spatial coverage of HW days were observed over CHZ in the El Nino case compared to the La Nina case (see Fig. 5). In case of El Nino, the areas of ≥ 8 HW days in the composite map extend northward along the Himalayan plains and eastward over Jharkhand and Bihar compared to that in the climatological map (Fig. 2a). However, in case of La Nina, the areas of ≥ 8 HW days in the composite map showed shrink over central and north-west India compared to that in the climatological map.

The composite maps of SHW days (figure not shown) showed that the composite areal coverage and frequency of SHW days during the El Nino (La Nina) case were more (less) than the climatology. In terms of persistency, it was observed that most of the longest HW/SHW spells were pertaining to the El Nino years.

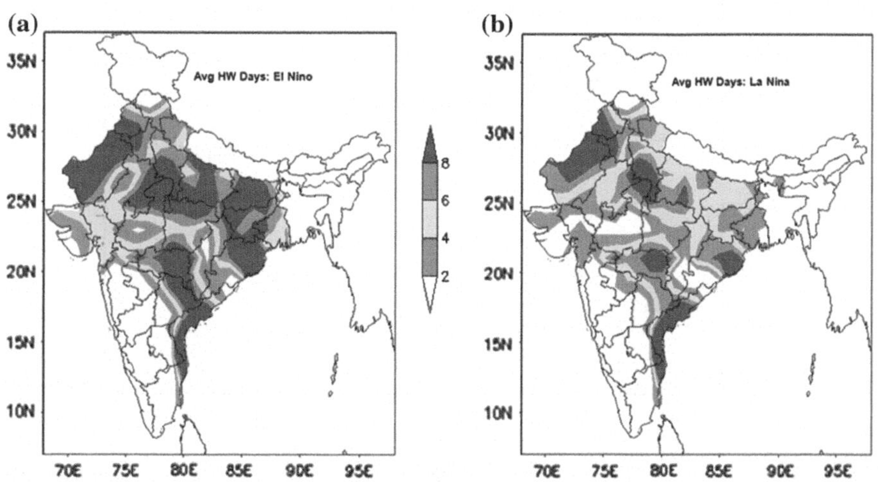

Fig. 5 Composite of average number of HW days during the AMJ season for the **a** 14 El Nino years and **b** 12 La Nina years

4.3 Trends in the Seasonal HW Days

Figure 6 shows the increasing (decreasing) trend in the station HW days using red rising arrow (blue falling arrow) with trends significant at 5 % or above level shown using filled arrows. About 20 stations from north India, north-west India, Central India and east coast showed significant increasing trends. These are also the stations that experience highest HWs during the season (Fig. 2a). However, 3 stations from the east coast (Kolkata Gopalpur, and Balasore) and 2 stations from North India (Ambala and Bhariach) showed significant decreasing trends. In case of SHW (Figure not shown), significant increasing trends were observed in 5 stations from north-west India (Amritsar, Hissar, Ganganagar, Phalodi and Jaipur). However, no stations showed significant decreasing trend. It was observed that even though the

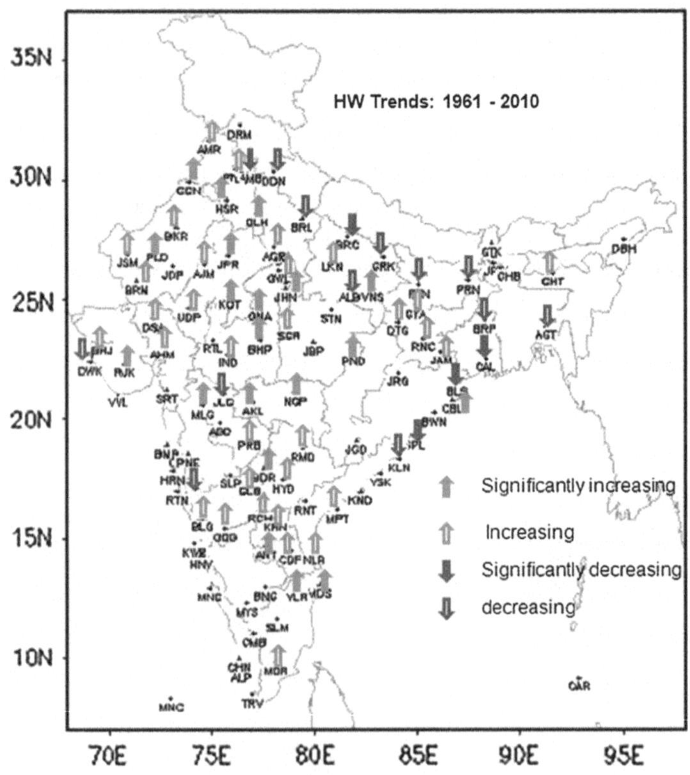

Fig. 6 Trends in the HW days of 103 stations during the AMJ season for the period 1961–2010. Nonparametric Mann–Kendall test was used to test the significance of the trends. *Red* rising (*blue* falling) *arrows* represent the increasing (decreasing) trends. *Filled arrows* represent the trends significant at 5 % level (color figure online)

day to day temperatures in the stations over the Peninsular India were not increasing above the HW/SHW thresholds, a general increasing trend was observed in the season averaged maximum temperatures over these stations.

4.4 Trends in the Total HW/SHW Days Over Core HW Zone (CHZ)

There are 71 stations in the CHZ. The time series of the total number of HW and SHW days over CHZ (Fig. 7a, b) for the period 1961–2010 shows slight but insignificant increasing trends. In Fig. 7, red (green) markers indicate El Nino (La Nina) years. The seasonal mean HW (SHW) days in the CHZ for the period 1961–2010 is 456 (39) days. In 9 (8) out of the 14 El Nino years, the HW (SHW) days in CHZ were more than its climatological value. Similarly, in 3 (1) out of the 12 La Nina years, the HW (SHW) days over CHZ were more than its climatological value. Decade wise, the HW (SHW) days over CHZ were less than normal during the first (middle) 3 decades with lowest decadal frequency of 393 (25) days in the central decade (1981–90). Among all the five decades, the highest decadal frequency of 536 (64) days was observed during the latest decade (2001–10).

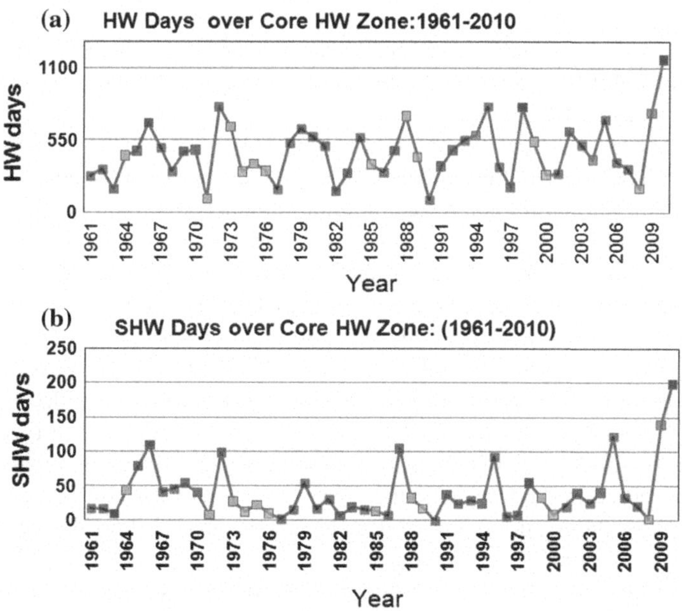

Fig. 7 Interannual variation **a** HW days and **b** SHW days in the core HW zone during the AMJ season for the period 1961–2010. The El Nino and La Nina years are indicated using *red* (*green*) *markers* (color figure online)

5 CW/SCW

5.1 Climatology and Decadal Variation

Frequency of the CW/SCW Days

Figure 8a, b shows the spatial variation in mean number (frequency) of CW and SCW days, respectively, during the cold weather season (DJF) over the country. It is seen (Fig. 8a) that on an average ≥ 2 CW days per season were experienced by most of the areas except southern Peninsula and north-east India. This region hereafter called the core CW zone (CCZ) and nearly same as CHZ expect that CCZ includes Jammu and Kashmir and excludes coastal Andhra Pradesh. Many areas over north and north-west of CCZ experienced ≥ 6 CW days and 1–3 SCW days (Fig. 8b).

Month wise distribution of CW days showed (figures not given) that during December, CW days of ≥ 2 were experienced by most areas of Rajasthan, some northern parts of Gujarat and Madhya Pradesh and western parts of Jammu and Kashmir. During January, CW days of ≥ 2 were experienced by many areas of Punjab, Rajasthan, Madhya Pradesh, western part of J & K, southern parts of UP, northern parts of Maharashtra and some areas of Bihar, Jharkhand, Chhattisgarh and north Telengana. During February, the case was nearly similar to December except for reduced frequency of CW days over west J & K. During in all the 3 months, most of the Rajasthan experienced CW days of ≥ 2 days with some areas of ≥ 4 days. Month wise, SCW days were experienced mainly during January and February (about 1–2 days in an average and mostly over north and north-west areas of CCZ).

The decade wise spatial distribution of season average of CW days (Fig. 9a–d) shows overall decrease in the average CW days over the country going from 1971–1980 to 1991–2000 with systematic and noticeable decrease over north, north-west and north-east India and slight increase over some southern parts of Central India. However, during recent decade (2001–2010), there was slight increase in the spatial coverage and frequency of CW days compared to the previous decade (1991–2000). During 1971–1980, many areas of north and north-west India reported ≥ 6 CW days. However, during the subsequent decade (1981–90), the areas of ≥ 6 CW days were mainly restricted to north-west India. During subsequent decades, 1991–2000 and 2001–2010, the areas of CW days of ≥ 6 days showed further decrease over north-west India and an increase over southern parts of Central India. Decrease was also observed in the CW frequency over eastern parts of north-east India from about 2–6 days during the first two decades to less than 2 days during the recent two days.

The decadal variation of the average SCW days/season showed (figures not presented) relatively higher seasonal frequency of SCW days than its long-term

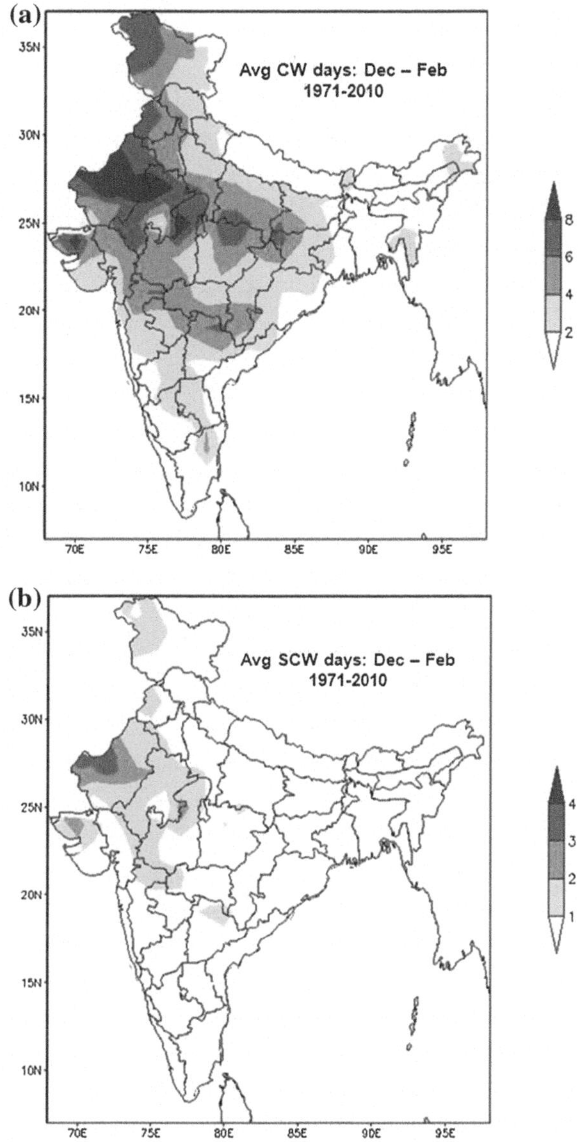

Fig. 8 **a** Average number of CW days over India during the DJF season computed using the CW information for the period of 1961–2010. **b** Same as **a**, but for SCW days

climatology during the first decade (1971–1980) over north and north-west area of CCZ. However, during the subsequent decades, an overall decrease in the average number of SCW days was observed over the region with lowest during 1991–2000 as in the case of CW days.

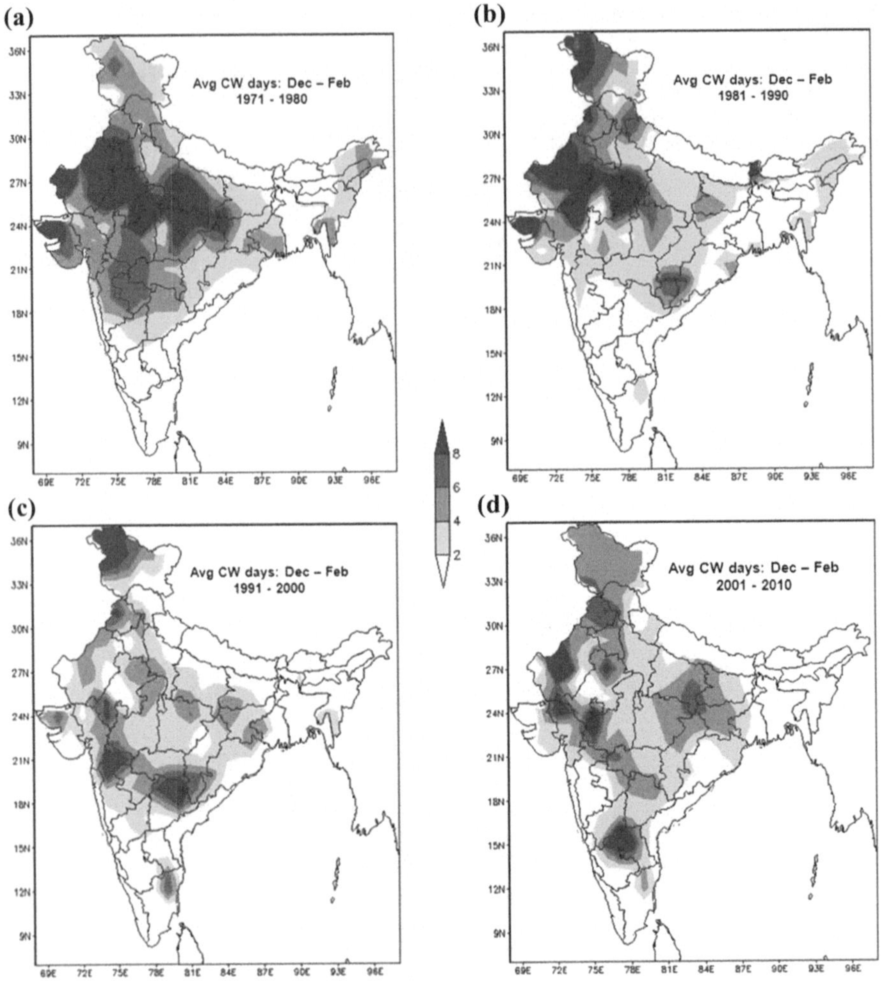

Fig. 9 Average number of CW days over India during the cold weather season (DJF) for the 4 decades viz. **a** 1971–80. **b** 1981–90. **c** 1991–2000 and **d** 2001–2010

Persistency of CW/SCW Events

For most of the stations, the most frequent CW/SCW spells were of about 1–2 days duration with some individual CW/SCW spells persisted over some stations for a very long period ($\geq 10/\geq 5$ days). Figure 10a, b shows the station-wise duration of the longest CW (SCW) spell in the country. The durations of longest CW (SCW) spells of ≥ 10 (≥ 5) days are shown in red. Stations that experienced longest CW (SCW) spells of duration ≥ 10 (≥ 5) days were mainly from north, north-west and central parts of CCZ. Gangtok from Sikkim, however, has experienced the longest CW spell of 17 days (10–26 December 1986).

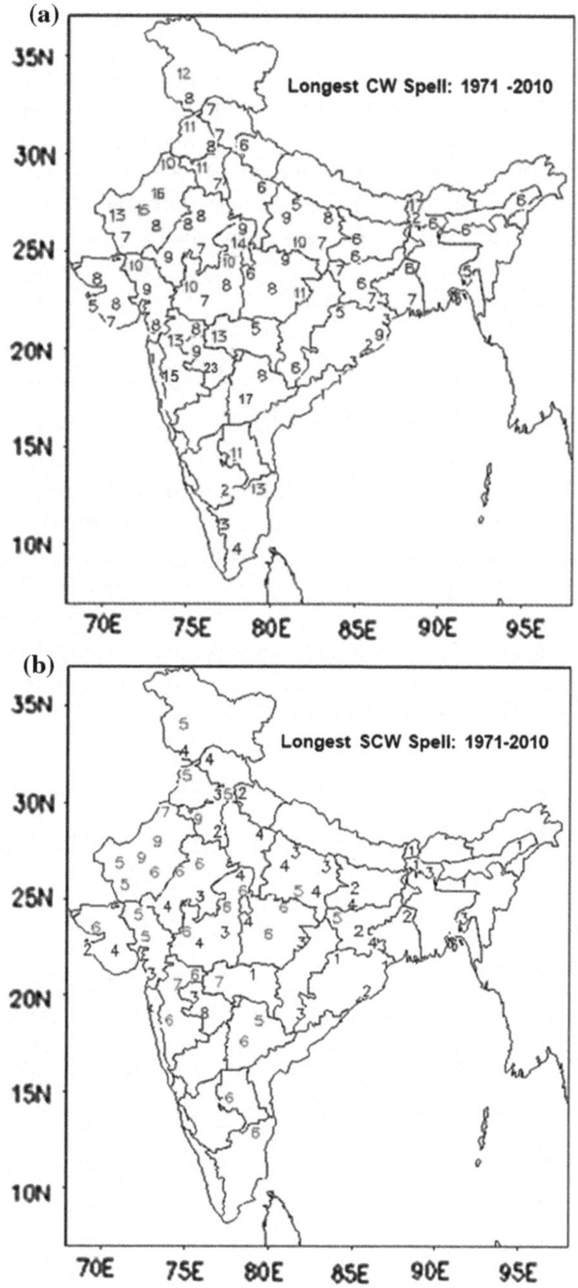

Fig. 10 **a** Duration of the longest HW spells during 1961–2010. The HW spells of ≥ 10-day duration are shown using *red colour*. **b** Same as **a**, but for the SCW spells. The duration of SCW spells ≥ 5 days are shown using *red colour* (color figure online)

Spatial Coverage of the CW/SCW

Number of stations affected by at least one spell of the CW (SCW) during a season was computed for the all the 40 years (1971–2010). During the period, on an average around 48 (17) out of 86 stations used in the study were affected by CW (SCW). The spatial coverage of CW was the highest (67 stations) in 1971 and 1972 and was the lowest (8 stations) in 2009. Highest and lowest spatial coverage of SCW (47 and 2 stations, respectively) was in 1984 and 2009, respectively. Decreasing trends significant at the 95 % confidence level was observed in spatial coverage of CW and SCW.

5.2 *ENSO and CW Days*

Composite seasonal spatial maps of mean CW days over India for the 13 El Nino years as well as that of 15 La Nina years for the DJF season during the data period are shown in Fig. 11a, b. In the El Nino case, the composite map showed relatively reduced frequency of CW days over most areas of CCZ. However, in the case of La Nina, significantly more number of CW days were seen in most parts of CCZ. In the case of SCW also (Figures not presented), significant increase (decrease) in the frequency of the SCW days were observed in the northern and north-western parts of CCZ during La Nina (El Nino) years.

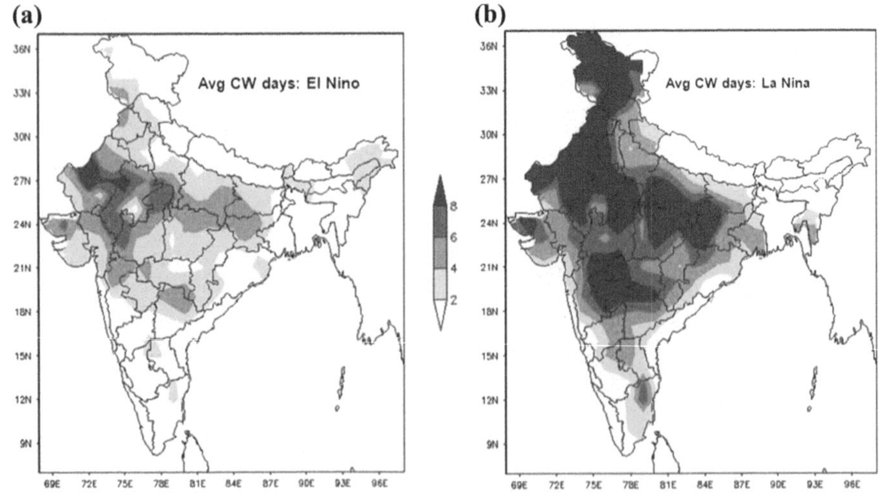

Fig. 11 Maps showing composite of average number of CW days during the AMJ season for the **a** 13 El Nino years and **b** 15 La Nina years

5.3 Trends in the Station-Wise Seasonal CW/SCW Days

In Fig. 12, red rising arrow (blue falling arrow) are used to represent stations with increasing (decreasing) trend in the CW days with filled arrows showing trends significant at the 95 % confidence level. Decreasing trends were observed in most of the stations (Fig. 12) north of about 18°N with significant trends observed in 23 stations (mostly from north and north-west parts of CCZ). On the other hand, two stations from south Peninsula (Ananthapur and Vellore) showed significant increasing trends. In case of SCW (figure not shown), decreasing trends were observed in many stations of CCZ with significant trends observed in 9 stations from north-western part and 4 stations from central part of CCZ. One station from south Peninsula (Ananthapur) showed significant increasing trend. The decreasing trends in the CW/SCW days was also corroborated by the increasing trends (figure not showed) in the season averaged (DJF) minimum temperature anomalies over

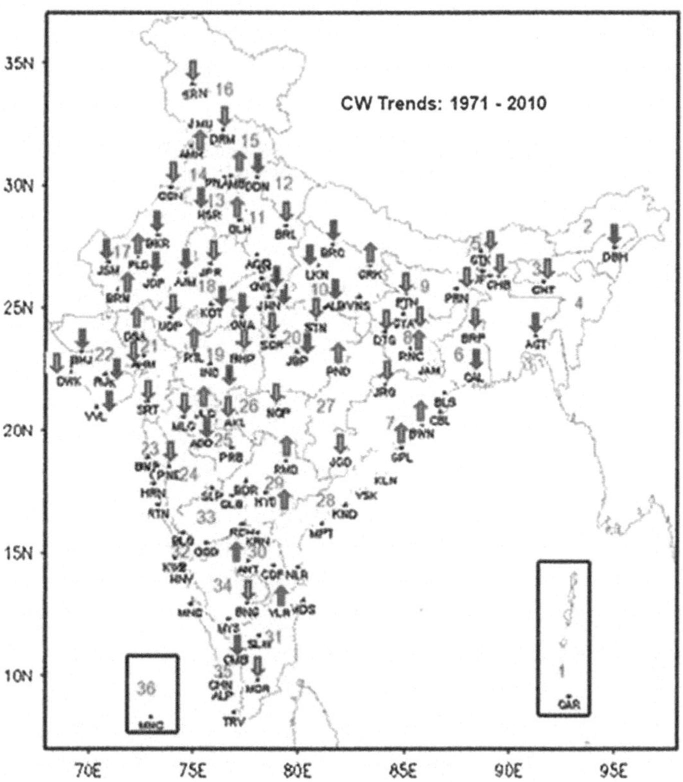

Fig. 12 Trends in the CW days of 86 stations during the DJF season for the period 1971–2010. Nonparametric Mann–Kendall test was used to test the significance of the trends. *Red* rising (*blue* falling) *arrows* represent the increasing (decreasing) trends. *Filled arrows* represent the trends significant at 5 % level (color figure online)

most of the stations from the country (Chap. 2) for the period 1971–2010. Of the total stations, 38 stations showed significant increasing trends in the seasonal minimum temperature anomaly.

5.4 *Trends in the Total CW/SCW Days Over CCZ*

There are 63 stations in the CCZ. Decreasing trends at significant level of 95 % can be seen in both the time series of the total number of CW and SCW days over CCZ during the DJF season for the period 1971–2010 (Fig. 13a, b). Red (green) markers indicate El Nino (La Nina) years. The mean CW (SCW) days over the CCZ for the period 1971–2010 was 300 (52) days. In 5(4) out of the 13 El Nino years, and 8(6) of the 15 La Nina years, the CW (SCW) days over CCZ were above its climatological value. Decade wise, the CW (SCW) days over CCZ during the first decade was more than the normal value with higher decadal frequency of 464 (91) days during the first decade (1971–80) and that during the last 3 decades were below the normal value with lower decadal frequency of 213 (28) days during the third decade (1991–2000).

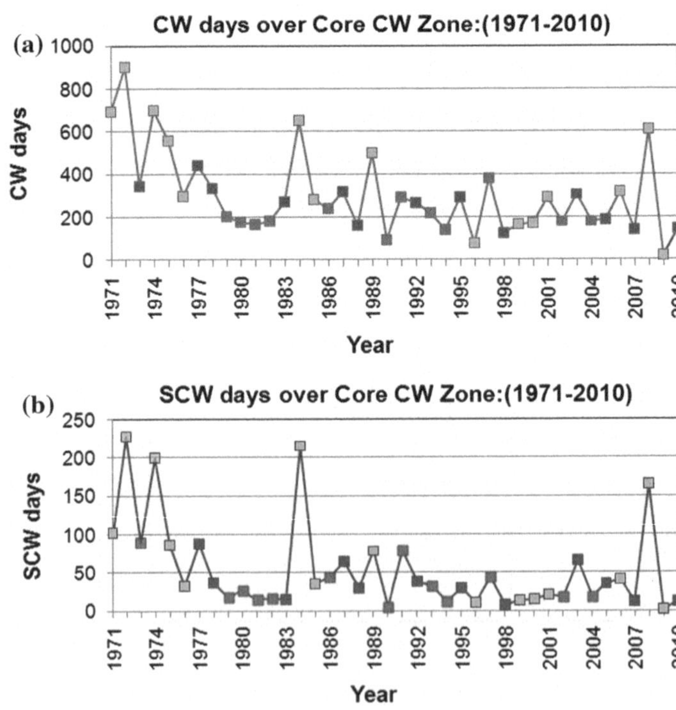

Fig. 13 Interannual variation **a** CW days and **b** SCW days in the core CW zone during the DJF season for the period of 1971–2010. The El Nino and La Nina years are indicated using *red* (*green*) *markers* (color figure online)

6 Conclusions

From the examination of various features of the HW (SHW) from 103 stations during the last five decades (1961–2010) and that of CW (SCW) from 86 stations during the last four decades (1971–2010), the following conclusions can be drawn.

1. During the hot weather season (AMJ), stations from the north, north-west, central, east India and north-east Peninsula (together called CHZ) are most prone for HW/SHW days with relatively highest frequency experienced during May. During the cold weather season (DJF), stations from CCZ that is nearly same as CHZ but includes Jammu and Kashmir and excludes coastal Andhra Pradesh are most prone for CW/SCW days with the highest frequency during January.
2. Appreciable increase (decrease) in the frequency and spatial coverage of HW/SHW days was observed over CHZ during the El Nino (La Nina) years. In case of CWs, the case was exactly opposite with decrease (increase) in their frequency and spatial coverage observed over the CCZ during the El Nino (La Nina) years.
3. Appreciable decadal variation was observed in the frequency, spatial coverage and area of maximum frequency both in the HW/SHW days and CW/SCW days. Decade wise, the all India frequency of HW days during the first 3 decades was less than normal with the lowest decadal frequency of 413 days during the middle decade (1981–90) and that during the last 2 decades were above normal value with the highest decadal frequency of 575 days during the latest decade (2001–10). The SHW days also showed similar decadal variation with the lowest (highest) frequency of 25 (67) during 1981–90 (2001–2010). However, above normal SHW days were also observed in the first decade (1961–70). In case of CWs, an overall decrease in the average CW/SCW days was observed from 1971–1980 to 1991–2000. But in the recent decade (2001–2010), there was slight increase in the area of coverage and frequency of CW days compared to the previous decade (1991–2000).
4. Significant increasing (decreasing) trends in the HW (CW) days were observed in many stations over CHZ (CCZ). However, insignificant increasing trend was observed in the total HW days of CHZ, and significant decreasing trend was observed in the total CW days of CCZ.
5. According to the latest IPCC report (IPCC 2013), the global averaged temperature during each of the last three decades has been warmer than any preceding decade since 1850 (Chap. 2). The observed increasing (decreasing) trend in the frequency of HW/SHW (CW/SCW) activity in recent decades, therefore, might be regional impact of the observed decadal scale global warming. Other possible reasons behind these changes are local factors such as deforestation and urbanization, which might have changed the exposure conditions of many of the stations used in this study. It may also be mentioned that the observed changes in the various characteristics of the HWs and CWs over India were in consistent with similar studies from various other parts of the world.

The latest projections from climate models (IPCC 2013) also suggest that increase (decrease) in the frequency, intensity and persistency of the HW (CW) are likely to continue due to the projected increase in the mean global temperatures. Therefore there is a need for proper operational service systems in the country for the advance warning of the heat and CWs, public information campaigns on dangers of these extreme temperature events, and social care networks to reach vulnerable sections of the populations.

References

Balbus J.M. and Malina C., 2009, Identifying vulnerable subpopulations for climate change health effects in the United States, J. Occup. Environ. Med., pp 51:33–37.

Bedekar, V. C., Dekate, M. V. and Banerajee, A. K., 1974, "Heat and cold waves in India", Forecasting Manual- Part IV-6, India Meteorological Department

Bedi, H.S. and Parthasarathy, B. 1967: 'Cold waves over northwest India and neighbourhood'. IJMG Vol. XV1I1 pp. 371-378.

Chaudhury, S.K., Gore, J.M. and Sinha Ray, K. C., 2000: "Impact of heat waves over India", Current Science, 79(2), 153–155.

Coumou, D. and Rahmstorf, S. 2012: A decade of weather extremes. Nature Climate, Change, DOI: 10.1038/NCLIMATE1452

De, U. S., 2001, "Climate change impact: Regional scenario", Mausam, 52, 201–212.

De, U.S., Dube, R.K. and Prakasa Rao, G.S.,2005, "Extreme Weather Events over India in the last 100 years", J. Ind. Geophys. Union, 9,3,173-187.

Donoghue, E.R., Nelson, N., Rudis, G., Sabogal, R.I., Watson, J.T., Huhn, G., and Luber, G., 2003. "Heat-related deaths—Chicago, Illinois, 1996–2001, and the United States, 1979–1999", Mortality and Morbidity Weekly Report, 52, 610-613.

Easterling, D. R., Meehl, G. A., Parmesan, C., Changnon, S. A., Karl, T. R., and Mearns, L. O., 2000, "Climatological extremes: observations, modeling and impacts", Science, 28, 2068-2074.

García-Herrera, R., Díaz, J., Trigo, R. M., Luterbacher, J. and Fischer, E. M.,2010 "A Review of the European Summer Heat Wave of 2003", Critical Reviews in Environmental Science and Technology, 40: 4, 267- 306.

Gilbert, R.O., 1987, Statistical methods for environmental pollution monitoring. Van Nostrand Reinhold, New York.

IPCC, 2013, Summary for Policymakers. In: Climate Change 2013: The Physical Science Basis. Contribution of Working Group I to the Fifth Assessment Report of the Intergovernmental Panel on Climate Change [Stocker, T.F., D. Qin, G.-K. Plattner, M. Tignor, S.K. Allen, J. Boschung, A. Nauels, Y. Xia, V. Bex and P.M. Midgley (eds.)]. Cambridge University Press, Cambridge, United Kingdom and New York, NY, USA.

Meehl, G. A. and Tebaldi, C., 2004. "More intense, more frequent, and longer lasting HWs in the 21st Century", Science,305,994–997.

Natarajan K K 1964 A note on the hot days of Madras (1875-1963); Indian J. of Meteorol. Geophys. 14 431-436.

Pai, D. S., Thapliyal, V. and Kokate, P. D., 2004, "Decadal variation in the heat and cold waves over India during 1971-2000", Mausam, 55, 2, 281-292.

Pai, D. S, Nair S.A, Ramanathan A.N., 2013, Long term climatology and trends of heat waves over India during the recent 50 years (1961-2010), Mausam, 64, 4, 585-604

Raghavan, K., 1966, "A climatological study of severe heat waves in India", Indian J. of Meteorol. Geophys., 17, 581-588.

Raghavan, K., 1967, Climatology of severe cold waves in India, Indian J. Met. Geophys., 18, 1,91-96

Rai Sircar, N.C. and Datar, S.V. 1963: Cold waves in northwest India. Ind. Jo rn. Met. Geophys. Vol. XIV, p. 315.

Subbaramayya, I. and Surya Rao, D. A., 1976, "Heat wave and cold wave days in different states of India", Indian J. Meteorol. Hydrol. Geophys., 27, 436-440.

WMO/WHO/UNEP, 1996, "In Climate and Human Health",Kalkstein, L.S., Maunder, W.J., Jendritzky, G. (eds), WMO No. 843, World Meteorological Organization, Geneva.

Chapter 5
Variability of Meteorological Droughts Over India

D.S. Pai, Pulak Guhathakurta, Ashwini Kulkarni and M.N. Rajeevan

1 Introduction

Drought is a complex, natural and recurrent feature of climate with significant impacts on various sectors like agriculture and water resources and the Indian economy. Droughts are generally observed in all the climatic zones. Drought is a relative, rather than absolute condition that should be defined for each region separately. Each drought differs in intensity, duration and the spatial extent. In general, drought over a geographic area represents a temporary condition of scarcity of water for an extended period of time caused by significantly low precipitation, high evapotranspiration and over-exploitation of water resources or a combination of all these (Bhuiyan et al. 2006; WMO 2006). A drought differs from aridity, which is a permanent feature of the climate over regions where climatological normal is low. In terms of the number of fatalities, drought ranks first among all natural hazards (Obasi 1994; Hewitt 1997).

Drought can be categorized into four major categories (Wilhite and Glantz 1985): meteorological, agricultural, hydrological and socioeconomic droughts. Meteorological drought is defined by the deficiency of precipitation from expected or "normal" amount over an extended period of time. The agricultural drought is characterized by a deficiency in water availability for crop or plant growth. It is usually associated with deficiencies in soil moisture, which is the most critical factor in defining crop production potential. Hot temperatures, low relative

D.S. Pai (✉) · P. Guhathakurta
India Meteorological Department, Earth System Science Organization, Pune, India
e-mail: sivapai@hotmail.com

A. Kulkarni
Indian Institute of Tropical Meteorology, Earth System Science Organization, Pune, India

M.N. Rajeevan
Earth System Science Organization, Ministry of Earth Sciences, New Delhi, India

M.N. Rajeevan and S. Nayak (eds.), *Observed Climate Variability and Change Over the Indian Region*, Springer Geology, DOI 10.1007/978-981-10-2531-0_5

humidity and strong winds often accompany the lack of rainfall which adds to further reduction in crop yield. It has been observed that there is substantial loss in agricultural output during the severe droughts (Parthasarathy et al. 1994). Gadgil and Gadgil (2006) have also shown that there is a significant reduction in the gross domestic product (GDP) of India during the drought years.

The hydrological drought is defined as the deficiency in surface and subsurface water supplies that lead to a lack of water availability to meet normal and specific water demands. Although climate is a primary contributor to hydrological drought, other factors such as changes in land use (deforestation), land degradation and dam construction also contribute. The socioeconomic droughts occur when physical water shortages start to affect the health, well-being and quality of life of people. Droughts put enormous demand on rural and urban water resources and immense burden on agricultural and energy production.

A drought index indicates the severity of a drought, useful in understanding the drought conditions over an area. There are several drought indices as seen in the literature (Palmer 1965, 1968; Bhalme and Mooley 1980; Rao et al. 1981; Sastri 1993; Heddinghaus 1991; Tate et al. 2000; Lloyd-Hughes and Saunders 2002; Kogan 1995, 1997). For more details, Heim (2002), Mishra and Singh (2010) and Sivakumar et al. (2011) can be referred. Rainfall is the primary factor that leads to the generation and maintenance of drought conditions. However, evapotranspiration also impacts the drought severity. The indices based on only rainfall data are simple to compute and perform better compared to more complex hydrological indices (Oladipio 1985).

One of the widely used drought indices is Palmer Drought Severity Index (PDSI). This index is the most prominent meteorological drought index used in the USA (Dai 2011). This index considers precipitation, evapotranspiration and soil water-holding capacity. However, this index cannot capture the multi-scalar nature of droughts. Also it may lag behind the emerging droughts by several months. Standardized Precipitation Index (SPI) (Mckee et al. 1993) considers the multi-scalar nature of droughts. However, SPI does not include temperatures which have substantial effects on severity of droughts. Standardized Precipitation-Evapotranspiration Index (SPEI) is a drought index that can account for the moisture demands of the atmosphere due to global warming (Begueria et al. 2010; Vicente-Serrano et al. 2010a, b).

In India, several studies have been carried out which are related to drought using drought indices based on rainfall data. These studies (Ramdas 1950; Banerji and Chabra 1964; Appa Rao 1991; Chowdhury et al. 1989; Sen and Sinha Ray 1997; Gore and Sinha Ray 2002; Sinha Ray and Shewale 2001; Guhathakurta 2003; Gore et al. 2010) mainly dealt with the droughts during the southwest monsoon season (June–September). Guhathakurta (2003) found the highest probability (>60 %) of droughts in some districts from northern India. Pai et al. (2011) examined the climatology droughts of different intensities over India using two most simple drought indices, PNP and SPI both based on rainfall only. Recently, Niranjan Kumar et al. (2013) examined observed variability of the droughts over India using the SPEI and showed that there is a general increase in percent area and intensity of droughts over India in the recent decades.

In this chapter, long-term climatology and interannual variability of the meteorological droughts over India for the southwest monsoon season (June to September) in the district-wide and all-India (country-wide) scales are discussed. For this purpose, district and all-India rainfall data for the period 1901–2010 have been used. Impact of sea surface temperatures (SSTs) over the equatorial Pacific and Indian Oceans on the variability of droughts has also been discussed.

2 Climatology of District-Wide Drought Frequency During Southwest Monsoon Season

Based on rainfall data of 458 districts in the country, Pai et al. (2011) examined district-wide drought climatology during the southwest monsoon season for the period 1901–2003 based on PNP and SPI. They suggested that SPI is a better index compared to PNP. The climatology of district-wide drought frequency over India based on PNP was found to be biased by the aridity of the region.

For this chapter, the district-wide SPI was recomputed using the rainfall data of 641 district data prepared by Guhathakurta et al. (2011) for the period 1901–2010. Districts with no data are shown as unshaded. The intensity of drought is classified as moderate drought for SPI from −1.0 to −1.49, severe drought for SPI from −1.5 to −1.99 and extreme drought for SPI of −2.0 and less. Figure 1a–c respectively shows the district-wide percentage frequency of moderate, severe and extreme droughts computed based on the SPI.

The highest frequency (>9 %) of moderate droughts is observed in many districts along the west and east coasts of Peninsular India, Madhya Pradesh, Maharashtra, Gujarat, Punjab, central Rajasthan, West Bengal, Sikkim, etc. (Fig. 1a). The highest frequency (>6 %) of the severe droughts is observed in few districts from southeast Rajasthan and neighboring areas of west Madhya Pradesh, Gujarat, Chhattisgarh, south interior Karnataka, Tamil Nadu, coastal Andhra Pradesh, Orissa, Gangetic West Bengal, etc. (Fig. 1b). The highest frequency of extreme droughts (≥3 %) is observed in many districts from Rajasthan, Uttarakhand, north Punjab and neighboring Himachal Pradesh, Gujarat and Saurashtra, north Madhya Pradesh, Chhattisgarh, Maharashtra, Tamil Nadu, north coastal Andhra Pradesh and some areas of northeast India.

3 All-India (Country-Wide) Droughts

In Fig. 2, the line plot presents the time series of all-India season (June–September) SPI and bar plot presents percentage area of the country affected by drought conditions for the period 1901–2010. The nation-wide season SPI time series was constructed from nation-wide season rainfall computed as the area-weighted

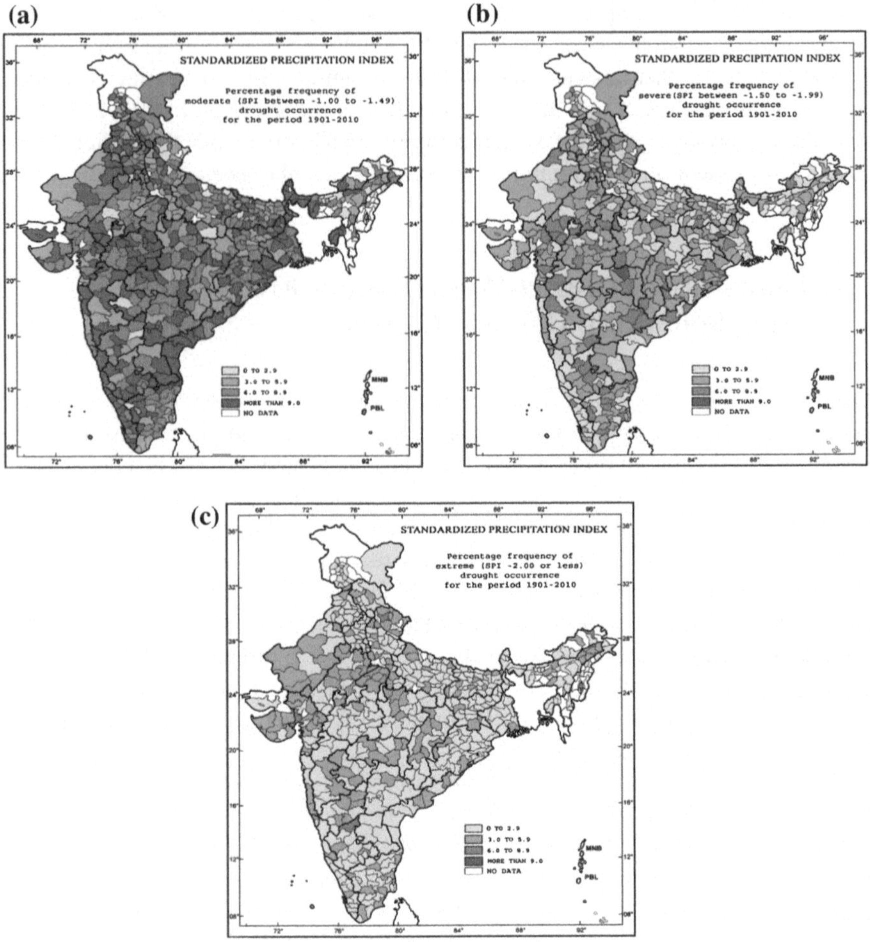

Fig. 1 Maps showing district-wide percentage of incidences (probability) of drought of **a** moderate, **b** severe and **c** extreme intensity for the southwest monsoon season (June to September). The season (June–September) Standardized Precipitation Index (SPI) was used to identify the drought incidences of various intensities. Data for the period 1901–2010 were used for the computation. Districts not used for this study are not shaded

average of all the 651 district season rainfall data. The percentage drought area for each year was computed as the total area of districts that experienced drought of moderate or higher intensity divided by the total area of all the districts for which data are available for that year. The correlation coefficient (CC) between the two time series (−0.84) is highly significant, indicating strong association between the all-India SPI and percent area under drought conditions. Therefore, the all-India drought has been defined by taking into account both the all-India SPI and area under drought conditions. When the all-India SPI is ≤ -1.0, the year is called

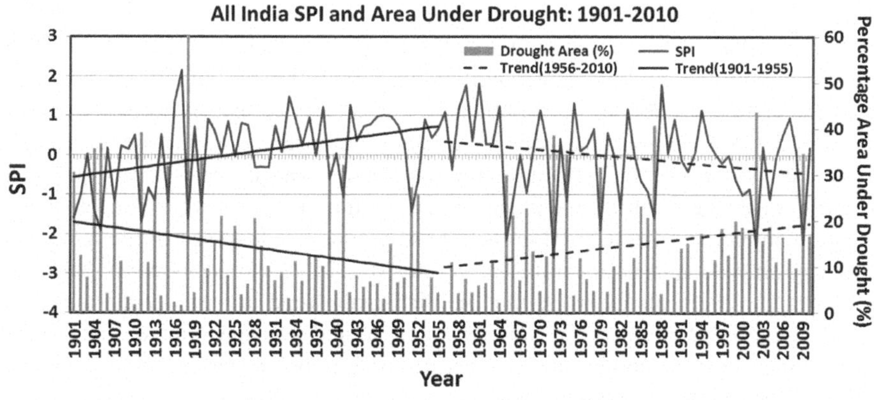

Fig. 2 Interannual variation and secular trends in the time series of all Standardized Precipitation Index (SPI) and all-India area under drought for the southwest monsoon season (June to September) rainfall for the period 1901–2010

deficient monsoon year. In addition to this, when the percentage drought area is 21–40 %, the situation is called all-India moderate drought. Similarly, in addition to deficient monsoon year, when the percentage of drought area is more than 40 %, the conditions are defined as severe all-India drought.

As per the above definition, there were 23 deficient monsoon years (all-India SPI ≤ -1.0) during the period 1901–2010 (Fig. 2). These 23 years arranged in ascending order of SPI values are 1972, 2002, 2009, 1965, 1905, 1979, 1911, 1918, 1901, 1987, 1951, 1904, 1982, 1920, 1915, 1907, 1974, 1913, 1966, 2004, 1941, 1902 and 2000. Out of these 23 years, except for 3 years (1902, 2000 and 2004), all other 20 years were moderate or severe drought years (percentage drought area ≥ 21 %). However, only 3 years (1972, 2002 and 1987) were severe drought years with drought areas of 63.4, 43.6 and 40.6 %, respectively. Thus even though all-India SPI value for 1987 is only 10th lowest among the 110 years, the area under drought in 1987 is the third highest among all the years resulting 1987 being classified as the severe drought year. It may be noted that 9 of the moderate or severe all-India drought years were experienced during the first 20 years of the data period (1901–1920). There were only two cases of two consecutive all-India drought years (1904–1905 and 1965–1966). During the recent 15 years (2001–2015), two consecutive drought years (2002 and 2009; 2014 and 2015) were observed. The study by Niranjan Kumar et al. (2013) using the SPEI observed significant interannual and decadal variability in the monsoon droughts of different time scales (6, 12, 18 and 24 months). They observed higher frequency (12 cases) of multi-year all-India droughts (of 24-month duration) during the period 1951–2010 in comparison with only 3 such long-lived droughts during the previous 50-year period 1901–1950.

On examining liner trends in both the country-wide SPI and drought area time series (Fig. 2) during the entire data period (1901–2010), no trends were observed. However, linear secular trends significant at the 95 % confidence level were observed in both the time series if we consider two different data periods (1901–1955 and 1956–2010). During first half of the data period (1901–1955), SPI showed significant positive trend and drought area showed significant negative trend indicating weakening in the intensity and area coverage of droughts during this period. On the other hand, opposite trends were observed during the recent half of the data period (1956–2010) indicting increasing trend in the intensity and area coverage of drought during the recent decades. Similar trends were also observed by Niranjan Kumar et al. (2013) in the intensity and area coverage of the monsoon drought events of various accumulated periods (6, 12, 18 and 24 months). Niranjan Kumar et al. (2013) also showed that the increasing trend in the intensity and area coverage of all-India droughts during the recent period was due to increasing annual temperatures over the country (associated with the global warming, Chap. 2) rather than decreasing precipitation.

4 Composite Spatial SPI Patterns Associated with the All-India Drought

Figure 3a, b shows the composite district-wide SPI over India during the 17 all-India moderate drought years and 2 all-India severe drought years. Both the figures show similar large-scale pattern with negative composite SPI values in most of the districts except many from east and northeast India where positive composite SPI values are observed. However, exception is in the moderate drought case where as shown in Fig. 3a, positive composite SPI (wet conditions) are observed in many districts in Tamil Nadu, central parts of interior Karnataka in addition to isolated districts from east and northeast India. In all the remaining districts, composite SPI is negative (dry) with SPI of ≤ -1.0 observed in few districts in Punjab and neighboring region, north Rajasthan, west Madhya Pradesh, east Madhya Pradesh, Chhattisgarh and Maharashtra.

In Fig. 3b, positive composite SPI (wet conditions) is observed in many districts of east and northeast India and some isolated districts in the central and north India. Negative composite SPI values are observed in all the remaining districts. Composite SPI values of ≤ -1.0 are seen in most of the districts from north and northwest India and along both the coasts (west and east) of Peninsular India with many districts from north and northwest India and southwest coast of Peninsular India having composite values of ≤ -2.0.

The drought is also generally characterized by long dry spells during the season (Krishnamurti and Bhalme 1976; Sikka 1980; Gadgil and Joseph 2003; Rajeevan et al. 2010), particularly over the core monsoon zone. Based on the break and active days data derived for the period 1901–2014 by Pai et al. (2015), it was observed that during the 20 all-India drought years (listed in Sect. 3), about 15 break days per year were experienced against the normal of 8 break days per/year. At the same time, during the drought years, active days were reduced to 5 days per year against the normal of 7 active days/year.

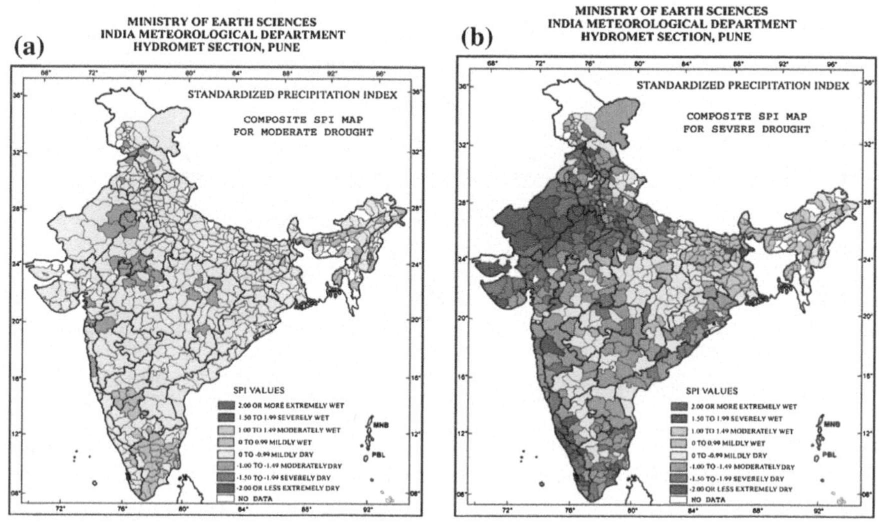

Fig. 3 Maps showing district-wide composite district-wide SPI for all-India **a** moderate and **b** severe drought years for the southwest monsoon season (June to September) for the period 1901–2010. Districts not used for this study are not shaded

5 Relationship of Sea Surface Temperatures (SSTs) Over the Tropical Pacific and the Indian Oceans with All-India Droughts

The physical mechanisms responsible for the meteorological droughts over India during the southwest monsoon season are natural variability of the Indian southwest monsoon and the slowly varying climate boundary conditions (Charney and Shukla 1981). SSTs have the most significant association with the Indian monsoon droughts. The association between anomalous warm phase of SSTs in the equatorial Pacific (El Nino) and the meteorological droughts over India during southwest monsoon season has been known for a long time (Sikka 1980; Angell 1981; Rasmussen and Carpenter 1983; Glantz et al. 1991; Webster and Yang 1992; Krishna Kumar et al. 1999; Pai 2004). This relationship is clearly visible in Fig. 4, which shows the CC map of global tropical SSTs with the all-India SPI for the southwest monsoon season (June–September) for the period 1901–2010. The SST data were derived from the NOAA extended reconstructed sea surface temperature (ERSST V3b) dataset (Smith et. al. 2008) at 2° × 2° resolution for the period 1901–2010. In Fig. 4, contour lines corresponding to CC values of ±0.19 and ±0.25 (Fig. 4) represent CC significant at the 95 and 99 % confidence level, respectively. As shown in Fig. 4, the strongest relationship of SSTs with all-India SPI is observed over the tropical Pacific with a distinct horseshoe pattern of negative CC areas over the equatorial Pacific east of 160°W wrapped around by positive CC areas extending to the north- and southeast from the western equatorial Pacific.

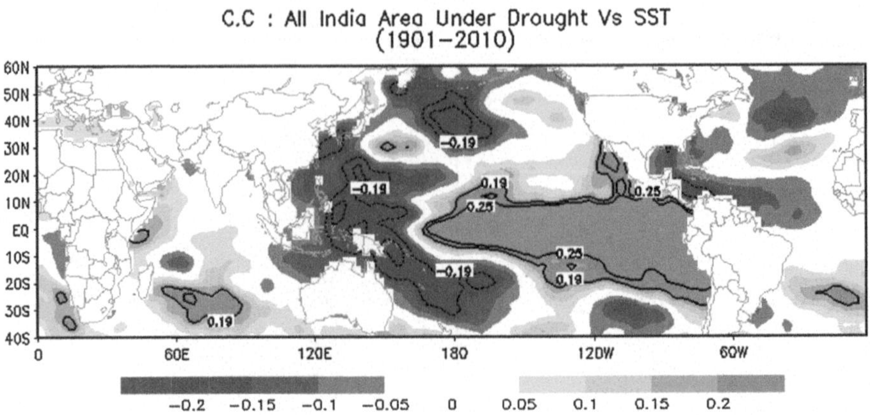

Fig. 4 Shaded map shows the correlation coefficient (CC) of tropical SSTs with active and break days (1901–2010). The CC values significant at 95 and 99 % level (±0.19 and ±0.25) are shown using *contour lines*

However, some studies have observed that in the recent years the characteristic properties of canonical ENSO have changed in their amplitude, frequency (An and Kang 2000) and the location of maximum anomalous SSTs (Latif et al. 1997; Ashok et al. 2007; Kao and Yu 2009; Kug et al. 2009). The anomalous warming over the central Pacific, also known as Modoki (Ashok et al. 2007), is associated with recent droughts after 1990s than the warming over the eastern Pacific (Krishna Kumar et al. 1999). In order to examine the secular variation in the relationship between Nino 3.4 and all-India SPI, moving 31 years CC was examined for the period 1901–2010 (Fig. 5). In Fig. 5, the horizontal dotted line represents the 99 % significant level of CC. As shown in Fig. 5, negative CC significant at the 99 %

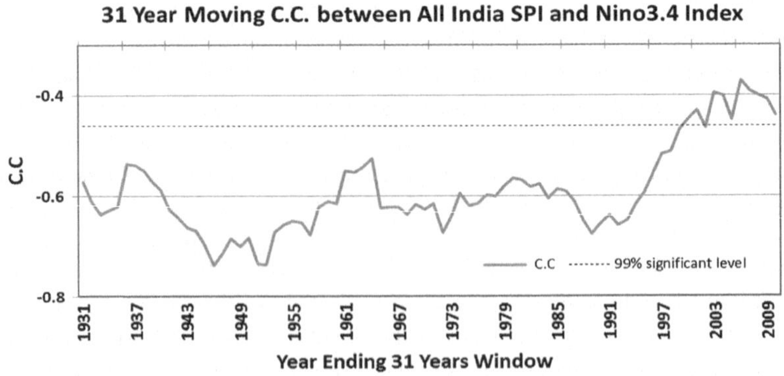

Fig. 5 Moving CC between the all-India SPI and Nino 3.4 SST index for the period 1901–2010 computed based on 31-year window period. The *horizontal dotted line* represents CC significant at 99 % level

confidence level was observed till the 31 years ending 1999 after which it weakened slightly though significant at the 95 % confidence level. This weakening is due to the 1997 El Nino, which was the strongest El Nino event during the period and Indian monsoon in that year was close to normal (102 % of long-period average).

In order to examine the impact of El Nino on regional droughts during the southwest monsoon season (JJAS), a map of CC between district-wide SPI and the Nino 3.4 SST was prepared (Fig. 6). Significant negative CCs are observed in most of the districts except those from east and northeast India and in some districts from north India and along the west coast region of the Peninsula, where insignificant

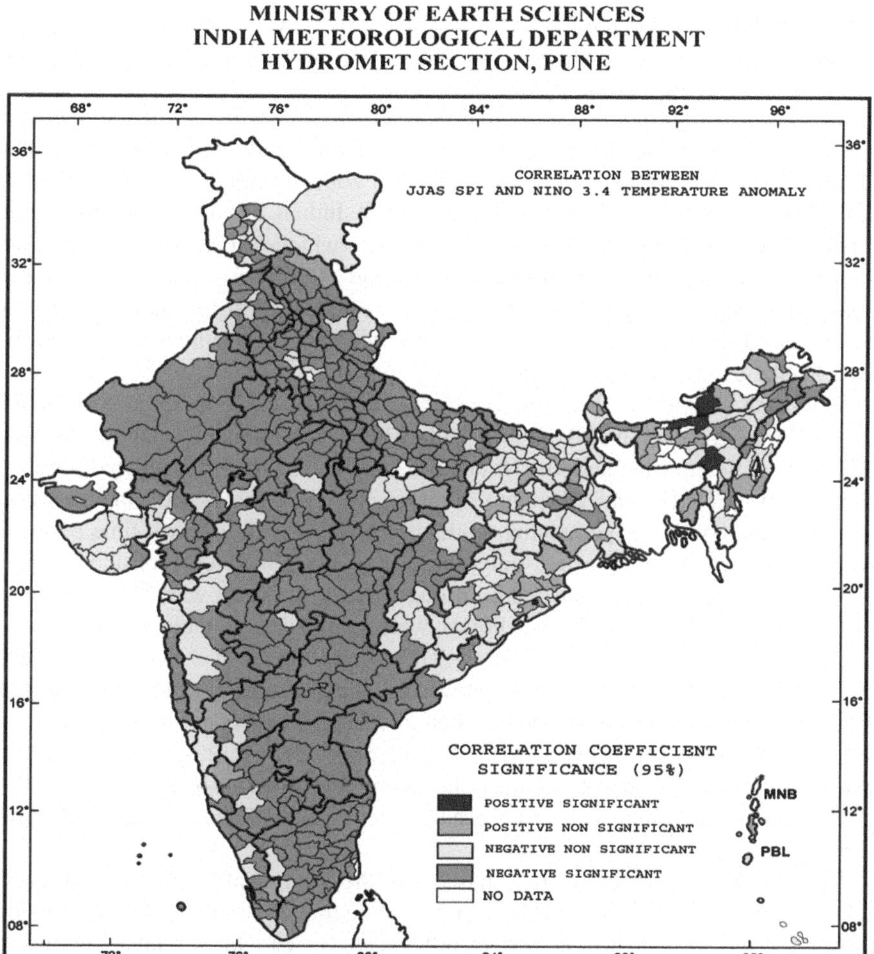

Fig. 6 Correlation coefficient (CC) of Nino 3.4 SST index with district-wide SPI over India

positive or negative CCs are observed. The large-scale pattern in Fig. 6 is nearly similar to the composite SPI pattern of drought years, indicating once again that the intensity and spatial distribution of the droughts over India are strongly related to the ENSO phenomena.

The variability in the monsoon droughts is also associated with the SSTs over other ocean basins such as the Indian Ocean (Joseph and Pillai 1984; Rao Kusuma and Goswami 1988; Krishnamurti et al. 1989; Webster et al. 1998; Saji et al. 1999; Behera et al. 1999; Ashok et al. 2001; Izumo et al. 2008; Niranjan Kumar et al. 2013), the western Pacific (Ju and Slingo 1995; Soman and Slingo 1997; Chandrashekhar and Kitoh 1998; Yun et al. 2008), and south China Sea (Ose et al. 1997). Gadgil et al. (2003) have shown that the Indian monsoon droughts are generally associated with the unfavorable mode of ENSO or the equatorial Indian Ocean Oscillation (EQUINOO) or both. The most severe droughts in the recent 50 years (e.g., severe droughts of 2002 and 2009) were linked to the unfavorable phases of both these modes (Gadgil 2009).

As shown in Fig. 4, over the Indian Ocean, some areas of significant negative (positive) CC are observed along the east coast off South Africa and along the latitudinal zone around 30°S (narrow latitudinal zone around 45°S). This suggests that when the SSTs over the subtropical south Indian Ocean are warmer than normal it weakens the Mascarene high pressure (which is one of the permanent components of the Indian monsoon system) and thereby weakens the low-level monsoon winds causing below normal moisture reaching in the Indian monsoon region and deficiency in the monsoon rainfall. This sometimes may lead to drought-like conditions over India.

Niranjan Kumar et al. (2013) analyzed the variability and long-term trends based on SPEI using the data of 1901–2004. With SPEI, it is possible to calculate drought indices for different time periods. Niranjan Kumar et al. (2013) considered drought events of 6-, 12-, 18- and 24-month duration. Figure 7 shows the time series of percent of the country experiencing moderate droughts with different time periods. To calculate these time series, only the plains of India were considered. For calculating percent of area affected by drought, the grid points with SPEI less than −1.0 were considered. Such grid points were counted, and their total area was considered to prepare the time series. These time series also show significant interannual and decadal variations. However, the most significant aspect is the observed increasing trends in percent area affected by moderate droughts. This is reflected in short-term drought-like SPEI (6) as well as multi-year drought-like SPEI (24). However, the significant increasing trends are observed from 1951 onwards, which are statistically significant at the 95 % confidence level for all time periods.

Niranjan Kumar et al. (2013) based on the canonical correlation analysis (CCA) of tropical SSTs and SPEI over Indian region have shown that the major portion of the drought variability is influenced by El Nino. Global warming, especially warming of the equatorial Indian Ocean, has played an important role in the occurrence of the monsoon droughts in the recent decades. The warming trend over the equatorial Indian Ocean (10°S–10°N, 50°–100°E) is shown in Fig. 8,

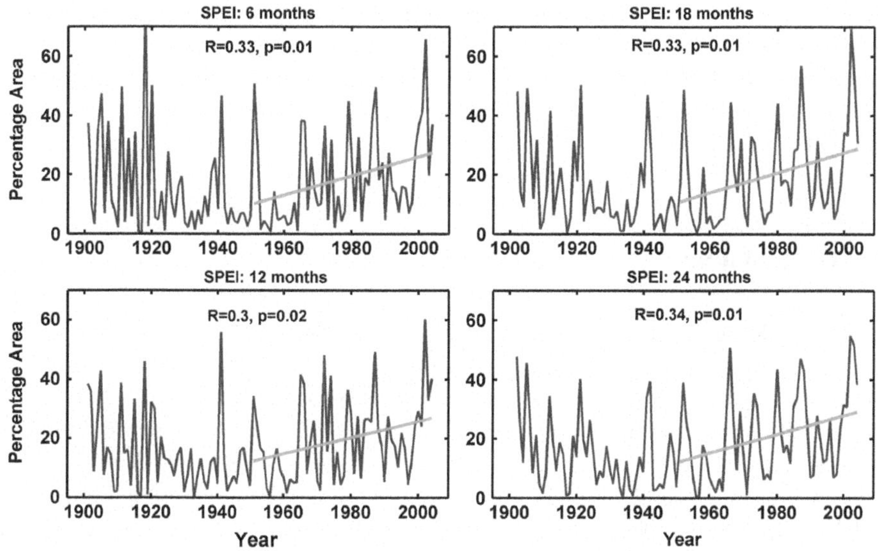

Fig. 7 Time series percent area affected by moderate drought (SPEI < −1.0) at different accumulated periods for the period, 1901–2004. The trend lines calculated using the data of 1951–2004 are shown as *red lines*. The statistical significance levels of trend values are shown as *p* values for each figure. All the trends are statistically significant at least at 95 % significance level (color figure online)

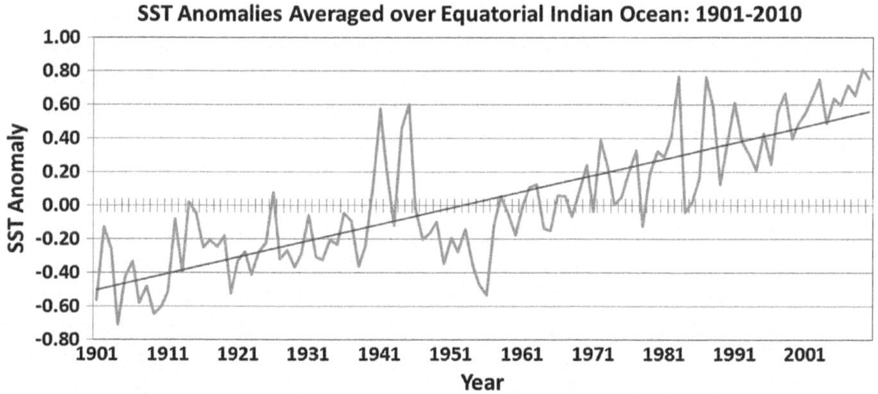

Fig. 8 Interannual variation and trend in the time series of SST anomaly averaged over the equatorial Indian Ocean (10°S–10°N, 50°–100°E) for the southwest monsoon season (June to September) rainfall for the period 1901–2010

which shows the time series of SST anomalies averaged over the region for the period 1901–2010 along with linear trend line. The trend is significant at the 99 % confidence level. However, a CC map (not presented here) similar to Fig. 4 but prepared using the data for the period 1971–2010 did not show any significant CCs in the equatorial region. This clearly indicates that though the year-to-year variation over the equatorial Indian Ocean may not have any impacts on the drought occurrences over India, the observed SST warming trend in the equatorial Indian Ocean may be impacting the increasing trend in the intensity and areal coverage of droughts.

6 Conclusions

District-wide and all-India SPI data for the period 1901–2010 were used to study the variability of meteorological droughts over India. The district-wise climatology of drought events over India showed that moderate droughts are equally likely in all parts of the country. However, relatively higher frequency of more intense droughts occurs in many districts from north, northwest and neighboring Central India. During the data period, drought was experienced in 20 years during the period 1901–2010, with 17 years being moderate drought years and 3 years being severe drought years. During the all-India drought years, most of the districts over the country except the districts from east and northeast India experience drier than normal conditions (negative SPI). In case of severe drought years, more intense drier conditions are experienced, particularly in districts from north and northwest India and southwest coast of peninsular India.

It was observed that warmer-than-normal SSTs over the equatorial Pacific and subtropical south Indian Oceans have significant association with the meteorological droughts over all-India scale as well as the spatial distribution of district-wise drought conditions over the country. As a fact, all the 3 severe all-India drought years and 9 out of the 17 all-India moderate drought years were the El Nino years. However, none of the 20 drought years were La Nino years establishing the strong relationship between El Nino and Indian monsoon droughts.

Significant secular trends were observed in the intensity and areal coverage of the droughts. During the later half of the period (i.e., 1956–2010), there is a significant increase of area affected by droughts and intensity over India. This is interesting because in spite of increasing trend in the intensity and areal coverage of the drought in recent half of the data period, the relationship between all-India SPI and Nino 3.4 was weakening in the recent years. As indicated by Niranjan Kumar et al. (2013), the significant warming trend in the SSTs over the equatorial Indian Ocean associated with global warming may be one of the reasons for the increasing trend in the intensity and areal coverage of droughts in the recent years. The warm SSTs over the equatorial Indian Ocean may cause drought conditions over India through weakening of the land–ocean temperature gradient and/or increased convection over the equatorial Indian Ocean causing anomalous subsidence over Central India and therefore reduced rainfall. However, more diagnostic and modelling studies are needed to establish the linkages with the Indian Ocean SSTs.

References

Angell JK, 1981, Comparison of variations in atmospheric quantities with SST variations in equatorial eastern Pacific. Mon Wea Rev 109: 230-243.

An, S.-I., and I.-S. Kang, 2000, Interdecadal Change of the Structure of the ENSO Mode and Its Impact on the ENSO Frequency, Journal of Climate, 13 , pp 321–340.

Appa Rao, G., 1991, Drought and southwest monsoon, Training course on Monsoon

Ashok K, Guan ZY, and Yamagata T, 2001, Impact of the Indian Ocean Dipole on the relationship between the Indian monsoon rainfall and ENSO. Geophys Res Lett.,28, pp 4499–4502.

Ashok, K., S. K. Behera, S. A. Rao, H. Weng, and T. Yamagata, 2007, El Niño Modoki and its possible Teleconnection, J. Geophys. Res., 112, C11007, doi:10.1029/2006JC003798.

Banerji, S. and Chabra, B.M, 1964, Drought Characteristics and Estimating Probabilities of Their Occurrences, in Surface Waters Symposium, WMO/IASH, Belgium, Publication No. 63, pp 189-192.

Begueria, S., Vicente Serrano, S.M.,Martínez., M.A., 2010. A multiscalar global drought dataset: the SPEI base: a new gridded product for the analysis of drought variability and impacts. Bull. Amer.Met.Soc., 10, pp 1351–1356.

Behera, S. K., S. Krishnan, and T. Yamagata, 1999, Unusual Ocean-Atmosphere conditions in the tropical Indian Ocean during 1994. Geophys. Res. Lett., 26, pp 3001-3004.

Bhalme, H.N. and Mooley, D.A., 1980, Large scale droughts/floods and monsoon circulation. *Mon. Wea. Rev.*, 108, pp 1197-1211.

Bhuiyan, C., R.P. Singh, and FN Kogan, 2006, Monitoring drought dynamics in the Aravalli region (India) using different indices based on ground and remote sensing data, Int. Journal of Applied Earth Observation and Geoinformation, 8, pp 289–302.

Chandrashekhar A, Kitoh A, 1998, Impact of localised sea surface temperature anomalies over the Equatorial Indian Ocean on the Indian summer monsoon. Journal of Meteorological Society of Japan,76, pp 841–853.

Charney, J. G. and Shukla, J., 1981, Predictability of monsoons, In monsoon dynamics (Ed. J. Lighthill), Cambridge University Press, pp 99–110.

Chowdhury, A., Dandekar, M.M., and Raut, P.S., 1989, Variability in drought incidence over India – A statistical approach, Mausam, 40, pp 207,-214.

Dai, A., 2011, Characteristics and trends in various forms of the Palmer Drought Severity Index (PDSI) during 1900–2008. J. Geophys. Res. 116, D12115.

Gadgil S and Gadgil S 2006, The Indian monsoon, GDP and agriculture. Econ. Political weekly, 41, pp 4887-1895

Gadgil S. 2009, The aberrant behaviour of the Indian monsoon in June 2009. Curr Sci, 97, 1291-1295

Gadgil, S. and P. V. Joseph, 2003, On breaks of the Indian monsson Proc. Indian Acad. Sci. (Earth Planet. Sci.), 112, 529–558.

Gadgil, S., P. N. Vinayachandran P. N. and P. A. Francis, 2003, Curr. Sci., 85, pp 1713–1719.

Glantz, M. H., R. W. Katz and N. Nicholls (Eds.), 1991, Teleconnections linking worldwide climate anomalies. Cambridge University Press, pp 535.

Gore, P. G., Thakur Prasad and H. R., Hatwar, 2010, Mapping of drought areas over India, NCC Research Report. No.12, India Meteorological Department, Pune.

Gore, P.G. and K.C. Sinha Ray, 2002, Variability in drought incidence over districts of Maharashtra, Mausam, 53, pp 533-542.

Guhathakurta P, Koppar AL, Krishan U, Menon P. 2011a. New rainfall series for the districts, meteorological sub-divisions and country as whole of India. National Climate Centre Research Report No: 2/2011, India Meteorological Department, Pune, India

Guhathakurta, P., 2003, Drought in districts of India during the recent all India normal monsoon years and its probability of occurrence. Mausam, 54, pp 542-545.

Heddinghaus, T.R., 1991, Monitoring and Dissemination of Drought Conditions at the Joint Agricultural Weather Facility, in D.A.Wilhite, D.A.Wood, and P.A.Kay (eds.), Proceedings of

the Seminar and Workshop on Drought Management and Planning, Institute of Agriculture and Natural Resources, University of Nebraska-Lincoln, pp 67- 72.
Heim, R.R., 2002, A review of twentieth century drought indices used in the United States, Bulletin of the American Meteorological Society, 83, pp 1149–1165.
Hewitt, K. 1997, Regions at Risk: A Geographical Introduction to Disasters, Addison-Wesley Longman, UK.
Izumo, T., C. B. Montégut, J.-J. Luo, et al., 2008: The role of the western Arabian sea upwelling in Indian monsoon rainfall variability, J. Climate, 21, pp 5603–5623.
Joseph, P. V. and Pillai, P. V., 1984, Air sea interaction on a seasonal scale over north Indian Ocean Part I: Interannual variation of sea surface temperature and Indian monsoon rainfall, Mausam, 35, pp 323–330.
Ju, J. and Slingo, J. M., 1995, The Asian summer monsoon and ENSO, Q. J. R. Meteorol. Soc., 121, pp 1133-1168.
Kao, H. -Y, and J. –Y. Yu, 2009, Contrasting Eastern-Pacific and Central-Pacific Types of ENSO, Journal of Climate, 22 , pp 615–632.
Kogan, F. N., 1997, Global drought watch from space, Bull.Amer. Meteor. Soc., 78, pp 21-636.
Kogan, F.N., 1995. Droughts of the late 1980 s in the United States as derived from NOAA polar-orbiting satellite data. Bulletin of the American Meteorological Society 76(5):655-668.
Krishnamurti, T. N. and Bhalme, H.N., 1976, Oscillation of a monsoon system. Part I: Observational aspects, J. Atmos. Sci., 133, pp 1937-1954.
Krishnamurti, T.N., H.S. Bedi, and M. Subramaniam, 1989, The summer monsoon of 1987, Journal of Climate, 2 , pp 321–340.
Krishna Kumar, K. Rajagopalan, B., and Cane, M. A., 1999, On the weakening relationship between the Indian monsoon and ENSO, Science, 284, pp 2156–2159.
Kug, J., F. Jin, and S. –I, An, 2009, Two Types of El Nino Events: Cold Tongue El Nino and Warm Pool El Nino, 22 (6), pp 1499–1515.
Latif, M., R. Kleeman, and C. Eckert, 1997, Greenhouse warming, decadal variability, or El Niño? An attempt to understand the anomalous 1990s, J. Climate, 10, pp 2221–2239.
Lloyd-Hughes, B., and M. A. Saunders, 2002: A drought climatology for Europe. Int. J. Climatol., DOI:10.1002/joc.846.
McKee T. B., Doesken N. J., Kliest, J. 1993. The relationship of drought frequency and duration to time scales, In Proceedings of the 8th Conference on Applied Climatology, 17–22 January, Anaheim, CA. American Meteorological Society: Boston, MA; 179–184.
Mishra, A.K., Singh, V.P., 2010, A review of drought concepts, Journal of Hydrology, 391, pp 202–216, doi:10.1016/j.jhydrol.2010.07.012.
Niranjan Kumar, K. Rajeevan, M. Pai, D.S. Srivastava, A.K. Preethi, B., 2013, On the observed variability of monsoon droughts over India, Weather and Climate Extremes, 42-50.
Obasi, G.O.P.,1994, WMO's role in the International Decade for Natural Disaster Reduction, Bulletin of the American Meteorological Association, 75(9): 1655-1661.
Oladipio EO. 1985. A comparative performance analysis of three meteorological drought indices. International Journal of Climatology, 5, pp 655–664.
Ose, T., Song, Y. and Kitoh, A., 1997, Sea surface temperature in the south China Sea and index for the Asian monsoon and ENSO system, J. Meteorol. Soc., Japan, 75,pp 1091-1107.
Pai, D.S., Latha Sridhar, and M.R., Ramesh Kumar, 2015, Active and break events of Indian summer monsoon during 1901-2014, Clim Dyn, DOI:10.1007/s00382-015-2813-9.
Pai, D.S., Latha Sridhar, P., Guhathakurta, Hatwar, H.R., 2011, District-wide drought climatology of the southwest monsoon season over India based on standardized precipitation index (SPI), Natural Hazards, 59, pp 1797–1813, doi:10.1007/s11069-011-9867-8.
Pai, D. S., 2004, A Possible Mechanism for the Weakening of ElNino-Monsoon Relationship during Recent Decade, Meteorol. Atmos. Physics, V86, pp 143–157.
Palmer WC., 1965, Meteorological drought. Research Paper No. 45, US Weather Bureau: Washington, DC.
Palmer, W.C., 1968, Keeping track of crop moisture conditions, nationwide: The new Crop Moisture Index, *Weatherwise*, 21, pp 156–161.

Parthasarathy B, Munot AA, Kothawale DR, 1994, All-India monthly and seasonal rainfall series; 1871—1993, Theor Appl Climatol,49, pp 217—224.

Rajeevan, M., Jyoti Bhate, and Sulochana Gadgil, 2010, Active and Break Spells of the Indian Summer Monsoon, J.Earth Science, 229-247.

Ramdas, D.A., 1950, Rainfall and Agriculture, Ind. J. Met and Geophys, 1, pp 262- 274.

Rao Kusuma, G. and Goswami, B. N., 1988, Interannual variation of sea surface temperature over the Arabian Sea and the Indian monsoon, A new perspective, Mon. Wea. Rev., 116, pp 558-568.

Rao, R., B.V., Sastri, A.S.R.A.S., and Rama Krishna, Y.S., 1981, An Integrated Scheme of Drought Classification as Applicable to Indian Arid Region, IDOJARAS, 85, pp 317-22.

Rasmussen E. M. and Carpenter T. H., 1983,. The relationship between eastern equatorial Pacific sea surface temperature and rainfall over India and Sri Lanka, Monthly Weather Review., 110, pp 354 – 384.

Saji, N.H., B.N. Goswami, P.N. Vinayachandran, and T. Yamagata, 1999, A dipole mode in the tropical Indian Ocean, Nature, 401, pp360–363.

Sastri, A.S.R.A.S., 1993, Agricultural Drought Management Strategies to Alleviate Impacts: Examples from the Arid and Sub humid Regions of the Indian Subcontinent, in D.A. Wilhite (ed.), Drought Assessment, Management, and Planning: Theory and Case Studies, Kluwer Academic Publishers, Dordrecht, pp 65-86.

Sen, A.K. and K.C. Sinha Ray. 1997. Recent trends in drought affected areas in India. Presented at International Symposium on Tropical Meteorology, INTROPMET-1997 at IIT, New Delhi.

Sikka DR, 1980, Some aspects of the large scale fluctuations of summer monsoon rainfall over India in relation to fluctuations in the planetary and regional scale circulation parameters. Proc Ind Acad Sci (Earth and Planetary Sci) 89: 179-195.

Sinha Ray, K.C. and M.P. Shewale, 2001, Probability of occurrence of drought in various subdivisions of India, Mausam, 52, pp 541-546.

Sivakumar, M. V. K., R. Motha, D. Wilhite, and J. Qu, Eds., 2011, Towards a compendium on national drought policy: Proceedings of an expert meeting. ES87, World Meteorological Organization AGM-12, WAOB-2011, pp 135.

Smith, T.M., R.W. Reynolds, Thomas C. Peterson, and Jay Lawrimore, 2008, Improvements to NOAA's Historical Merged Land-Ocean Surface Temperature Analysis (1880–2006), Journal of Climate, 21, 2283–2296.

Soman, M.K., and Slingo, J.M., 1997, Sensitivity of the Asian summer monsoon to aspects of sea surface temperature anomalies in the tropical Pacific Ocean, Q. J. R. Meteorol. Soc., 123, pp 309-336.

Tate, E.L., J.R. Meigh, C. Prudhomme and M.P. McCartney., 2000, Drought assessment in Southern Africa using river flow data, DFID report 00/4, London, UK.

Vicente-Serrano, Sergio, M., Beguería, Santiago, López-Moreno, JuanI., 2010a. A multi scalar drought index sensitive to global warming: the standardized precipitation evapotranspiration index. Journal of Climate, 23, pp 1696–1718.

Vicente-Serrano, Sergio,M., etal., 2010b. A new global 0.5 gridded dataset (1901– 2006) of a multi scalar drought index: comparison with current drought index datasets based on the Palmer Drought Severity Index, Journal of Hydrometeorology, 11, pp 1033–1043.

Webster, B. C. and Yang, S., 1992, Monsoon and ENSO: Selectively interactive systems, Q. J. Roy. Meteorol. Soc., 118, pp 877-926.

Webster, P. J., Magana, V. O., Palmer, T. N., Shukla, J., Tomas, R. A., Yanai, M. and Yasunari, T., 1998, Monsoons: Processes, Predictability, and the prospects for prediction, J. Geophys. Res., 103, pp 14451-14510.

Wilhite DA, Glantz MH., 1985, Understanding the drought phenomenon: the role of definitions. Water International, 10, pp 111–120.

WMO, World Meteorological Organization, 2006, Drought monitoring and early warning: Concepts, progress and future challenges, WMO-No. 1006.

Yun, K.-S., K.-H. Seo, and K.-J. Ha, 2008,, Relationship between ENSO and northward propagating intraseasonal oscillation in the east Asian summer monsoon system, J. Geophys. Res., p 113, D14120, doi:10.1029/2008JD009901.

Chapter 6
Inter-annual Variation and Trends in Tropical Cyclones and Monsoon Depressions Over the North Indian Ocean

M. Mohapatra, A.K. Srivastava, S. Balachandran and B. Geetha

1 Introduction

Tropical cyclones (TCs) are large synoptic scale weather systems which originate over the warm oceans of the world and develop into massive vortices composed of swirling winds, intense clouds and torrential rains by drawing energy from the oceans. While moving over land, they cause large-scale destruction to life and property over the coastal areas of the world. The east and west coasts of India are prone to the destructive features of TC activity (Mohapatra et al. 2012a) over the North Indian Ocean (NIO) comprising of the Bay of Bengal (BOB) and the Arabian Sea (AS).

The climatological features of the cyclonic disturbances (CDs) forming over the NIO, BOB and AS are given in Tables 1 and 2. On an average, 11 CDs with maximum sustained wind speed (MSW) of 17 knots or more including depressions and TCs develop over the NIO every year (based on data of 1961–2010). Over the BOB and AS, 9 and 2 CDs, respectively, form as per the data of satellite period (1961–2010) (Mohapatra et al. 2014). Out of these, about five CDs intensify into TCs (MSW ≥ 34 knots), including about 4 over the BOB and 1 over the AS. About 3 severe TCs (MSW ≥ 48 knots) form over the NIO (2 over BOB and 1 over the AS). The frequency of very severe TCs (64 knots or more) over the NIO is about 2.

M. Mohapatra (✉)
India Meteorological Department, Earth System Science Organization, New Delhi, India
e-mail: mohapatraimd@gmail.com

A.K. Srivastava
India Meteorological Department, Earth System Science Organization, Pune, India

S. Balachandran · B. Geetha
India Meteorological Department, Earth System Science Organization, Chennai, India

M.N. Rajeevan and S. Nayak (eds.), *Observed Climate Variability and Change Over the Indian Region*, Springer Geology, DOI 10.1007/978-981-10-2531-0_6

Table 1 Mean, standard deviation and linear trend in frequency of CDs, TCs and severe TCs over the NIO based on data of 1901–2010

Basin	Annual/season	CDs			TCs			Severe TCs		
		Mean	SD	Trend/decade	Mean	SD	Trend/decade	Mean	SD	Trend/decade
NIO	Annual	12.6	3.8	−0.188	5.1	1.9	**−0.193**	2.5	1.6	0.068
	Pre-monsoon	1.4	0.9	0.003	1.0	0.8	0.001	0.7	0.7	0.002
	Monsoon	6.4	2.8	**−0.254**	1.5	1.3	**−0.173**	0.4	0.7	−0.009
	Post-monsoon	4.5	1.7	0.062	2.6	1.2	−0.02	1.3	1.2	**0.074**
BOB	Annual	9.8	3.2	**−0.299**	4.1	1.7	**−0.196**	1.9	1.3	0.057
	Pre-monsoon	1.1	0.8	−0.002	0.8	0.7	0	0.5	0.6	0.014
	Monsoon	4.9	2.3	**−0.299**	1.1	1.1	**−0.176**	0.2	0.5	−0.021
	Post-monsoon	3.7	1.5	0.02	2.2	1.2	−0.019	1.1	1.1	**0.064**
AS	Annual	1.8	1.4	**0.117**	1.0	1.1	0.011	0.6	0.8	0.017
	Pre-monsoon	0.4	0.6	0.004	0.3	0.5	0.002	0.2	0.4	−0.012
	Monsoon	0.6	0.7	**0.054**	0.3	0.6	0.009	0.2	0.4	0.011
	Post-monsoon	0.8	0.9	**0.060**	0.4	0.6	0.003	0.2	0.5	0.018

Trends significant at the 95 % confidence level are emphasised in bold

Table 2 Linear trend in frequency of CDs, TCs and severe TCs over NIO during the satellite era (1961–2010)

Basin	Annual/season	CDs (1961–2010)			TCs (1961–2010)			Severe TCs (1961–2010)		
		Mean	SD	Trend/decade	Mean	SD	Trend/decade	Mean	SD	Trend/decade
NIO	Annual	11.9	3.9	**−1.818**	4.8	1.8	**−0.683**	3.0	1.6	**−0.524**
	Pre-monsoon	1.4	0.9	−0.085	1.1	0.8	−0.078	0.8	0.8	−0.065
	Monsoon	5.5	2.8	**−1.183**	0.9	0.9	**−0.243**	0.4	0.7	−0.090
	Post-monsoon	4.8	1.8	**−0.507**	2.7	1.3	**−0.377**	1.8	1.2	**−0.370**
BOB	Annual	8.6	3.1	**−1.467**	3.6	1.5	**−0.627**	2.3	1.3	**−0.513**
	Pre-monsoon	1.0	0.7	−0.065	0.8	0.6	−0.084	0.6	0.6	−0.070
	Monsoon	3.7	2.2	**−0.931**	0.5	0.7	**−0.188**	0.2	0.4	**−0.090**
	Post-monsoon	3.7	1.5	**−0.429**	2.3	1.2	**−0.369**	1.5	1.1	**−0.355**
AS	Annual	2.2	1.5	−0.118	1.1	1.0	−0.020	0.7	0.8	0.001
	Pre-monsoon	0.4	0.6	−0.025	0.3	0.5	0.007	0.2	0.4	0.005
	Monsoon	0.8	0.8	−0.001	0.4	0.6	0.003	0.2	0.5	0.012
	Post-monsoon	1.0	1.0	−0.091	0.5	0.6	−0.028	0.3	0.5	−0.016

Trends significant at the 95 % confidence level are emphasised in bold

Development of TCs is in general seasonal in nature, with most tropical ocean basins having maximum frequency of formation during the late summer-to-early autumn period. This is associated with the period of maximum sea surface temperature (SST), although other factors, such as the seasonal variation in the Inter-Tropical Convergence Zone (ITCZ)/monsoon trough location, are also important (Gray 1968). However, unlike other ocean basins, the TCs frequency over the NIO shows bimodal character with primary peak during October to December (OND, post-monsoon season) followed by the secondary peak during the pre-monsoon season (March–May, MAM) (Li et al. 2013). During the south-west monsoon season (SWM) of June to September (JJAS), intense low-pressure systems usually do not develop due to the northward shift of the convergence zone (monsoon trough) over the land and high vertical wind shear over the region (Rao 1976). About 5–6 CDs including about one TCs form over the NIO during the monsoon season (Tables 1 and 2). The TCs in monsoon season usually occurs during the onset phase (month of June) and withdrawal phase (month of September) of south-west monsoon (Rao 1976).

India Meteorological Department (IMD), the official operational weather forecaster of India, is responsible for the detection, tracking and forecasting the movement and intensity of TCs over the NIO and has meticulously archived records of CDs over the NIO for more than a century. The data have been initially archived and presented month-wise in chart/map form from 1877 to 1990 [IMD (1979) and IMD (1996)]. In the present form, this archive is available as a more versatile electronic tool, *Cyclone eAtlas—IMD* (IMD 2008, 2011), which could generate tracks and statistics of CDs over the NIO (www.rmcchennaieatlas.tn.nic.in). An extensive climatology of TCs over the NIO covering statistical aspects of CDs formation, intensification, movement, landfall, dissipation, etc. has been documented by IMD using these data and products (Raj 2011).

Regarding the trends in CDs/TCs activity over the NIO, detailed review is available in Sikka (2006), Mohapatra et al. (2012b) and Niyas et al. (2009). Sikka (2006) has noted a drastic decrease in the number of total CDs in the decade of 1991–2000 to less than even 50 % of the average of earlier 10 decades (1891–1990). Not only the total number of CDs has drastically decreased, but the number of TCs and severe TCs on the decadal basis has also shown significant decrease in the last two decades. Considering all CDs over the NIO (excluding short-lived CDs with life period of less than a day), Mohapatra et al. (2012b) noted that there is no significant trend in frequency of CDs over the BOB and AS during the period 1891–2010. Tyagi et al. (2010) have shown that there is no significant trend in frequency of land-falling CDs over the east and west coasts of India during 1891–2007.

Considering only TCs, based on data of 1891–2008, Niyas et al. (2009) have observed a significant decreasing trend in the frequency of TCs over the NIO, and the rate of decrease in frequency is maximum for the monsoon season. Considering the frequency of (i) TCs, (ii) severe TCs and (iii) very severe TCs during satellite era (1961–2010), there are significant decreasing trends in all these frequencies over the BOB and NIO as a whole (Mohapatra et al. 2014). The frequency of TCs has decreased at the rate of about 0.7 and 0.6 per decade, respectively, over the NIO and BOB.

It has been documented by Webster et al. (2005) that TCs frequency has decreased over all the oceanic basins of the world except the North Atlantic, but the number of severe TCs has increased over all the oceanic basins. Knutson et al. (2010) examined the impact of climate change on TCs forming over different oceanic basins. The study concluded that it was uncertain whether past changes in TCs activity have exceeded the variability expected from natural causes. A review of multi-decadal scale TCs variations and possible "greenhouse warming" effects has been covered by Landsea (1998). According to Landsea (1998) and Chan (2006), the recent trend in TCs genesis frequency and intensity cannot be attributed to trends in SST, rather, to large-scale oscillations such as ENSO. However, TCs over various oceanic basins do not respond identically to ENSO. Some show changes in frequency of genesis, while others show shifts in the genesis locations or in intensity due to coupled interaction of local parameters and large-scale interaction of oceanic and atmospheric circulation features, including ocean thermal energy, vertical wind shear, relative vorticity embedded in large-scale oscillations such as ENSO, Indian Ocean Dipole (IOD), Pacific Decadal Oscillation (PDO), Atlantic Multi-decadal Oscillation (AMO). Detailed review is available in Girish Kumar and Ravichandran (2012), Rajeevan et al. (2013) and Girish Kumar et al. (2014).

Patnaik (2005) examined variability of storm activity over the NIO during the pre-monsoon, post-monsoon and monsoon seasons for a period of 113 years (1891–2003) and found that large-scale atmospheric circulation is the main cause of the observed inter-decadal variability of the storm activity over the Indian region rather than the variability of SSTs over the region. Ng and Chan (2011) have examined the inter-annual variations in TCs activity over the NIO during 1983–2008 and found an evidence of the influence of ENSO and have shown that instead of local SSTs, such variations, at least over the BOB during the post-monsoon season, can be attributed to similar variations in the atmospheric flow patterns and moist static energy that are apparently forced largely by the ENSO.

Monsoon lows and depressions are the major synoptic scale systems that contribute a large quantum of south-west monsoon rainfall (Mooley and Shukla 1989). As the frequency of monsoon depression and depression days has decreased, the SW monsoon rainfall has been compensated by an increase in the number of monsoon lows and low days in the recent years, and therefore the all-India monsoon rainfall did not show any significant trend. However, the relationship between monsoon rainfall and frequency of CDs shows secular changes (Mooley and Shukla 1989; Mohapatra and Mohanty 2004, 2007). Similarly, the north-east monsoon rainfall over southern peninsular India during the post-monsoon season also shows secular variations with frequency of CDs. Land-falling CDs during OND contribute about 10 % of the seasonal rainfall over the peninsular India (Geetha and Raj 2014).

Considering all the above, trends in seasonal and annual frequency of CDs, TCs and severe TCs over NIO in inter-annual and decadal scales are presented in this chapter based on long-period data of 1901–2010 as well as for the period of 1961–2010, corresponding to the satellite era. Impacts of these trends in the monsoon and post-monsoon seasonal rainfall over various meteorological sub-divisions are also brought out. The trends are also analysed from the point of view of climate forcing such as the ENSO.

2 Data and Methodology

The primary data used for the analysis are IMD's ***Cyclone eAtlas*** (IMD 2008, 2011). Time series analysis of frequency of tropical disturbances is carried out based on intensity stratification into (a) CDs, (b) TCs and (c) severe TCs for the pre-monsoon, monsoon and post-monsoon seasons for the NIO as well as for the BOB and the AS separately. Long-term linear trend for each of these categories is determined based on the data of 1901–2010 in the inter-annual and decadal scales and tested for significance using Mann–Kendall trend test. Trends significant at the 95 % confidence level are discussed. In the view of less reliability of the data during the pre-satellite era (Mohapatra et al. 2012b) when observing and monitoring tools were inadequate, trends are also analysed separately using data of the last five decades (1961–2010) of satellite era.

It is of great scientific interest to understand the atmospheric features (vorticity and windshear) associated with the trends in CDs/TCs/severe TCs frequency despite increasing SST and ocean thermal energy. Using the NCEP data and Mann–Kendall test, the trend in the vertical shear of the horizontal wind between 850 and 200 hPa levels and low-level relative vorticity (at 850 hPa level) on monthly and seasonal scales for the period 1951–2010 is also examined. To examine the ENSO influence, correlation coefficients (CC) of SST anomaly over Nino 3.4 (5°N–5°S, 170°W–120° W) region (https://climatedataguide.ucar.edu/climate-data/nino-sst-indices) with frequency of (a) CDs, (b) TCs and (c) severe TCs over the NIO in both antecedent and concurrent modes for pre-monsoon, monsoon and post-monsoon seasons are computed and analysed. The ENSO index is taken from the climate data sets of Climate Prediction Centre, NOAA (www.esrl.noaa.gov/psd/data/climateindices/).

IMD's monsoon seasonal rainfall series for all-India and various homogeneous sub-regions of India available in IMD website (www.imd.gov.in/) as well as the monthly rainfall data for various meteorological sub-divisions of India available in IITM website (www.tropmet.res.in) are used. The analysis is based on (i) calculation of CC between rainfall over all-India/homogeneous regions/meteorological sub-divisions and the CDs frequency during the monsoon and post-monsoon seasons of 1901–2010 and (ii) 30-year sliding CCs for the same period to find out the secular variations in their relationships.

The results are presented and discussed in Sect. 3 and broad conclusions drawn are presented in Sect. 4.

3 Results and Discussions

3.1 *Linear Trends in CDs, TCs and Severe TCs Frequencies*

Table 1 presents the seasonal and annual linear trend in the frequency of CDs, TCs and severe TCs over the NIO, BOB and AS during the period 1901–2010. Time series of significant trends (95 % confidence level) over the BOB and the AS is presented in Fig. 1. Considering the entire NIO basin, on a seasonal scale, significant decreasing trend (−0.25 and −0.17 per decade) is observed in the frequencies of CDs and TCs during the monsoon season. However, there is a significant increasing trend in severe TCs (0.07 per decade) during the post-monsoon season. In the annual scenario, there is a significant decreasing trend (−0.19 per decade) in the frequency of TCs over the NIO, and there is no trend in severe TCs. Considering the BOB alone, similar significant decreasing trends are observed for the monsoon and annual CDs and TCs. Trend values of −0.30 (−0.30) and −0.17 (−0.19) per decade are obtained for frequencies of CDs and TCs during the monsoon season (year as a whole). A significant increasing trend of 0.06 per decade is observed for severe TCs over the BOB during the post-monsoon season. For the AS, there is a significant increase in the frequency of CDs during the monsoon and the post-monsoon seasons (0.05 and 0.06 per decade), which is reflected in the trend for the year as a whole (0.12 per decade).

3.2 *Trends During Satellite Era (1961–2010)*

Table 2 presents the annual and seasonal trends in CDs, TCs and Severe TCs over the NIO, BOB and AS during the satellite era, 1961–2010. Frequencies of CDs, TCs and severe TCs over the BOB during the last five decades show significant decreasing trends in the monsoon and post-monsoon seasons as well as for the year as a whole (Fig. 2). Similar significant decreasing trends are observed in the frequencies of CDs, TCs and severe TCs over the NIO except for the severe TCs during the monsoon season. There are no significant trends in the pre-monsoon season over both BOB and NIO. It is interesting to note that contrasting trends emerge for post-monsoon severe TCs over BOB and NIO for the two periods 1901–2010 and 1961–2010 with significant positive trends in the former period and significant negative trends in the later period. Over the AS, there are no significant trends in CDs, TCs and severe TCs frequencies in the last five decades for any season and year as a whole unlike long-term significant increasing trend in frequency of CDs for the monsoon and post-monsoon seasons and a year as whole based on data of 1901–2010. It may be mentioned that these results are consistent with findings of Srivastava et al. (2000), Singh et al. (2000), Mandal and Premkrishna (2009) and Mohapatra et al. (2014).

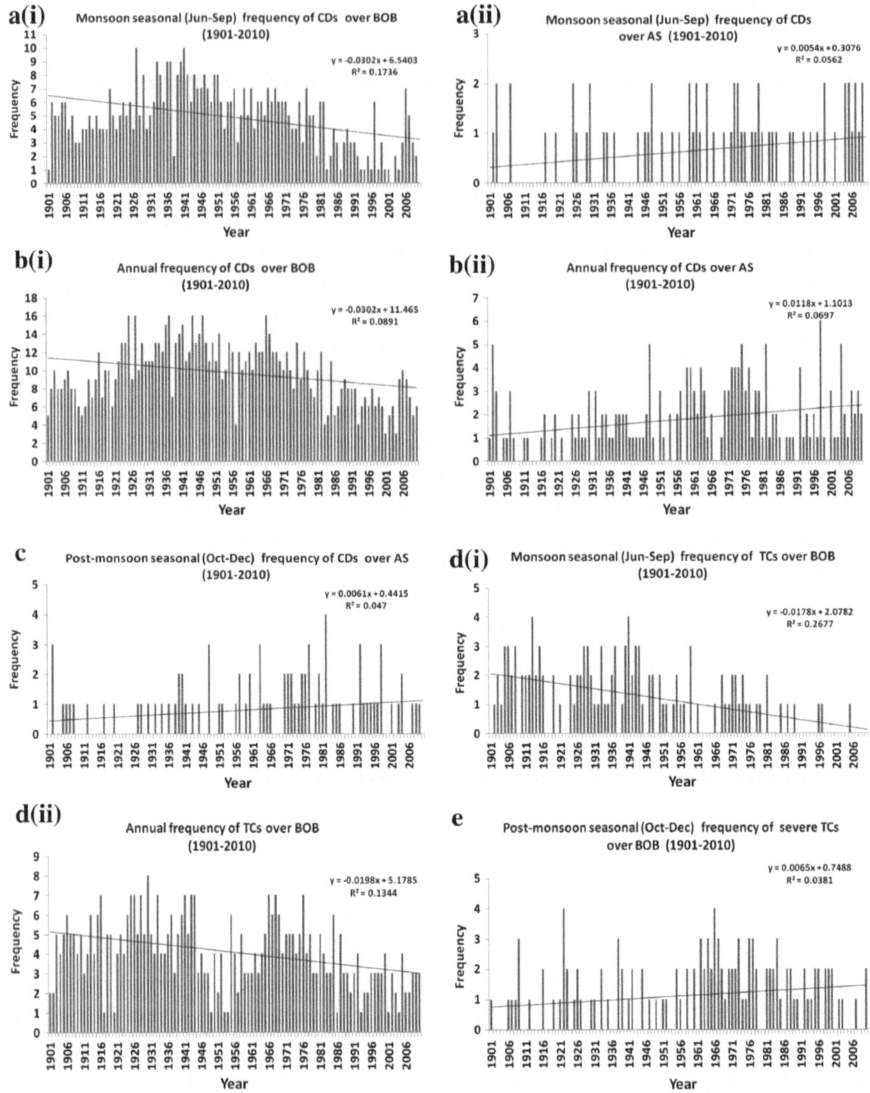

Fig. 1 Frequency of CDs in **a** monsoon season and **b** year as a whole over the (*i*) BOB and (*ii*) AS; **c** frequency of CDs in post-monsoon season over the AS, **d** frequency of TCs over the BOB during (*i*) monsoon season and (*ii*) year as a whole and **e** frequency of severe TCs over the BOB during post-monsoon season for the period of 1901–2010. Linear trend lines are also shown

3.3 Trends in Intensification of CDs and TCs

Table 3 presents the trends in ratio of frequency (30 year running total) of TCs to total CDs and that of severe TCs to total TCs over the NIO, BOB and AS for different seasons and year as a whole for the period 1961–2010. There are

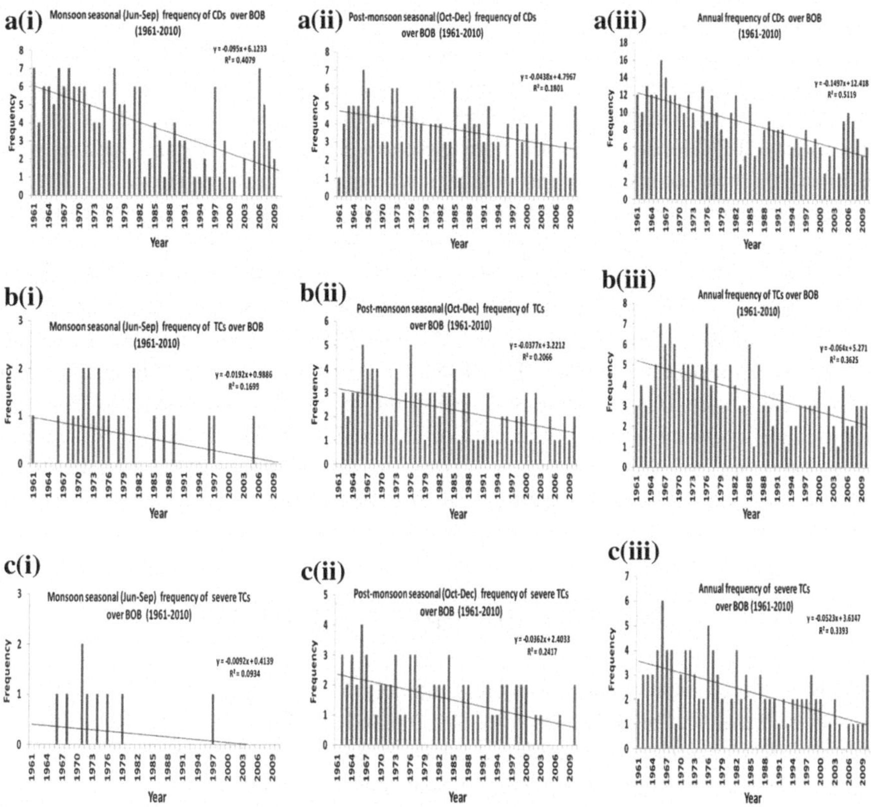

Fig. 2 Frequency of **a** CDs, **b** TCs and **c** severe TCs in (*i*) monsoon, (*ii*) post-monsoon and (*iii*) year as a whole over the BOB during 1961–2010. Trend lines are also shown

Table 3 Linear trend in ratio of TCs to CDs and severe TCs to TCs frequencies over North Indian Ocean during 1961–2010

Basin	Annual/season	Ratio of 30-year running total TCs to CDs frequency			Ratio of 30-year running total severe TCs to TCs frequency		
		Mean	SD	Trend/10 years	Mean	SD	Trend/10 years
NIO	Annual	0.40	0.01	−0.003	0.62	0.02	−0.009
	Pre-monsoon	0.74	0.05	**−0.051**	0.71	0.04	**0.033**
	Monsoon	0.17	0.02	**−0.017**	0.46	0.03	**0.017**
	Post-monsoon	0.56	0.01	**−0.017**	0.65	0.03	**−0.036**
BOB	Annual	0.44	0.01	−0.011	0.62	0.02	**−0.026**
	Pre-monsoon	0.77	0.07	**−0.084**	0.78	0.03	0.008
	Monsoon	0.15	0.03	**−0.033**	0.31	0.09	**−0.117**
	Post-monsoon	0.62	0.02	**−0.030**	0.64	0.03	**−0.035**
AS	Annual	0.45	0.03	**0.045**	0.65	0.05	**0.039**
	Pre-monsoon	0.77	0.06	**0.064**	0.49	0.12	**0.139**
	Monsoon	0.43	0.05	**0.052**	0.71	0.08	0.058
	Post-monsoon	0.39	0.03	**0.044**	0.70	0.05	**−0.031**

Trends significant at the 95 % confidence level are emphasised in bold

significant decreasing trends in the ratio of frequency of TCs to total CDs during pre-monsoon, monsoon and post-monsoon seasons over the BOB and NIO, but significant positive trends during all the three seasons and year as a whole over the AS. Also, there is a significant decreasing trend in the ratio of frequency of severe TCs to total TCs during post-monsoon season over the BOB, AS and NIO as well as during the monsoon season and year as a whole over the BOB. However, there is an increasing trend over the AS during pre-monsoon season and year as a whole as well as over the NIO during pre-monsoon and monsoon seasons.

Figure 3 depicts the long-period trends in the above ratios based on the data of 1901–2010. Considering the long period of 1901–2010, there are significant decreasing trends in intensification of CDs into TCs during the monsoon and the post-monsoon seasons over the NIO. However, once formed, rate of intensification of TCs into severe TCs has increased during the recent years. It is evident from the significant increasing trends in the ratio of severe TCs to TCs frequency during all the seasons and the year as a whole during the period 1901–2010 over the NIO. Similar results have been reported by Singh (2001) also.

3.4 Trends in Associated Atmospheric Parameters

Trends in the environmental parameters, namely SST, vertical wind shear (VWS), low-level vorticity at 850 hPa (VOR850) and mid-tropospheric humidity at 500 hPa (RH500) have been analysed for BOB and AS by Mohapatra et al. (2015) from the time series of area averaged seasonal and annual mean values of the parameters during the period 1951–2010. It has been shown that there is a significant increasing trend in the SST and a significant decreasing trend in the RH500 during all the three seasons (pre-monsoon, monsoon and post-monsoon seasons) over both BOB and AS. The vertical wind shear has decreased significantly over AS during the pre-monsoon season, over both BOB and AS during the monsoon season and over AS and southern parts of BOB during the post-monsoon season. It has increased over the northern parts of BOB. The low-level cyclonic vorticity has increased over the BOB during the pre-monsoon season and over northern parts of the BOB during the post-monsoon season in the recent decades.

3.5 Impact of CDs Trends on Seasonal Rainfall

Table 4 presents the CCs between the CDs frequency and monsoon seasonal rainfall of all-India and four homogeneous regions of India during the period 1901–2010. It endorses the earlier findings of Mooley and Shukla (1989) that CDs influence monsoon rainfall over Central India. As it can be seen, all-India monsoon rainfall does not have a significant relationship with seasonal CDs frequency. The rainfall over the Central India is directly related to the frequency of CDs during the

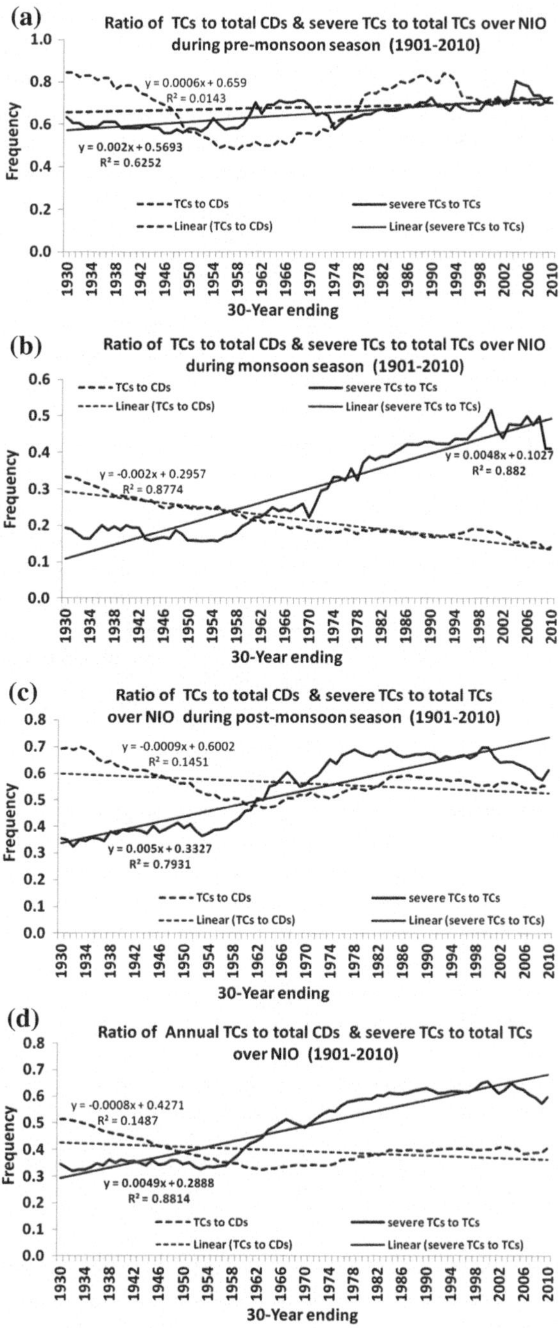

Fig. 3 Linear trends in ratio of 30-year moving total TCs to CDs and severe TCs to TCs frequencies during **a** pre-monsoon, **b** monsoon, **c** post-monsoon and **d** year as a whole over the NIO for the period of 1901–2010

Table 4 Correlation coefficient (CC) between monsoon rainfall over various regions of India and CDs frequency based on data of 1901–2010

Region of rainfall	CC between rainfall and CDs frequency
All India	0.13
North-west India	0.05
North-east India	−0.02
Central India	**0.21**
Southern Pen. India	0.03

CC significant at the 95 % confidence level is emphasised in bold

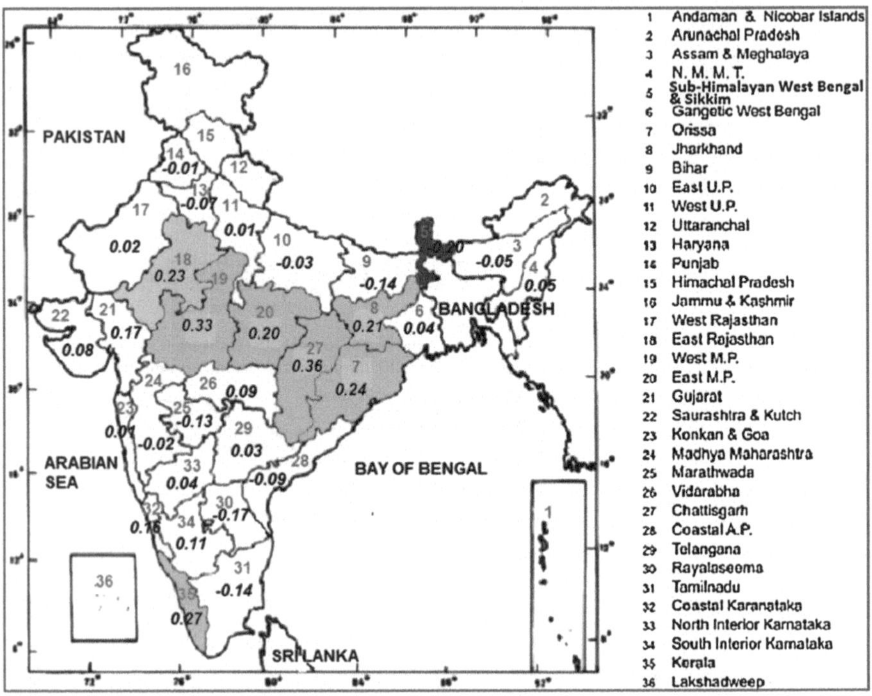

Fig. 4 Correlation coefficient (CC) between sub-divisional rainfall and CDs frequency during the monsoon season based on data of 1901–2010. Sub-divisions showing significantly positive CCs are *shaded red* and negative CCs, *blue* (*whole numbers* indicate the meteorological sub-division as listed on the *right side*; *Real numbers* (with two decimal) are the CC values; |CC| > 0.19: significant at the 95 % confidence level) (color figure online)

monsoon season. Figure 4 presents the relation between seasonal CDs frequency and monsoon rainfall on sub-divisional scale. Monsoon rainfall over the sub-divisions of Central India, viz. Odisha, Jharkhand, Chhattisgarh, East Madhya Pradesh, West Madhya Pradesh, East Rajasthan and Kerala significantly increases with the increase in CDs frequency during the monsoon season. All the above 6 sub-divisions of Central India lie along the tracks of monsoon depressions (IMD 2008, 2011). Higher rainfall over Kerala in association with monsoon CDs may be due to increased cross-equatorial flow and hence stronger winds over south-east AS enhancing rainfall activity through orographic lifting by the Western Ghats. Significantly negative CC is obtained between the monsoon rainfall over sub-Himalayan West Bengal and Sikkim and the monsoon CDs frequency, indicating a decrease in seasonal rainfall over this region due to frequent formation and movement of CDs along the monsoon trough. Figure 5a presents 30-year moving CCs between monsoon seasonal rainfall of all-India/Central India and CDs frequency. It is noted that the significantly positive relation that existed between

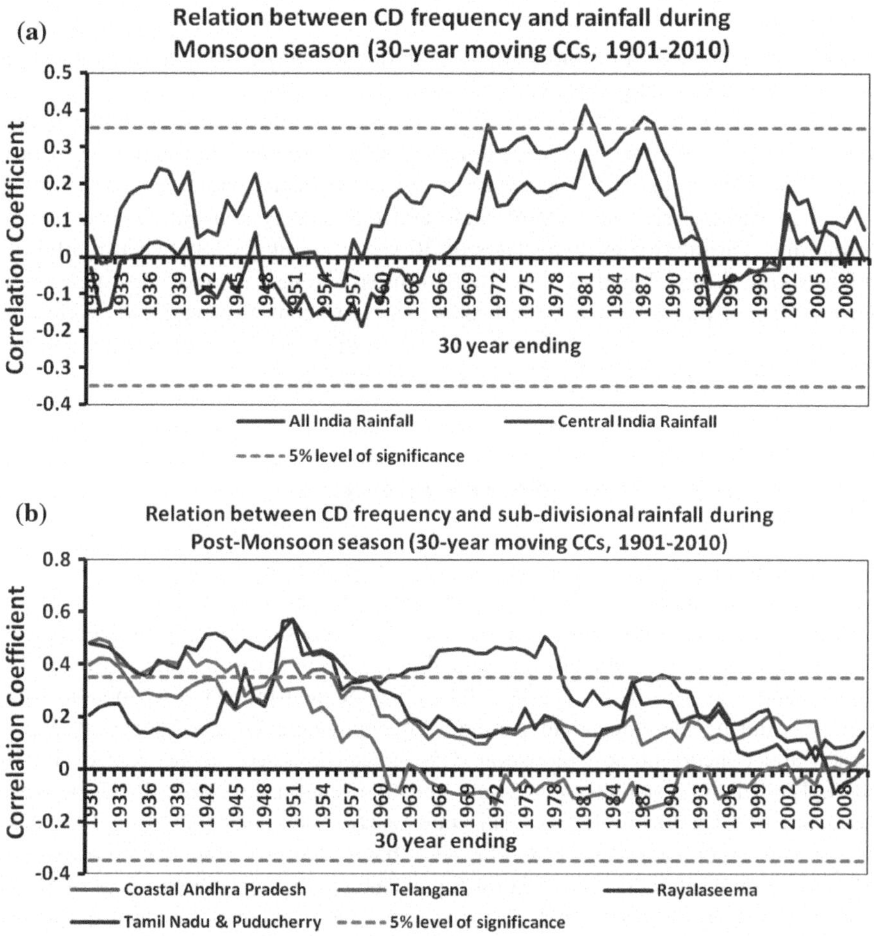

Fig. 5 30-year moving CCs of frequency of CDs with **a** All India and Central India rainfall during monsoon season and **b** sub-divisional rainfall over south-eastern India during post-monsoon season (CCs significant at the 95 % confidence level are indicated by *dotted lines*)

Table 5 Correlation coefficient between post-monsoon rainfall over sub-divisions of southern Peninsular India and CDs frequency based on data of 1901–2010

Sub-division	CC between rainfall and CDs frequency
Coastal Andhra Pradesh	0.19
Telengana	0.15
Rayalaseema	**0.24**
Tamil Nadu and Puducherry	**0.20**
Coastal Karnataka	0.10
North Interior Karnataka	0.14
South Interior Karnataka	0.06
Kerala	0.06

CCs significant at the 95 % confidence level are emphasised in bold

monsoon rainfall of Central India and CDs frequency during the period 1940–1990 has decreased drastically subsequently.

Table 5 presents the CCs between the post-monsoon seasonal rainfall of sub-divisions of southern peninsular India and CDs frequency during post-monsoon season. Significant positive relationship exists between post-monsoon seasonal rainfall of Rayalaseema and Tamil Nadu and Puducherry and the CDs frequency during this season. However, analysis of 30-year moving CCs between rainfall of major sub-divisions benefitted by the north-east monsoon and CDs frequency during the post-monsoon season indicates that significant positive relationship that existed between rainfall of Rayalaseema and Tamil Nadu and Puducherry and the CDs frequency up to 1980 has decreased subsequently.

3.6 *Role of ENSO on CDs/TCs Frequency*

Relationships between the frequency of CDs, TCs and severe TCs over NIO and the Nino 3.4 index during the pre-monsoon, monsoon and post-monsoon seasons in both the antecedent and concurrent modes have been studied by Mohapatra el al. (2015). Figure 6(a–c). shows the 30-year sliding CCs between frequency of (a) CDs (b) TCs and (c) severe TCs over the NIO during the post-monsoon season and Nino 3.4 index during the antecedent winter(JF)/pre-monsoon (MAM)/monsoon (JJAS) season and concurrent post-monsoon (OND) season based on the data of 1951–2010. There is a persistently increasing negative relationship, after 1960, between frequency of CDs during the post-monsoon season and concurrent Nino 3.4 index, though not significant (Fig. 6a). The CCs have changed sign from positive to negative around 1960. Similar relationship exists with antecedent Nino 3.4 of JJAS. However, a positive relation is building up since mid-seventies between frequency of CDs during post-monsoon season and Nino 3.4 of winter/pre-monsoon seasons.

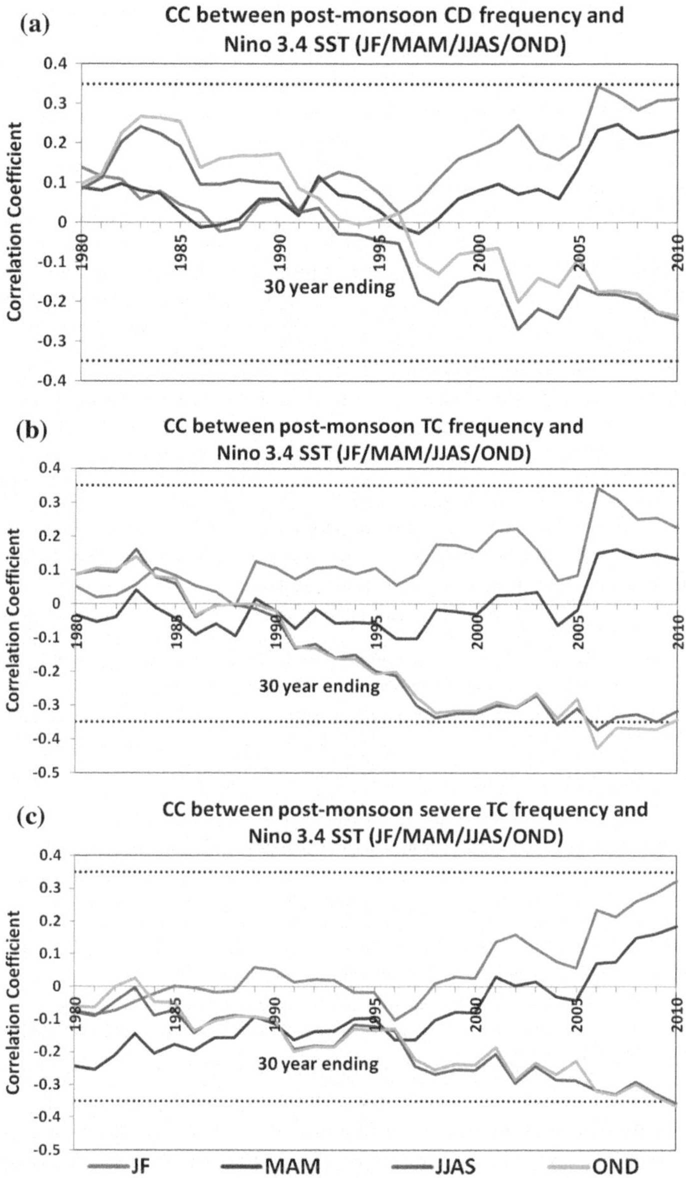

Fig. 6 The 30-year sliding correlation coefficients (CC) between frequency of **a** CDs, **b** TCs and **c** severe TCs over NIO during the post-monsoon season and Nino 3.4 SST during the antecedent winter (JF)/pre-monsoon (MAM)/monsoon (JJAS) season and concurrent post-monsoon (OND) season based on data of 1951–2010 (CCs significant at the 95 % confidence level are indicated by *dotted lines*)

In the case of TCs, significant negative relation has emerged since mid-seventies with concurrent Nino 3.4 and antecedent Nino 3.4 of monsoon season (Fig. 6b). Similar relation has emerged for severe TCs also (Fig. 6c).

4 Conclusion

Long-term trends in annual frequency and intensity of CDs and TCs over the NIO, BOB and AS are examined for various seasons using the data of 1901–2010. Significant decreasing trends in the frequencies of CDs as well as TCs are observed for the monsoon season and year as a whole over the BOB and hence over the NIO. There is a decrease of about 3 CDs, including 2 TCs in a year from the beginning of twentieth century to the beginning of twenty-first century over the BOB and hence the NIO. However, there is a significant increasing trend in severe TCs (about 0.7 per 100 years) during the post-monsoon season over BOB and hence NIO during the same period. For the AS, there is a significant increase in the frequency of CDs during the monsoon and the post-monsoon seasons (0.5 and 0.6 per 100 years) and hence in the year as a whole (1.1 per 100 years).

Trends are also examined during the satellite period (1961–2010). The CDs, TCs and severe TCs frequencies during the last five decades over the BOB and NIO show a significant decreasing trend for the monsoon and post-monsoon seasons and the year as a whole with maximum decrease in CDs frequency during the monsoon season.

There are significant decreasing trends in rate of intensification of CDs into TCs during the pre-monsoon, monsoon and the post-monsoon seasons over the BOB and NIO. There are also increasing trends of intensification during the pre-monsoon, monsoon and post-monsoon seasons and the year as a whole over the AS during the period 1961–2010. However, once formed, more TCs intensify into severe TCs during the pre-monsoon and monsoon seasons over the NIO and during the pre-monsoon season and the year as a whole over the AS.

Considering the impacts of trends in CDs on south-west monsoon rainfall, significant positive relationship existed between monsoon rainfall of sub-divisions over Central India and CDs frequency during the period 1940–1990. The relationship weakened subsequently and became insignificant. Significant positive relationship that existed between north-east monsoon rainfall over Rayalaseema and Tamil Nadu and Puducherry up to 1980 has weakened and become insignificant during the recent years. This decreasing relationship is attributable to decrease in frequency of CDs over BOB, along with various other factors. However, in the case of post-monsoon TCs and severe TCs, significant negative relationship has emerged since mid-seventies with Nino 3.4 SST, which could be in association with the climate shift over the Pacific Ocean in mid-1970s. However, it needs further investigation.

References

Chan J.C.L., 2006, Comment on "Changes in Tropical Cyclone Number, Duration and Intensity in a Warming Environment", *Science*, **311**, 1713.

Gray W.M., 1968, "Global view of the origin of tropical disturbances and storms", *Mon. Wea. Rev.*, **96**, 669-700.

Geetha B., and Y.E.A. Raj, 2014, Spatial patterns of Northeast Monsoon rainfall over sub-regions of Southern Peninsular India and Sri Lanka as revealed through Empirical Orthogonal Function analysis, *Mausam*, **65**, 185-204

Girish Kumar, M.S., and M. Ravichandran, 2012; The influence of ENSO on tropical cyclone activity in the Bay of Bengal during October to December, *Journal of Geophysical Research*, **117**, c02033, DOI.10.1029/2011 JC 007417.

Girish Kumar, M.S., V.P. Thanga Prakash, M. Ravichandran, 2014, Influence of Pacific Decadal Oscillation on the relationship between ENSO and tropical cyclone activity in the Bay of Bengal during October–December, *Climate Dynamics*, DOI 10.1007/s00382-014-2282-6

India Met. Dep., 1979, Tracks of storms and depressions in the Bay of Bengal and the Arabian Sea (1877-1970)

India Met. Dep., 1996, Tracks of storms and depressions in the Bay of Bengal and the Arabian Sea 1971-1990: An Addendum to storm track atlas (1877-1970)

India Met. Dep., 2008, Cyclone eAtlas-IMD (Version 1.0/2008), Tracks of Cyclones and Depressions over North Indian Ocean

India Met. Dep., 2011, Cyclone eAtlas-IMD (Version 2.0/2011), Tracks of Cyclones and Depressions over North Indian Ocean (www.rmcchennaieatlas.tn.nic.in).

Knutson T., J.L. McBride, J. Chan, K. Emanuel, G. Holland, C. Landsea, I. Held, J.P. Kossin, A. K. Srivastava, and M. Sugi, 2010, Tropical cyclones and Climate change, *Nat. GeoSci.*, **3**, 157-163.

Landsea, C.W., 1998, Climate variability of tropical cyclones: Past, Present and Future, Storms (Ed. A. Pielke, Sr. and R. A. Pielke, Jr).

Li Z., W. Yu, T. Li, V.S.N. Murty, and F. Tangang, 2013, Bimodal Character of Cyclone Climatology in the Bay of Bengal Modulated by Monsoon Seasonal Cycle, *Jl. Climate*, **26**, 1033–1046.

Mandal, G. S. and Prem Krishna, 2009, Global warming, climate change and cyclone related destructive winds- Discussion of results from some selected studies with emphasis on the north Indian Ocean, *Global Environmental Research*, **13**, 141-150.

Mohapatra, M and U.C. Mohanty, 2004, Some characteristics of low pressure systems and summer monsoon rainfall over Orissa, *Current Science*, **87**, 1245-1255.

Mohapatra, M. and U. C. Mohanty, 2007, Inter-annual variability of summer monsoon rainfall over Orissa (India) in relation to cyclonic disturbances, *Natural Hazards*, 42(2); 301-315; DOI:10.1007/s11069-006-9090-1

Mohapatra, M., B.K. Bandyopadhyay, and A. Tyagi, 2014, Construction and quality of Best Tracks Parameters for Study of Climate Change Impact on Tropical Cyclones over the North Indian Ocean during Satellite era, Monitoring and Prediction of Tropical Cyclones in the Indian Ocean and Climate Change (Eds. Mohanty et al), Springer, Netherlands and Capital Publishing Company, New Delhi

Mohapatra, M., G. S. Mandal, B. K. Bandyopadhyay, A. Tyagi and U. C. Mohanty, 2012a, Classification of cyclone hazard prone districts of India, *Natural Hazards*, **63**, 1601-1620.

Mohapatra, M., B.K. Bandyopadhyay, and A. Tyagi, 2012b, Best Track parameters of tropical cyclones over the North Indian Ocean: A review, *Natural Hazards*, **63**, 1285-1317.

Mohapatra, M., B. Geetha, S. Balachandran, and L.S. Rathore, 2015, On the Tropical Cyclone activity and associated environmental features over North Indian Ocean in the context of Climate Change, *Jl. of Cli. Change*, **1**, 1-26.

Mooley, D.A., and J. Shukla, 1989, Main features of the westward moving low pressure systems which form over the Indian region during the summer monsoon season and their relation to the monsoon rainfall, *Mausam*, **40**, 137-152.

Ng, E.K.W., and J.C.L. Chan, 2011, Inter-annual variations of tropical cyclone activity over the North Indian Ocean, *Int. J. Climatol.*, Wiley online library, DOI:10.1002/joc.2304

Niyas, N.T., A.K. Srivastava, and H.R. Hatwar, 2009, Variability and trend in the cyclonic storms over North Indian Ocean, IMD, Met. Monograph No. Cyclonic Warning -3/2009.

Patnaik, D.R., 2005, Variability of oceanic and atmospheric conditions during active and inactive periods of storms over the Indian region, *Int. J. Climatol.*, **25**, 1523-1530.

Rao, Y.P., 1976, Southwest monsoon, IMD, Met. Monograph No. Synoptic Meteorology – 1/76, 335 pp.

Raj, Y.E.A., 2011, Climatology and Forecasting based on climatology of tropical cyclones of Indian seas, IMD, Met. Monograph No. Cyclone Warning 9/2011.

Rajeevan, M., J. Srinivasan, K. NiranjanKumar, C. Gnanaseelan, M.M. Ali, 2013, On the epochal variation of intensity of tropical cyclones in the Arabian Sea, *Atmospheric Science Letters*, **14**, 249-255.

Sikka, D.R., 2006, Major advances in understanding and prediction of tropical cyclones over north Indian Ocean: a perspective, *Mausam*, **57**, 165-196.

Singh, O.P., T.M.A. Khan, and M.S. Rahman, 2000, Changes in the frequency of tropical cyclones over the North Indian Ocean, *Meteorology and Atmospheric Physics,* 75, 11-20.

Singh, O.P., 2001, Long term trends in the frequency of monsoonal cyclonic disturbances over the North Indian Ocean, *Mausam*, **52**, 655-658.

Srivastava, A.K., K.C. Sinha Ray, and U.S. De, 2000, Trends in the frequency of cyclonic disturbances and their intensification over Indian seas, *Mausam,* **51**, 113-118.

Tyagi, A., M. Mohapatra, B. K. Bandyopadhyay and Naresh Kumar, 2010, Inter-annual variation of frequency of cyclonic disturbances landfalling over WMO/ESCAP Panel Member Countries, WMO/TD-No. **1541** on Ist WMO International conference on Indian Ocean Tropical Cyclones and climate change, Muscat, Sultanate of Oman, 08-11 March 2009', WWRP-**2010/2**, pp. 1-7

Webster, P.J., G.J. Holland, J.A. Curry, and H.R. Chang, 2005, "Changes in Tropical Cyclone Number, Duration and Intensity in a Warming Environment", *Science*, **309**, 1844-1846.

Chapter 7
Variability and Changes in Cloud Cover Over India During 1951–2010

A.K. Jaswal

1 Introduction

Clouds have an enormous impact on weather and climate by reflecting sunlight, blocking outgoing long-wave radiation and producing precipitation. In addition, clouds also play an important role in the recycling of water vapour from the earth's surface to the atmosphere and back again. Evaporation of water removes heat from the surface and represents an important cooling process. Clouds generally form in rising air, which expands and cools in the presence of condensation nuclei in the supersaturated air. The actual form of clouds so created depends on local conditions such as the strength of the uplift and air stability. In unstable atmospheric conditions, convection dominates, creating vertically developed clouds. Stable air produces horizontally homogeneous clouds. Due to high albedo, low-level clouds have a cooling effect, whereas high-level clouds trap outgoing long-wave radiation contributing to warming of the earth's surface (Mace et al. 2006; Zelinka and Hartmann 2010).

Different cloud types contribute to total cloud amount and are associated with a wide variety of thermal and dynamic processes in the climate system. Clouds cover about two-thirds of the earth's surface and exert great effects on earth's radiation budget (Wild et al. 2004) and climate change by producing precipitation, reflecting short-wave solar radiation coming from the Sun and returning outgoing long-wave radiation from the surface (IPCC 2007). The effect of clouds on the earth's radiation budget, the "cloud radiative forcing" (Ramanathan et al. 1989; Harrison et al. 1990), is generally negative in the daytime but positive at night. Therefore, knowledge of variations in cloud cover may improve our comprehension of the role of clouds in contemporary climate change (Warren et al. 2007).

A.K. Jaswal (✉)
India Meteorological Department, Earth System Science Organization (ESSO), Pune 411005, India
e-mail: jaswal4@gmail.com

M.N. Rajeevan and S. Nayak (eds.), *Observed Climate Variability and Change Over the Indian Region*, Springer Geology, DOI 10.1007/978-981-10-2531-0_7

2 Review of Past Studies

There have been a large number of studies from specific countries or regions documenting spatial and temporal variations and trends in cloud cover (Henderson-Sellers 1992; Sun and Groisman 2000; Kaiser 2000; Groisman et al. 2004; Dai et al. 2006; Warren et al. 2007). These studies have focused mostly on long-term trends for Europe (Henderson-Sellers 1986; Warren et al. 2007; Sanchez-Lorenzo et al. 2012), North America (Henderson-Sellers 1989; Milewska 2004; Warren et al. 2007), Australia (Jones and Henderson-Sellers 1992; Jovanovic et al. 2011), the United States (Sun 2003; Sun and Groisman 2004; Groisman et al. 2004; Dai et al. 2006) and the former Soviet Union (Sun and Groisman 2000; Sun et al. 2001). Analyses of cloud data suggest increased total cloud cover over the United States (Karl and Steurer 1990; Sun 2003; Groisman et al. 2004; Dai et al. 2006; Warren et al. 2007), Canada (Milewska 2004), the former Soviet Union (Sun and Groisman 2000; Sun et al. 2001), Western Europe, mid-latitude Canada, Russia (Chernokulsky et al. 2011; Sanchez-Lorenzo et al. 2012) and Australia (Jones and Henderson-Sellers 1992; Jovanovic et al. 2011). In contrast, a decreasing trend in the total cloud cover has been revealed over much of China during 1951–1994 (Kaiser 1998, 2000) and 1954–2005 (Xia 2010), over the United States since the 1980s (Sun 2003; Sun and Groisman 2004), over India during 1961–2007 (Jaswal 2010), most of South Africa during 1960–2005 (Kruger 2007) and Italy during 1951–1996 (Maugeri et al. 2001). On the other hand, total cloud cover trends are also period-specific over some regions. Several authors have found significant positive trends in the United States total cloud for periods beginning in or after 1947 and ending before 1996 (e.g., Angell 1990; Plantico et al. 1990; Sun et al. 2001), but Elliott and Angell (1997) found no significant trend in cloudiness for 1973–1993. Sun (2003) indicates increasing trends from about 1950–1990 but declining cloud cover from the 1980s to 2000, suggesting that the increasing trend might be limited to the period from about 1950 to the 1980s. On continent scale, Warren et al. (2007) found a large decrease for South America, small decreases for Eurasia and Africa and no trend for North America in total cloud cover during 1971–1996. Warren et al. (2007) have attributed decreasing trends in cloud cover on many continents, and that some of these trends are linked to ENSO variations. However, cloud feedback remains an uncertain source in global climate change (IPCC 2007).

Published studies about long-term trends in total cloud cover over India have reported decreasing trends (Rao et al. 2004; Warren et al. 2007; Biggs et al. 2007; Jaswal 2010). Rao et al. (2004) studied long-term trends in annual total cloud amount over 15 stations in India and found decreasing trends at 11 stations. Biggs et al. (2007) have reported decrease in annual total cloud amount by 0.09 % per year during 1952–1997. Decreasing trends in total cloud cover over India during 1961–2007 were reported by Jaswal (2010). While on monthly scale, statistically significant decrease has occurred during 1961–2007 for April (3 % per decade), June to September (2 % per decade) and December (5 % per decade); decline in total cloud cover is significant for annual, summer and monsoon (Jaswal 2010). However, Roy and Balling (2005) have found significant increase in cloud cover during summer over Jammu and Kashmir. In our neighbouring country China, total cloud cover decreased up to the middle of 1990s and then levelled off at much of synoptic stations (Xia 2010; Zong et al. 2012).

3 Data and Methodology

3.1 Method of Observations

Clouds are observed visually by trained observers at surface meteorological stations in units of eighths of sky (oktas) following the guidelines given by the World Meteorological Organization (WMO). The surface observations of clouds are made in the "synoptic" code, which is exchanged globally through Global Telecommunication System (GTS). Clouds are classified as high, middle or low according to the altitude of the cloud base above the surface of earth. The information about clouds in the synoptic weather report consists of total cloud amount, individual cloud layer's cloud amount and cloud type (high, medium and low) and base height of the lowest layer of cloud. Total cloud amount recorded in the observations are the amount of sky estimated to be covered by all cloud types. The surface cloud observations have provided the long-term cloud climatologies (Warren et al. 1986, 1988; Hahn and Warren 1999), evaluation of satellite cloud observations (Rossow and Schiffer 1999) and analyzing decadal and long-term changes in cloud cover during the last 100 years or so. Since weather reports of clouds are available for several decades with no change in official observing methods, long-term variations and trends in cloud cover can be studied.

3.2 Data Used

A total 195 surface meteorological stations under the network of India Meteorological Department were selected for analyses for the period 1951–2010. Geographical locations of these 195 meteorological stations with numbers of year data available are shown in Fig. 1. Total cloud amount, maximum temperature, minimum temperature and numbers of rainy days data for these 195 stations were taken from the archives of the India Meteorological Department (IMD). For the entire study period, the observation practice of cloud observation in IMD has remained consistent with the WMO standards. Total cloud amount refers to the portion of the sky in eights (okta) covered by clouds at any height. Surface observations of total cloud amount are recorded at every synoptic hour and observations recorded at 0300 and 1200 UTC were used for preparing mean total cloud cover (TCC), which is expressed in per cent by dividing by 8 and multiplying

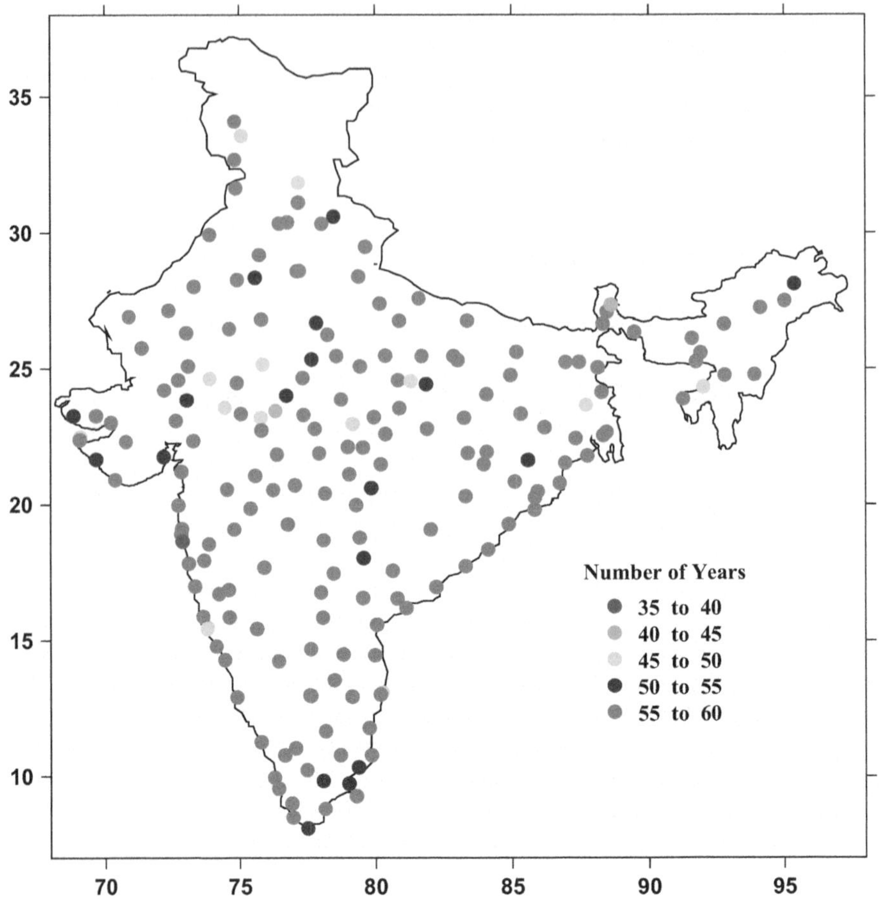

Fig. 1 Geographical locations of stations used for the study

by 100 in this study. Additionally, total cloud cover relationship with climate variables, diurnal temperature range (DTR—the difference between the daily maximum and daily minimum temperature) and number of rainy days is also studied. A rainy day is defined as a day when total precipitation is 2.5 mm or more, as such number of rainy days (NRD) in a month is count of such days. From monthly values, time series of mean annual (January–December), winter (December, January, February), summer (March–May), monsoon (June–September) and postmonsoon (October–November) TCC, DTR and NRD was prepared for 1951–2010. The all India averaged time series for annual and seasonal TCC, DTR and NRD was prepared by averaging all 195 stations.

3.3 Estimation of Long-Term Trends

Trends are determined using a nonparametric Mann–Kendall test to assess the probability that there is a trend statistically different from zero and to evaluate increasing or decreasing slope of trends in the time series of temperature and rainfall by using Sen's method (Sen 1968). The Mann–Kendall test consists of comparing each value of the time series with the others remaining always in sequential order. The number of times that the remaining terms are greater than that under analysis is counted. The calculated TCC trends are tested for significance at the 95 % confidence level by Mann–Kendall and Sen's method.

Temporal variations in annual and seasonal average TCC are shown in Fig. 2. Annual and seasonal spatial patterns of long-term mean TCC are shown in Fig. 3, where coefficient of variation (CV) is shown as contour lines in foreground and variations in mean TCC are shaded in colour in the background. Spatial distributions of annual and seasonal trends in TCC are shown in Fig. 4, where trends significant at 95 % level are marked by an outer circle in black. Scatter plot of India averaged TCC against DTR and NRD is shown in Figs. 5 and 7, respectively. Spatial patterns of correlation coefficients between TCC, DTR and NRD are shown in Figs. 6 and 8, respectively.

4 Results and Discussion

4.1 Temporal Variations in TCC

Figure 2 shows temporal variations in the all India averaged mean annual and seasonal TCC for the period 1951-2010. Mean annual TCC for India is 43.5 %, which is showing a significantly decreasing trend at the rate of −0.44 % per decade. Figure 2a suggests that annual mean TCC has been decreasing steadily since the late 1970s. Time series of annual mean TCC indicates the highest TCC (48.9 %) in 1961 and the lowest (39.2 %) in 1992 as shown in Fig. 2a. Seasonal mean TCC values are 25.8, 32.8, 69.1 and 34.5 % for winter, summer, monsoon and postmonsoon seasons, respectively. The highest cloudy winter season in India was in 1998 with mean TCC 35.4 % and lowest was in 1951 having mean TCC 18.7 %. Summer season total cloud cover was the highest in 1951 (38.8 %) and lowest was 28.1 % in 1973. In monsoon season, the highest obtained TCC was in 1961 (76.2 %), while the lowest cloud cover was in 2009 (61.3 %). It is evident from Fig. 2d that mean TCC for monsoon has been steadily declining since the mid-1980s. For postmonsoon season, TCC was highest in 2010 (43.5 %) and lowest in 1988 (26.3 %). Seasonal TCC trends are decreasing for all seasons, but it is significant for the monsoon season only, where rate of decline is −0.93 % per decade as shown in Fig. 2d. Annual and seasonal temporal variations as shown in Figs. 2a–e indicate consistent lower TCC during last two decades of the data period.

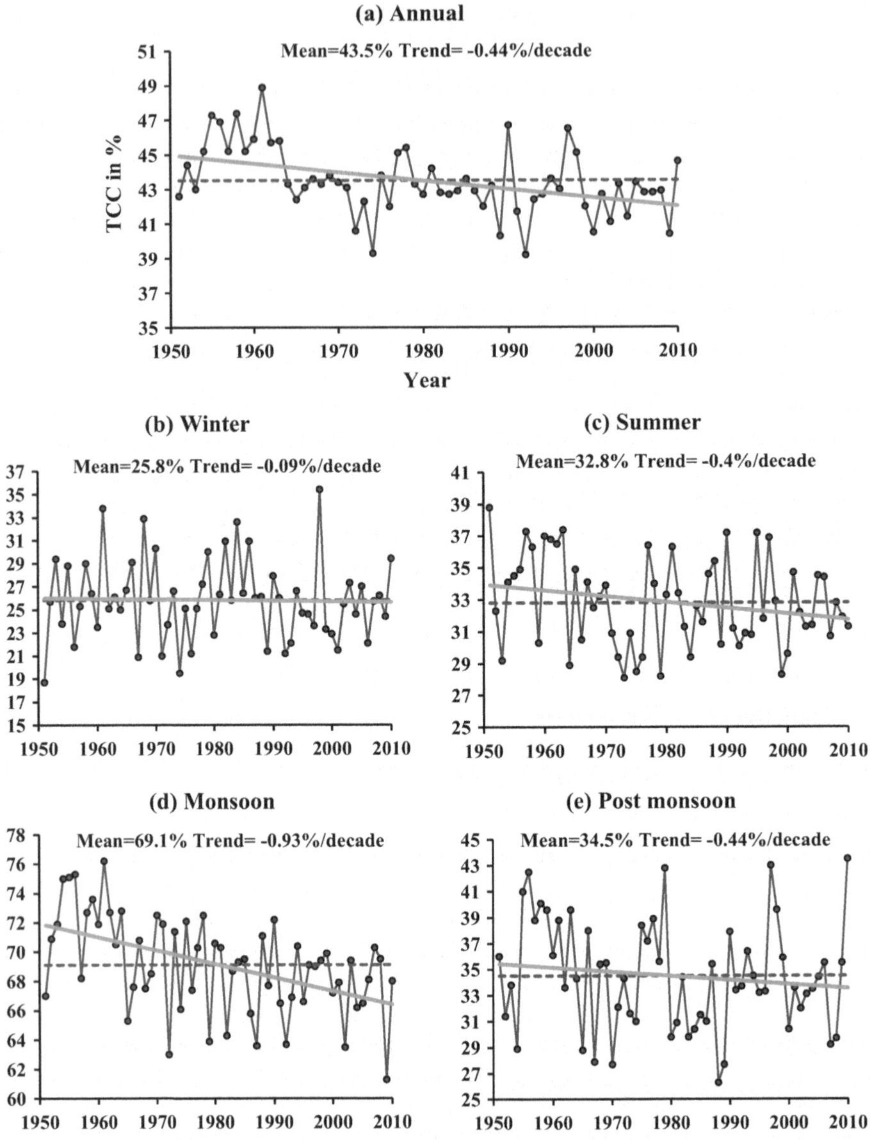

Fig. 2 Temporal variations in all India total cloud cover (TCC) for **a** annual, **b** winter, **c** summer, **d** monsoon and **e** postmonsoon during 1951–2010. *Dashed line in blue* indicates long-term mean, while linear trend line is shown in *red*. TCC trends are significant at the 95 % confidence level for annual and monsoon season only (color figure online)

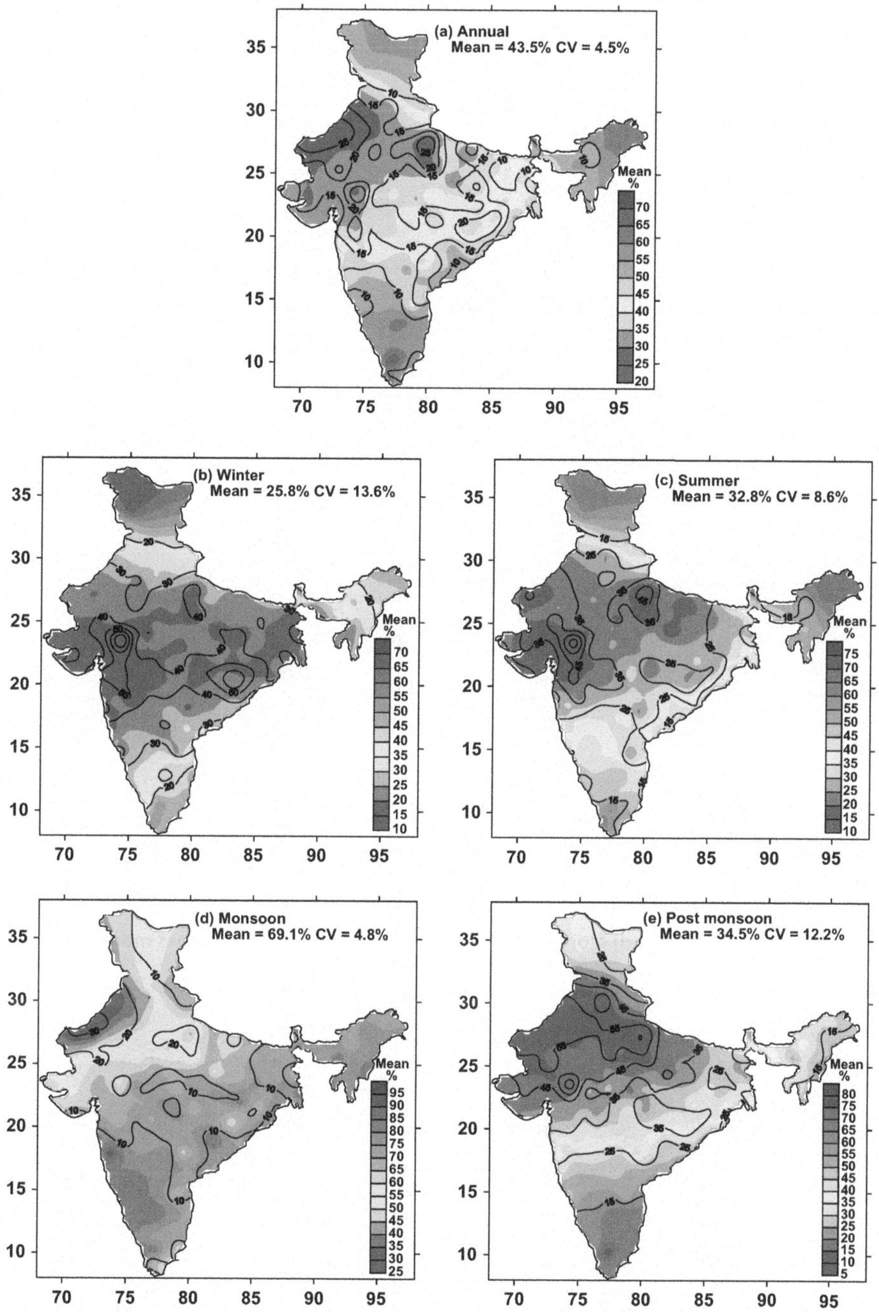

Fig. 3 Spatial pattern of mean total cloud cover (TCC) for **a** annual, **b** winter, **c** summer, **d** monsoon and **e** postmonsoon. Contours are coefficient of variations CV (%), while long-term mean TCC is *shaded*

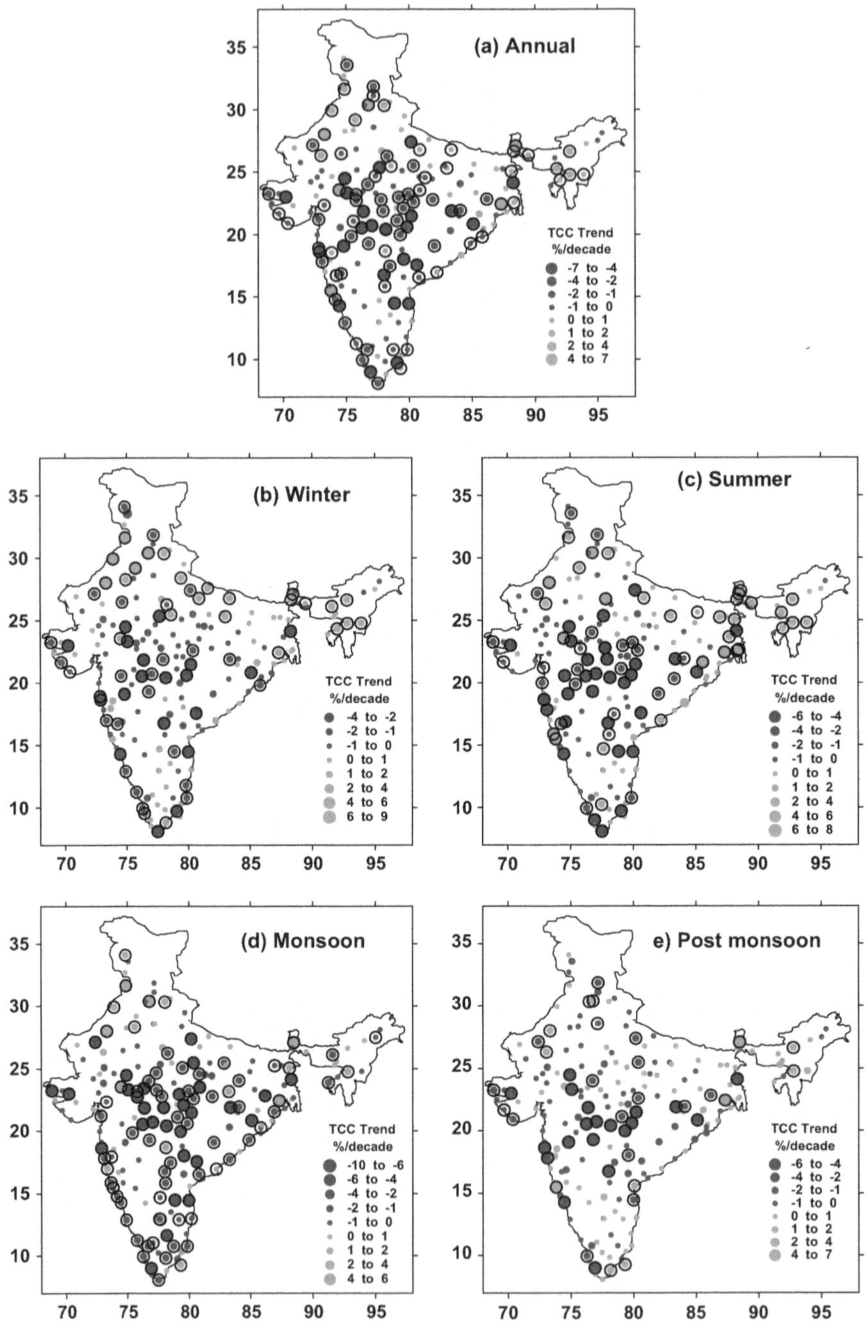

Fig. 4 Spatial pattern of trends in mean total cloud cover (TCC) for **a** annual, **b** winter, **c** summer, **d** monsoon and **e** postmonsoon. Stations having trends significant at the 95 % confidence level are marked by an *outer circle in black*

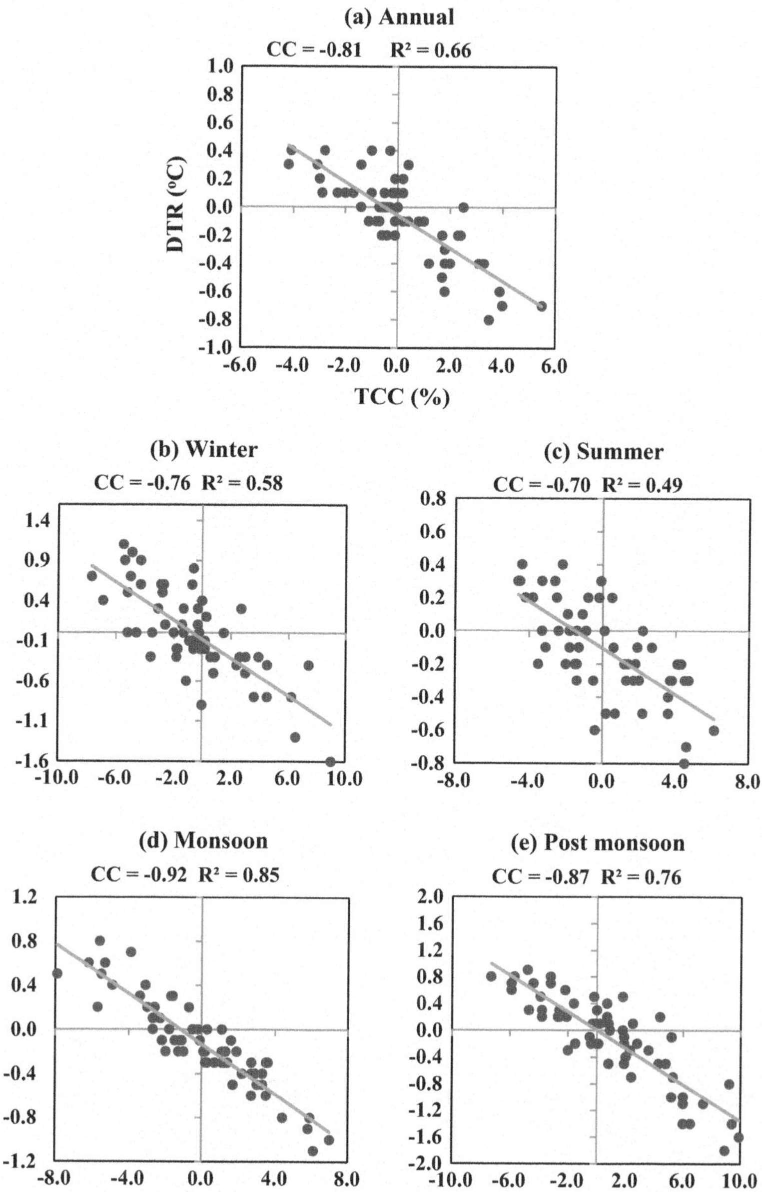

Fig. 5 Scatter plots of all India mean total cloud cover (TCC) and diurnal temperature range (DTR) anomalies for **a** annual, **b** winter, **c** summer, **d** monsoon and **e** postmonsoon. Data series are anomalies from the 1961–1990 period

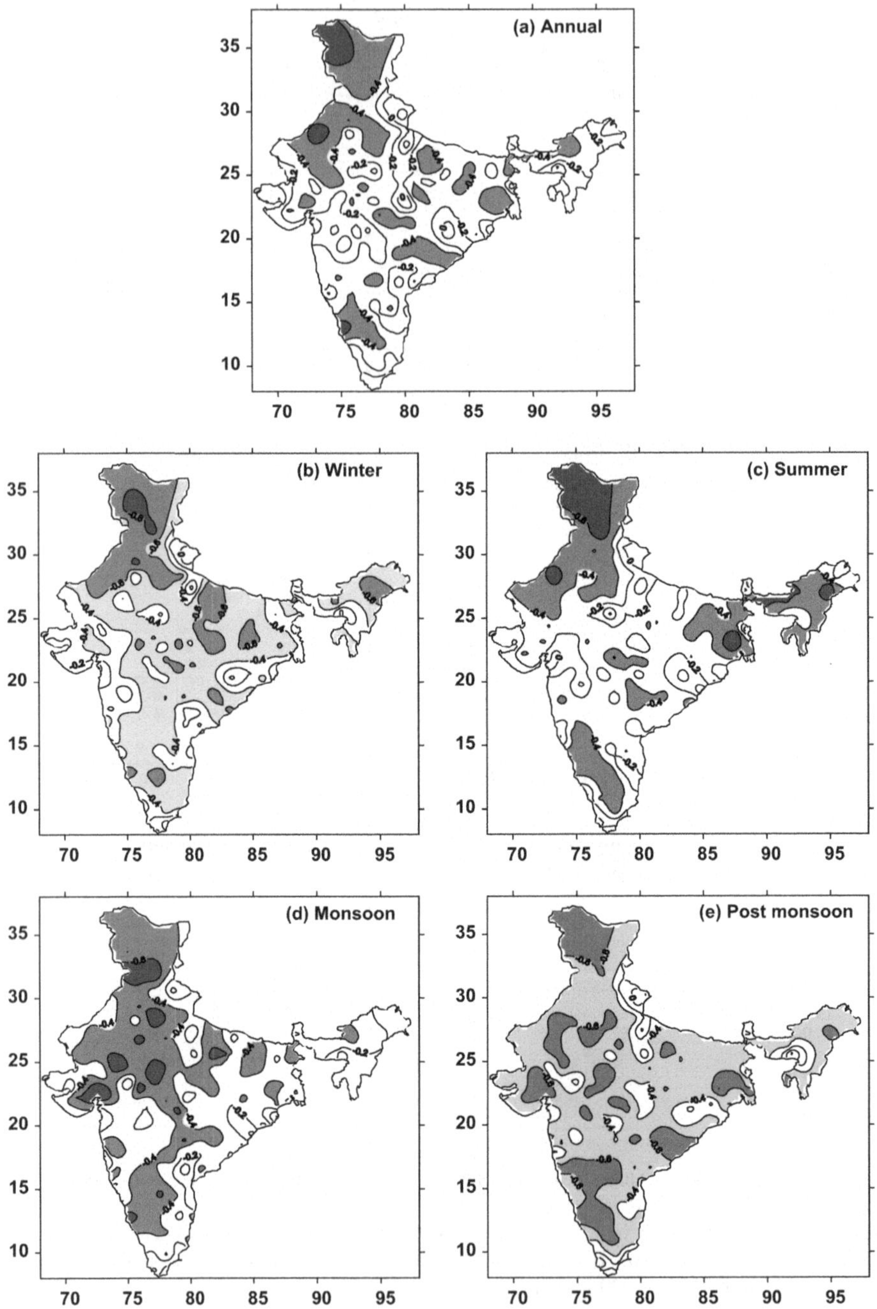

Fig. 6 Spatial pattern of correlation coefficients between total cloud cover (TCC) and diurnal temperature range (DTR) for **a** annual, **b** winter, **c** summer, **d** monsoon and **e** postmonsoon

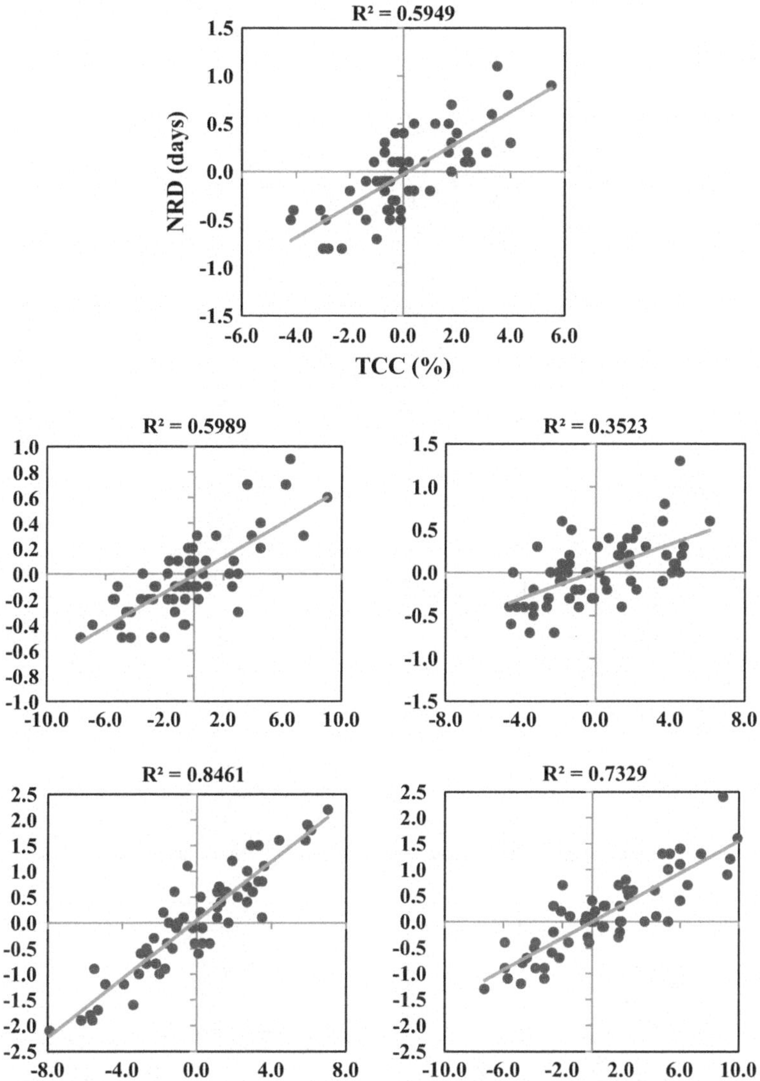

Fig. 7 Scatter plots of all India mean total cloud cover (TCC) and number of rainy days (NRD) anomalies for **a** annual, **b** winter, **c** summer, **d** monsoon and **e** postmonsoon. Data series are anomalies from the 1961–1990 base period

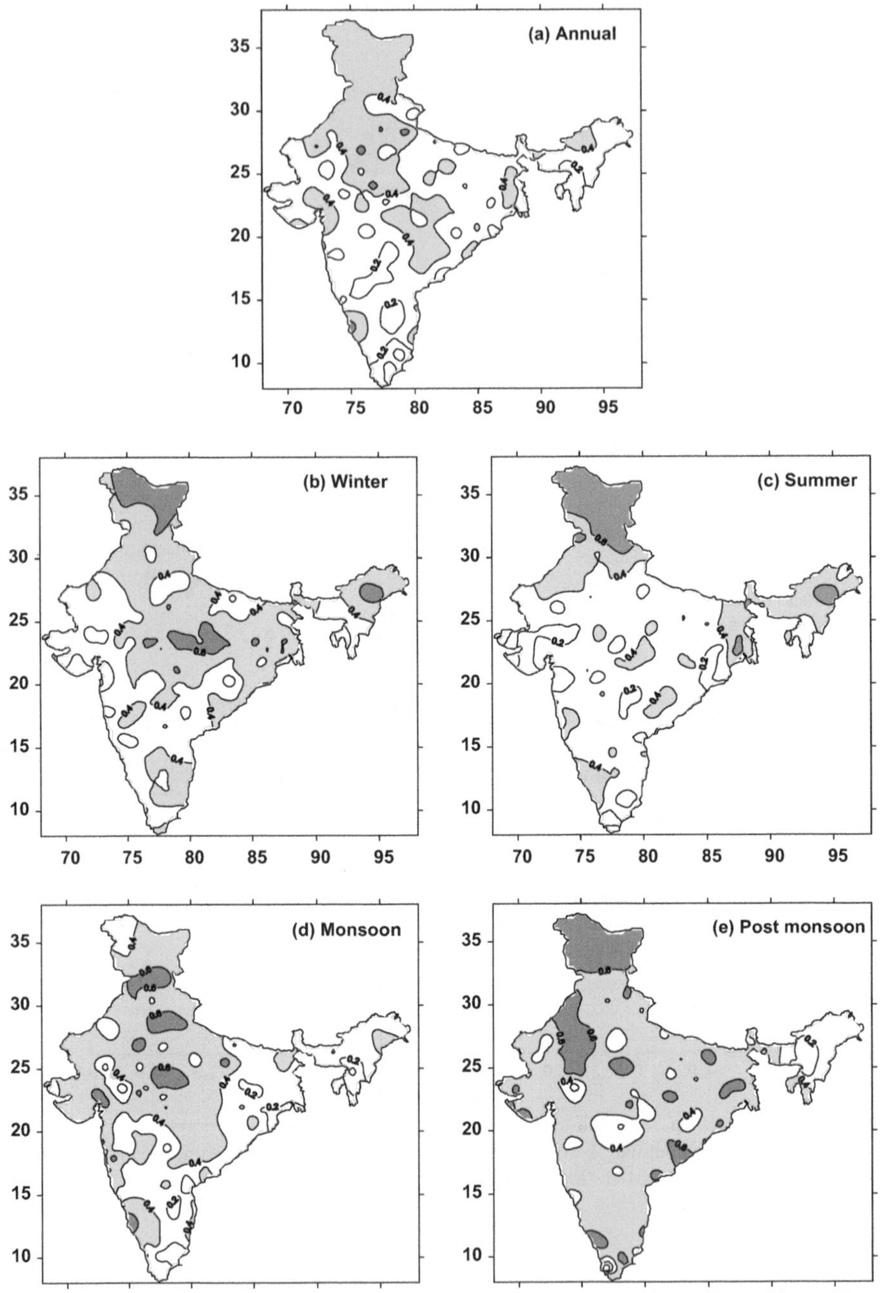

Fig. 8 Spatial pattern of correlation coefficients between total cloud cover (TCC) and number of rainy days (NRD) for **a** annual, **b** winter, **c** summer, **d** monsoon and **e** postmonsoon

4.2 Annual and Seasonal Mean TCC Variation

Spatial distribution of long-term mean TCC over India based upon 1951–2010 period is shown in Fig. 3. The patterns of annual mean TCC indicate region of the highest clouding and lower CV over the Western Himalayas, north-east and over south peninsula (Fig. 3a), where some stations such as Gangtok, Kodaikanal, Alleppey, Thiruvananthapuram have >65 % mean cloud cover. The Western and Central India are having the lowest annual mean clouding and the highest coefficient of variation, where some stations such as Phalodi, Hardoi, Jaisalmer and Sri Ganganagar have <25 % mean cloud cover. Similar to annual CV, variations in seasonal CV are the highest (the lowest), where TCC is the lowest (highest). Winter season TCC is the highest in the Western Himalayas, extreme north-east and extreme south peninsula, while rest of the country is having the lowest TCC as shown in Fig. 3b. Summer season TCC is the highest over the Western Himalayas, north-east, south peninsula and along the eastern coastline as shown in Fig. 3c. Highest TCC over India is during the monsoon season when entire country except western Rajasthan, parts of Punjab and Jammu and Kashmir is having cloud cover above 60 %. Stations in Western Ghats, Chhattisgarh and north-east India have the highest clouding. The highest monsoon season mean TCC is in Mahabaleshwar (94.7 %), while the lowest TCC is in Sri Ganganagar (28.6 %). Postmonsoon season TCC is highest in the south peninsula and north-east India. While Kodaikanal, Thiruvananthapuram and Nagapattinam are having >70 % TCC, stations in the west and Central India such as Hardoi, Sri Ganganagar, Jaisalmer and Phalodi have lower than 10 % mean TCC for the postmonsoon season.

4.3 Spatial Patterns of Annual and Seasonal TCC Trends

Annual

Out of 195 stations selected for study, 125 stations are exhibiting decrease in TCC out of which trends for 74 stations are significant at the 95 % level of confidence as given in Table 1. However, the total cloud cover is significantly increasing at 28 stations, which are mainly located in the north-west and north-east India. Spatial

Table 1 Number of stations having decreasing/increasing trends of total cloud cover (TCC) during 1951–2010 for annual, winter, summer, monsoon and postmonsoon

	Number of stations				
	Annual	Winter	Summer	Monsoon	Postmonsoon
Decreasing	125	118	112	141	112
Significantly decreasing	74	45	54	82	37
Increasing	70	77	83	54	83
Significantly increasing	28	24	32	18	11

distribution of annual TCC trends indicates overall decrease in cloud cover over the country except in the north-west and the north-east India as shown in Fig. 6a. The annual TCC trend values range between −6.2 % per decade and +6.9 % per decade over the country. Stations showing significant decreasing trends are mostly concentrated over the Central India, extreme south peninsula and along western and eastern coasts of India, where rate of decline is in the range of −4 % per decade to −7 % per decade. In contrast, many stations in the north-west India, Indo-Gangetic plains and north-east India have significant increasing trends in the range of +4 % per decade to +7 % per decade. The patterns of annual TCC trends obtained are consistent with trends reported earlier by Jaswal (2010).

Winter

Winter season TCC trends are decreasing at 118 stations out of which trends are significant for 45 stations. Twenty-four stations show significantly increasing trends (Table 1). Spatial distribution of winter TCC trends is similar to the annual patterns, wherein large parts of the country show decline in cloud cover as is shown in Fig. 6b. The trend values range between −3.3 % per decade and +8.5 % per decade. Stations showing significant decrease are more coherent over Central India, south-east peninsula and along the Indian coastline, where trends are in the range of −2 % per decade to −4 % per decade. Stations in hilly regions of the Western Himalayas are also having significant decreasing trends for the winter season. Winter TCC trends are increasing over Indo-Gangetic plains and north-east India, where stations with significant increase in the range of +6 % per decade to +9 % per decade are located.

Summer

Summer total cloud cover is decreasing at 112 stations out of which trends are significant for 54 stations, while 32 stations are having significant increasing trends (Table 1). Spatial distribution of trends suggests decrease over the Western Himalayas, Central India, south peninsula and along the western coast of India as shown in Fig. 6c. The magnitude of trend values ranges between −5.3 % per decade and +7.5 % per decade. Similar to annual and winter season, stations showing significant decrease are more coherent over Central India and south peninsula, where large number of stations is having trends in the range of −4 % per decade to −6 % per decade. Summer season TCC trends are increasing over the Indo-Gangetic plains and north-east India, where many stations having significant increasing trends in the range of +6 % per decade to +8 % per decade are concentrated.

Monsoon

Out of 195 stations, 141 stations are showing decreasing trend in the monsoon season cloud cover. Numbers of stations having significantly decreasing trends are 82, while 18 stations are showing significantly increasing trends (Table 1). Spatial patterns of trends indicate a general decline in cloud cover over the country except over north-west India, where there are increasing trends (Fig. 6d). The magnitude of the monsoon season TCC trend values ranges between −9.3 % per decade and +5.8 % per decade. Stations showing significant decrease in monsoon TCC are more coherent over west, east, Central and Peninsular India, where large numbers of stations are having higher magnitude (−6 % per decade to −10 % per decade) of decreasing trend. In contrast, many stations in the north-west India and hilly regions of the Western Himalayas are having significant increasing trends in the range of +4 % per decade to +6 % per decade in monsoon season.

PostMonsoon

Postmonsoon season cloud cover is decreasing at 112 stations out of which trends are significant at 37 stations, while trends are increasing significantly at 11 stations as given in Table 1. Spatial distribution of trends suggests a decrease over Central India and along the western coast as shown in Fig. 6e. The magnitude of trend values of postmonsoon season TCC is between −5.5 % per decade and +6.8 % per decade. Stations showing significant decrease are more coherent over Central India and along the west coast, where trends are in the range of −4 % per decade to −6 % per decade. Stations showing significant increasing trends in postmonsoon TCC are over north-eastern states and coastal Tamil Nadu, where rising trends are in the range of +4 % per decade to +7 % per decade.

4.4 Relationship Between TCC and Associated Climatic Variables DTR and NRD

Scatter plot of all India averaged time series of annual and seasonal TCC, DTR and NRD anomalies from 1961 to 1990 base years (Figs. 5 and 7) shows relationship between these variables. Reliable cloud observation values should usually have a significant negative relationship with diurnal temperature range and a significant positive relationship with the number of rainy days. In our study, the regions having significantly negative correlations between TCC and DTR have significantly positive correlations between TCC and NRD suggesting a closer relationship between them.

Correlation Between TCC and DTR

The relationship between mean TCC and DTR for annual and four seasons during the period 1951–2010 is shown in Fig. 5. The notable feature of this scatter plot is the opposite relationship between TCC and DTR during all periods indicating significant negative correlation between these two atmospheric parameters as shown in Fig. 5a–e. Upon comparing the two time series, strong correlations were found in the annual means (−0.81) and also for the different seasons: winter (−0.76), summer (−0.70), monsoon (−0.92) and postmonsoon (−0.87). All correlations are significant at the 95 % level of confidence. With coefficient of determination (R^2) in the range of 0.49–0.85, there is strong relationship between these two variables explaining the DTR variability. It is clear that the higher the correlations, the bigger the influence of TCC on DTR; indeed the highest values can be found in monsoon and postmonsoon followed by winter and summer. The relationship between TCC and DTR obtained here is similar to what was reported by Dai et al. (1997). Nonetheless, the relatively strong correlation between observed decreases in all India averaged TCC with increase of DTR supports the notion that the increase in DTR is in response to these physical changes.

Spatial patterns of correlation between TCC and DTR are shown in Fig. 6, where regions having correlation significant at the 95 % level of confidence are shaded. Correlation between mean TCC and DTR is negative over most of the country. However, the observed coincident decrease in TCC and DTR or increase in TCC and DTR over some regions suggests that other mechanisms may have been involved in changing the DTR. The magnitude of correlation of annual mean TCC and DTR time series is between −0.82 and +0.38. Shaded region in Fig. 6a indicates significantly negative correlation between annual mean TCC and DTR, which are primarily located in north-west India and over east, north-east, east coast, west coast of India. Correlation coefficient of winter mean TCC and DTR time series is between −0.88 and +0.23. Spatial patterns of winter mean TCC and DTR correlations (Fig. 6b) suggest regions with moderate-to-strong negative correlation over north-west India, east Uttar Pradesh, Jharkhand and Assam. Figure 6c shows spatial patterns of summer mean TCC and DTR correlation, which have moderate-to-strong correlation over north-west India, West Bengal, Karnataka and north-east India. Monsoon season mean TCC and DTR correlations are between −0.79 and +0.25. Spatially, regions of moderate-to-strong correlations are located over north-west, central, west and south peninsula as shown in Fig. 6d. Postmonsoon season correlation coefficient of TCC and DTR time series varies between −0.81 and +0.29.

Correlation Between TCC and NRD

The notable feature of Fig. 7a–e is the closer relationship between TCC and NRD during all periods indicating significant positive correlation between these two atmospheric parameters. All India NRD is decreasing for annual and four seasons but significant for annual (−0.07 days per decade) and monsoon season (−0.23 days per decade) only. However, the coefficient of correlation between annual TCC and NRD is 0.77, which is highly significant. The coefficient of correlation between seasonal TCC and NRD is 0.77, 0.59, 0.92 and 0.86 for winter, summer, monsoon and postmonsoon, respectively, which is significant at the 95 % level.

Spatial patterns of correlation between TCC and NRD are shown in Fig. 8, where regions having significant correlation at the 95 % level of confidence are shaded. Correlation between mean TCC and NRD is positive over the entire country as 98 % stations are positively correlated for all periods as given in Table 2. Annual mean TCC and NRD correlations are between −0.17 and +0.74. Spatial patterns in Fig. 8a indicate regions of moderate-to-strong correlation between annual mean TCC and NRD, which are primarily located in the north-west, Central, west and north-east India. Winter mean TCC and NRD correlations are between −0.10 and +0.78. Figure 8b shows region of moderate-to-strong positive correlations between winter mean TCC and NRD over north-west, Central, east and north-east India. Summer season mean TCC and NRD correlation coefficients are between −0.18 and +0.73 while spatially, regions of moderate-to-strong correlations are over the Western Himalayas, extreme north-west India, West Bengal, parts of Karnataka and north-east India as shown in Fig. 8c. Monsoon season correlation coefficients between mean TCC and NRD are between −0.10 and +0.77. Spatial patterns of monsoon season mean TCC and NRD correlations suggest regions of moderate-to-strong positive correlation are located over the entire north-west, west, Central India and some pockets over south peninsula as shown in Fig. 8d. Postmonsoon season mean TCC and NRD correlations coefficients range between −0.37 and +0.78, while almost entire country has moderate-to-strong positive correlation as shown in Fig. 8e.

About 60 % of the world is covered by cloud and cloud is an important factor for weather prediction and climate change. Due to the profound influence of clouds on both the water balance and global radiation budget, even small variations can alter the climate response. The response of clouds to global warming represents an area of uncertainty in our understanding of climate change (Cess et al. 1996; Weare et al. 1996). At present, it is not known what will be the effect of cloudiness on global temperatures. Another uncertainty is the magnitude of radiative forcing by

Table 2 Number of stations having positive/negative correlation between total cloud cover (TCC) and diurnal temperature range (DTR) and number of rainy days (NRD) for annual, winter, summer, monsoon and postmonsoon

	Sign of correlation	Number of stations				
		Annual	Winter	Summer	Monsoon	Postmonsoon
TCC and DTR	Positive	36	9	22	18	10
	Negative	159	186	173	177	185
TCC and NRD	Positive	192	194	191	193	193
	Negative	3	1	4	2	2

anthropogenic aerosols. It is also very important for deep understanding related to climate changes, for example global dimming and brightening, global warming and global water cycling (Wild 2009; Wang and Dickinson 2013; Xia 2013).

Annual and seasonal trends in total cloud cover over India and their relationship with climatic variables, diurnal temperature range and rainy days are investigated for the period1951–2010. For country as a whole, statistically significant decrease in TCC has occurred during annual (0.44 % per decade) and monsoon (0.93 % per decade). The TCC trends obtained for India are similar to earlier those reported by Jaswal (2010). TCC trends over India are similar to global trends reported by Dai et al. (2006) and Warren et al. (2007) and trends for China reported by Kaiser (2000) and Qian et al. (2006). Spatially, the strongest and the most consistent evidence for decreasing TCC over India during 1951–2010 is seen over Central India and along the west coast of India, where rate of decrease is in the range of −4 % per decade to −6 % per decade. However, Indo-Gangetic plains are having significant increase in TCC, where some stations show a rate of increase in the range of +4 % per decade to +8 % per decade.

Further, it is found that for most cases, mean annual and seasonal TCC are negatively correlated with annual and seasonal mean DTR, and positively correlated with annual and seasonal NRD. Durre and Wallace (2001) reported that the spatial patterns of DTR trends are physically consistent with the pattern of trends in cloud cover areas. In our study also, areas of moderate-to-strong negative correlations between TCC and DTR have significant decreasing trends in TCC. Similarly, areas of moderate-to-strong positive correlations between TCC and NRD have significant increasing trends in TCC. Almost all the cases of TCC decreases were in moderate and negative relationship with DTR (correlation coefficients varied from −0.37 to −0.86) and in moderate and positive relationship with NRD (correlation coefficients varied from 0.36 to 0.88). As is shown in Fig. 8a–e, time series of number of days with precipitation are well correlated with those of total cloud cover over India during 1951–2010.

Clouds affect surface temperature which in turn also affect cloud development (Warren et al. 2007). The decreased cloud cover enhances incoming solar radiation during the day and accelerates outgoing long-wave radiation at night, which can lead to an increase in daily maximum temperature and reduction in daily minimum temperature. Similarly, decreased cloud cover reduces chances of precipitation. The decrease in total cloud cover may have a large bearing on agriculture and water availability. On the other side, high daytime temperatures and low night-time temperatures are a major constraint to crop productivity and human health. Crop yields would be negatively affected owing to increased insect damages and plagues by all kinds of pathogens, which are likely to occur with increasing temperatures.

5 Conclusions

In summary, the mechanism contributing to the increasing/decreasing total cloud cover over India is unknown and uncertain at present. One factor causing decrease in total cloud cover may be the direct effect of aerosols. As aerosols can cool the earth's surface by reflecting sunlight and warm the aerosol layer by absorbing downward long-wave radiation, the lapse rate will decrease and atmospheric stability will increase, suppressing cloud formation and reducing the cloudiness (Dai et al. 1997; IPCC 2007). Therefore, further investigations are needed for attributing changes in total cloud cover over India.

The analysis of total cloud cover carried out in this investigation provides spatially and temporally detailed results for India based upon long-term quality checked data. The results of this study are summarized as follows:

(a) The data analysis indicates that there has been a general decline in mean daily total cloud cover over most parts of the country during the past 60 years (1951–2010), except over the Indo-Gangetic plains and north-east India.
(b) All India averaged annual and seasonal total cloud cover trends are decreasing for all periods but statistically significant for annual (0.44 % per decade) and monsoon (0.93 % per decade) only.
(c) Out of 195 stations considered, annual total cloud cover has decreased at 64 % of the stations out of which 59 % are significant at the 95 % confidence level. On the seasonal scale, the maximum spatial extent for decreased cloud cover has occurred during the monsoon (at 72 % stations) followed by winter (at 61 % stations), summer and postmonsoon (at 57 % stations each) seasons.
(d) Seasonally, the declining trends are statistically significant at 38, 48, 58 and 33 % stations during winter, summer, monsoon and post-monsoon, respectively.
(e) Areas of moderate-to-strong negative correlations between total cloud cover and diurnal temperature range have significant decreasing trends in total cloud cover for all periods. Similarly, areas of moderate-to-strong positive correlations between total cloud cover and number of rainy days have significant increasing trends in total cloud cover for all periods.

References

Angell, J. K., 1990, Variation in United States cloudiness and sunshine duration between 1950 and the drought year of 1988, Journal of Climate, **3,** 296–308.
Biggs, T. W., Scott, C. A., Rajagopalan, B. and Turral, H. N., 2007, Trends in solar radiation due to clouds and aerosols, southern India, 1952-1997, Int. J. Climatol., **27**, 1505-1518.
Cess, R. D. and Coauthors, 1996,Cloud feedback in atmospheric general circulation models: An update, J. Geophys. Res., **101**, 12791-12794.

Chernokulsky, A. V., Bulygina, O. N., Mokhov, I. I, 2011, Recent variations of cloudiness over Russia from surface daytime observations, Environ. Res. Lett., **6,** 3, 035202.

Dai, A., Del Genio, A. D. and Fung, I. Y., 1997, Clouds, precipitation and temperature range. Nature, **386**, 665–666.

Dai, A., Karl, T. R, Sun, B. E. and Trenberth, K. E., 2006, Recent trends in cloudiness over the United States: A tale of monitoring inadequacies, Bull. Amer. Meteor. Soc., **87**, 597-606.

Durre, I., Wallace, J. M., 2001, Factors influencing the cold-season diurnal temperature range in the United States, Journal of Climate, **14,** 3263–3278.

Elliott, W. P., and Angell, J. K., 1997, Variations of cloudiness, precipitable water, and relative humidity over the United States: 1973–1993, Geophys. Res. Lett., **24**, 41–44, doi:10.1029/96GL03616.

Groisman, P. Y., Knight, R. W., Karl, T. R., Easterling, D. R., Sun, B. and Lawrimore, J. H., 2004, Contemporary changes of the hydrological cycle over the contiguous United States: Trends derived from in situ observations, J. Hydromet., **5**, 64-85.

Hahn, C. J., and Warren, S. G., 1999, Extended edited synoptic cloud reports from ships and land stations over the globe, 1952–1996, Report NDP026C. Oak Ridge, TN: Carbon Dioxide Information Analysis Center, Oak Ridge National Lab.

Harrison, E. F., Minnis, P., Barkstrom, B. R., Ramanathan, V., Cess, R. D. and Gibson, G. G., 1990, Seasonal variation of cloud radiative forcing derived from the Earth Radiation Budget Experiment, J. Geophys. Res., **95**, 18687–18703.

Henderson-Sellers, A., 1986, Cloud changes in a warmer Europe, Clim. Change, **8**, 25-52.

Henderson-Sellers, A., 1989, North American total cloud amount variations this century, Global Planet Change, **1**, 175–194.

Henderson-Sellers, A., 1992, Continental cloudiness changes this century, Geo. J., **27**, 255-262.

IPCC, 2007, Climate change 2007: the physical science basis, contribution of working group I to the fourth assessment report of the IPCC, Cambridge University Press, Cambridge

Jaswal, A. K., 2010, Changes in total cloud cover over India based upon 1961–2007 surface observations, Mausam, **61,** 4, 455–468.

Jones, P. A. and Henderson-Sellers, A., 1992, Historical records of cloudiness and sunshine in Australia, Journal of Climate, **5**, 260-267.

Jovanovic, B., Collins, D., Braganza, K., Jakob, D., Jones, D. A., 2011, A high-quality monthly total cloud amount dataset for Australia, Clim. Change, **108,** 3, 485–517.

Kaiser, D. P., 1998, Analysis of total cloud amount over China, 1951-1994, Geophys. Res. Lett., **25,** 19, 3599-3602.

Kaiser, D. P., 2000, Decreasing cloudiness over China: An updated analysis examining additional variables, Geophys. Res. Lett., **27**, 15, 2193-2196.

Karl, T. R. and Steurer, P. M., 1990, Increased cloudiness in the United States during the first half of the twentieth century Fact or fiction?, Geophys. Res. Lett., **17**, 1925–1928, doi:10.1029/GL017i011p01925.

Kruger, A. C., 2007, Trends in cloud cover from 1960 to 2005 over South Africa, Water SA, **33,** 5, 603–608.

Mace, G. G., Benson, S., and Kato, S., 2006, Cloud radiative forcing at the Atmospheric Radiation Measurement Program Climate Research Facility: 2. Vertical redistribution of radiant energy by clouds, J. Geophys. Res., **111**, D11S91, doi:10.1029/2005JD005922.

Maugeri, M., Bagnati, Z., Brunetti, M., Nanni, T., 2001, Trends in Italian total cloud amount, 1951–1996, Geophys. Res. Lett., **28,** 24, 4551–4554.

Milewska, E., 2004, Baseline cloudiness trends in Canada 1953–2002, Atmos. Ocean, **42**, 267–280.

Plantico, M. S., Karl, T. R., Kukla, *G.* and Gavin, J., 1990, Is recent climate change across the United States related to rising levels of anthropogenic greenhouse gases? J. Geophys. Res., 95, 16617–16637, doi:10.1029/JD095iD10p16617.

Qian, Y., Kaiser, D. P., Leung, L. R. and Xu, M., 2006, More frequent cloud-free sky and less surface solar radiation in China from 1955 to 2000, Geophys. Res. Lett., **33,** L01812, doi:10.1029/2005GL024586.

Rao, G. S. P., Jaswal, A. K. and Kumar, M. S., 2004, Effects of urbanization on meteorological parameters, Mausam, **55**, 3, 429-440.

Roy, S. S., and Balling Jr., R. C., 2005, Analysis of trends in maximum and minimum temperature, diurnal temperature range, and cloud cover over India, Geophys. Res. Lett., **32**, L12702, doi:10.1029/2004GL022201.

Ramanathan, V., Cess, R. D., Harrison, E. F., Minnis, P., Barkstrom, B. R., Ahmad, E. and Hartmann, D., 1989, Cloud-radiative forcing and climate: Results from the Earth Radiation Budget Experiment, Science, **243**, 57–63.

Rossow, W. B., and Schiffer, R. A., 1999, Advances in understanding clouds from ISCCP, Bull. Amer. Meteor. Soc, **80,** 2261– 2287.

Sanchez-Lorenzo, A., Calbo, J., Wild, M., 2012, Increasing cloud cover in the 20th century: review and new findings in Spain, Climate Past, **8,** 4, 1199–1212.

Sen, P. K., 1968, Estimates of regression coefficient based on Kendall's tau, J. Am. Stat. Assoc, **63,** 1379–1389.

Sun, B., 2003, Cloudiness over the contiguous United States: Contemporary changes observed using ground-based and ISCCP D2 data, Geophys. Res. Lett., **30**, 1053, doi:10.1029/2002GL015887.

Sun, B. and Groisman, P. Y., 2000, Cloudiness variations over the former Soviet Union, Int. J. Climatol., **20,** 1097-1111. Sun, B., Groisman, P. Y. and Mokhov, I. I., 2001, "Recent changes in cloud-type frequency and inferred increases in convection over the United States and the former USSR", J. Climate, **14,** 1864–1880.

Sun, B. and Groisman, P. Y, 2004, Variations in low cloud cover over the United States during the second half of the twentieth century, J. Climate, **17**, 1883–1888.

Sun B.,Groisman P.Y. and Mokhov I.I. 2001 Recent changes in cloud type frequency and inferred increases in convection over the United States and the former USSR J Climate 14 1864-1880.

Wang, K., Dickinson, R. E., 2013, Contribution of solar radiation to decadal temperature variability over land, Proc. Natl. Acad. of Sci., **110**, 37, 14877–14882.

Warren, S. G., Hahn, C. J., London, J., Chervin, R. M. and Jenne, R. L., 1986, Global distribution of total cloud cover and cloud type amounts over land, NCAR Tech. Note TN-273_STR, 229 pp.

Warren, S. G., Hahn, C. J., London, J., Chervin, R. M. and Jenne, R. L., 1988, Global distribution of total cloud cover and cloud type amounts over the ocean, NCAR Tech. Note TN-317_STR, 212 pp.

Warren, S. G., Eastman, R. M. and Hahn, C. J., 2007, A survey of changes in cloud cover and cloud types over land from surface observations, 1971-1996, J. Clim., **20,** 717-738.

Weare, B. C. and AMIP modeling groups, 1996 Evaluation of the vertical structure of zonally averaged cloudiness and its variability in the Atmospheric Model Intercomparison Project, J. Climate, **9**, 3419-3431.

Wild, M., Ohmura, A., Gilgen, H. and Rosenfeld, D., 2004, On the consistency of trends in radiation and temperature records and implications for the global hydrological cycle, Geophys. Res. Lett., **31,** 11, L11201.

Wild, M., 2009, Global dimming and brightening: a review, J. Geophys. Res., **114**, D00D16, doi:10.1029/2008JD011470.

Xia, X., 2010, Spatiotemporal changes in sunshine duration and cloud amount as well as their relationship in China during 1954–2005, J. Geophys. Res., **115**.

Xia, X., 2013, Variability and trend of diurnal temperature range in China and their relationship to total cloud cover and sunshine duration, Ann. Geophys., **31**, 5, 795–804.

Zelinka, M. D., and Hartmann, D. L.,2010, Why is longwave cloud feedback positive?, J. Geophys. Res., **115**, D16117.

Zong, X., Wang, P. and Xia, X., 2012, Variability and long term trend of total cloud cover in China derived from ISCCP, ERA-40, CRU3, and ground station Datasets, Atmos. Ocean. Lett., **6**, 133–137.

Chapter 8
Variability and Trends of Atmospheric Moisture over the Indian Region

P. Mukhopadhyay, A.K. Jaswal and Medha Deshpande

1 Introduction

Atmospheric water vapour is one of the most important factors in determining weather and climate as it plays a dominant role in the radiative budget of the troposphere. The main sources of water vapour in the lower atmosphere are evaporation from the Earth's surface and transpiration by plants. Since surface humidity and temperature regulate evaporation and transpiration processes, these are connected to both the hydrological cycle and the surface energy budget. Water vapour is one of the most important greenhouse gasses and exceeds carbon dioxide several times in terms of its greenhouse contribution (Trenberth et al. 2007). As temperature rises, the atmosphere's water holding capacity increases leading to increase in atmospheric water vapour (Philipona et al. 2005).

Long-term mean distributions of surface specific humidity show large seasonal and spatial variations (Oort 1983), while the variations in surface relative humidity are relatively small over the oceans (Peixoto and Oort 1996) but considerable over the USA (Gaffen and Ross 1999). Increased surface specific humidity and dew point temperature during the second half of the twentieth century have been reported over the contiguous USA by Robinson (2000), Sun et al. (2000) and Groisman et al. (2004). Increase in surface water vapour is also found over Europe (Schönwiese and Rapp 1997; New et al. 2000; Philipona et al. 2004), the former Soviet Union and China (Sun et al. 2000; Wang and Gaffen 2001), Canada (Vincent et al. 2007) and Japan (New et al. 2000). Investigating changes in global surface

P. Mukhopadhyay (✉) · M. Deshpande
Indian Institute of Tropical Meteorology, Earth System Science Organization,
Pune 411008, India
e-mail: parthasarathi64@gmail.com

A.K. Jaswal
India Meteorological Department, Earth System Science Organization,
Pune 411005, India

M.N. Rajeevan and S. Nayak (eds.), *Observed Climate Variability and Change Over the Indian Region*, Springer Geology, DOI 10.1007/978-981-10-2531-0_8

humidity, Dai (2006) has reported increase in both specific and relative humidity over USA, India and China. A 4 % increase in atmospheric moisture has been observed, consistent with a warming climate (Trenberth et al. 2007). Dee et al. (2001) have found increasing trends in air temperature and relative humidity at most of the airport stations in India in winter season. Rao et al. (2004) have reported increasing trends in annual mean relative humidity at 11 out of 15 urban stations over India. Singh et al. (2008) have studied annual trends in relative humidity over nine river basins in north-west and central India and have found increasing trends in six river basins and decreasing trends in three river basins.

Along with atmospheric water vapour, surface soil moisture is also an important variable in the climate system. Understanding the changes and variability in temperature, precipitation and soil moisture is important for monitoring drought and floods, planning and management of water resources in crop-growing seasons and ensuring food security. In situ soil moisture measurements are costly and labour intensive to perform and are thus only available in limited regions of the world. Furthermore, historical records of soil moisture content (Robock et al. 2000) are sparse but still trends and dynamics in soil moisture have been studied from ground-based measurements (Robock et al. 2005). As in situ measurements commonly lack global coverage and representativeness, recent studies have mostly relied on model estimates (Sheffield and Wood 2008; Zhu and Lettenmaier 2007).

There is a close dynamical and thermos-dynamical link among the processes related to changes in precipitation, water vapour in the atmosphere, evapotranspiration and soil moisture, etc., which in turn manifests to produce the effect of global warming and climate change (Koster et al. 2004). In the context of changes in precipitation with climate change, Trenberth (2011) suggested a direct link between global warming and precipitation. Global warming leads to more evaporation and also increases the water holding capacity of air (by about 7 % per 1 °C warming). Increased water vapour increases intensity of weather systems such as thunderstorms or tropical cyclones.

There are other articles based on observations where the fact of changes in moisture content of the atmosphere has been reported (Albergel et al. 2013) and its link with other processes, namely the evapotranspiration (Jung et al. 2010). Jung et al. (2010) have attempted to test whether a soil moisture shortage could be the reason behind the declining evapotranspiration (ET) trend since 1998 and they found strong coherence between TRMM-based soil moisture data and ET trend (1998–2008) particularly for those regions where moisture supply controls the soil moisture. A detailed evaluation of global trends of surface soil moisture using multi satellite remote sensed data have been reported by Dorigo et al. (2012).

Recently, Calla et al. (2013) analysed surface soil moisture data from Soil Moisture and Ocean Salinity Satellite (SMOS) from 6 a.m. to the evening of the following day, i.e. 6 p.m., particularly to understand the subsurface soil water behaviour over arid region of Rajasthan. Using micro-meteorological tower data from four sites over the Indian region, Eswar et al. (2013) developed a model for disaggregation of evaporative fraction and evaluated the model based on available satellite data, e.g. MODIS. However, they mentioned that it will be worth to explore similar evaluation based on INSAT-3D and INSAT-3A NDVI data. Using SMOS soil moisture product, Calla et al. (2013) also mentioned the possibility of monitoring and analysing overnight surface soil moisture.

In a recent study by Jung et al. (2010), and references there in a significant negative trend is reported coherently in the soil moisture and evapotranspiration anomaly over the global tropics (28°S–38°N) and also over different land regions. Although this study has been carried out in a data-driven way using TRMM data for the period of 1998–2008, the limitation of remote-sensed data prompts towards a need of such analyses based on direct measurement over the global region in general and the Indian region in particular.

Although relatively fewer studies have been carried out over the Indian region in respect of analysing moisture/humidity trends over different parts, Jaswal and Koppar (2011) reported significant increasing trend of surface specific humidity, relative humidity and dry bulb temperature during summer, winter and monsoon months. The increasing trend of specific and relative humidity is particularly noted over north, north-west, central and south-east India. They have also mentioned that the increasing trend of surface moisture and temperature is contributing to enhanced human discomfort over the country.

Keeping above studies in mind and the gap areas that need to be addressed in the context of the Indian region, this article attempts to bring out the variability of atmospheric moisture over the Indian region.

2 Data and Methodology

Humidity is a general term used to indicate moisture in the atmosphere. The specific humidity (g kg^{-1}) is the mass of water vapour contained within a unit mass of moist air which means that the higher the amount of water vapour, the higher the specific humidity. The commonly used term relative humidity is defined as the ratio (%) of the actual vapour pressure of the air to the saturation vapour pressure at the same pressure and temperature. In other words, it is a measure of the actual amount of water vapour in the air compared to the total amount of vapour that can exist in the air at its current temperature.

In this study, we have analysed data from 244 surface meteorological stations for surface humidity and 27 agro-meteorological stations for surface moisture. The geographical locations of these stations in the network of IMD are shown in Fig. 1, where six homogeneous regions of India, viz., north-west (NW), east (E), north-east (NE), central (C), west (W) and south (S), are also marked. These data records are processed and quality checked before archival at the National Data Centre of India Meteorological Department. Daily observations recorded at 0300 and 1200 UTC are used for calculating mean monthly specific humidity and relative humidity. Since daily surface data are available from 1969 onwards, the period of study is restricted to 1969–2012. Daily specific humidity values in g kg^{-1} are calculated by

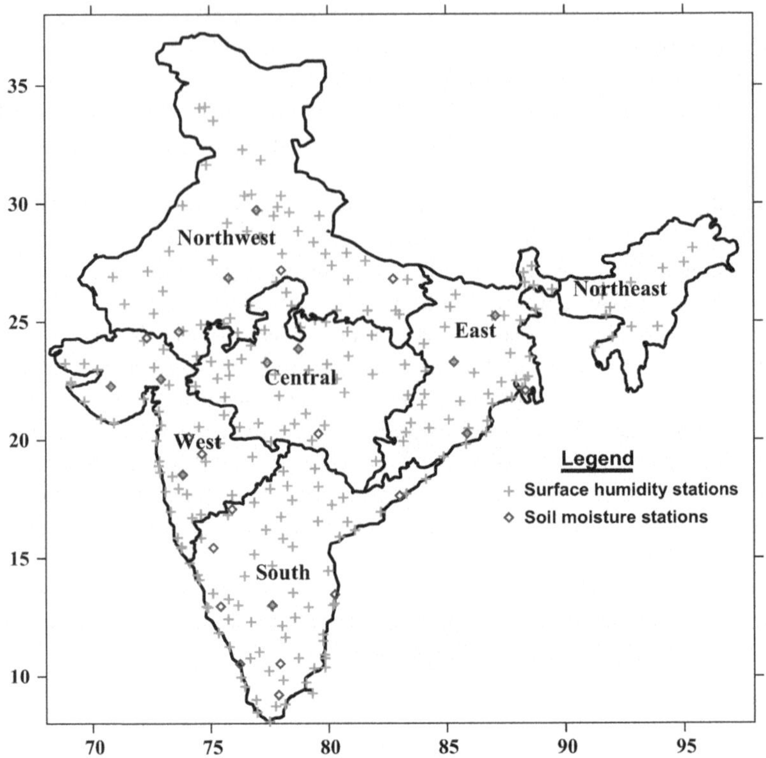

Fig. 1 Geographical locations of 244 stations measuring surface humidity and 27 stations measuring soil moisture distributed over six regions of India

using the following relation originally given by Tetens (1930) and simplified by Murray (1967).

$$e_w = 6.1078 \times \exp^{\frac{17.269388 \times (\mathrm{dpt} - 273.16)}{(\mathrm{dpt} - 35.86)}}$$

Here e_w is the vapour pressure in hPa and dpt is the dew point temperature in °K. The specific humidity sph in g kg^{-1} is calculated by the following relation:

$$\mathrm{sph} = \frac{622 \times e_w}{\mathrm{slp} - 0.378 \times e_w}$$

where slp is the station-level pressure in hPa at the time of observation.

From daily values, monthly means of specific humidity and relative humidity were obtained which are then used to prepare time series for annual (January–December), winter (December–February), summer (March–May), monsoon (June–September) and post-monsoon (October–November). Based upon 244 stations considered in this study, averaged all India and regional (six regions) time series of

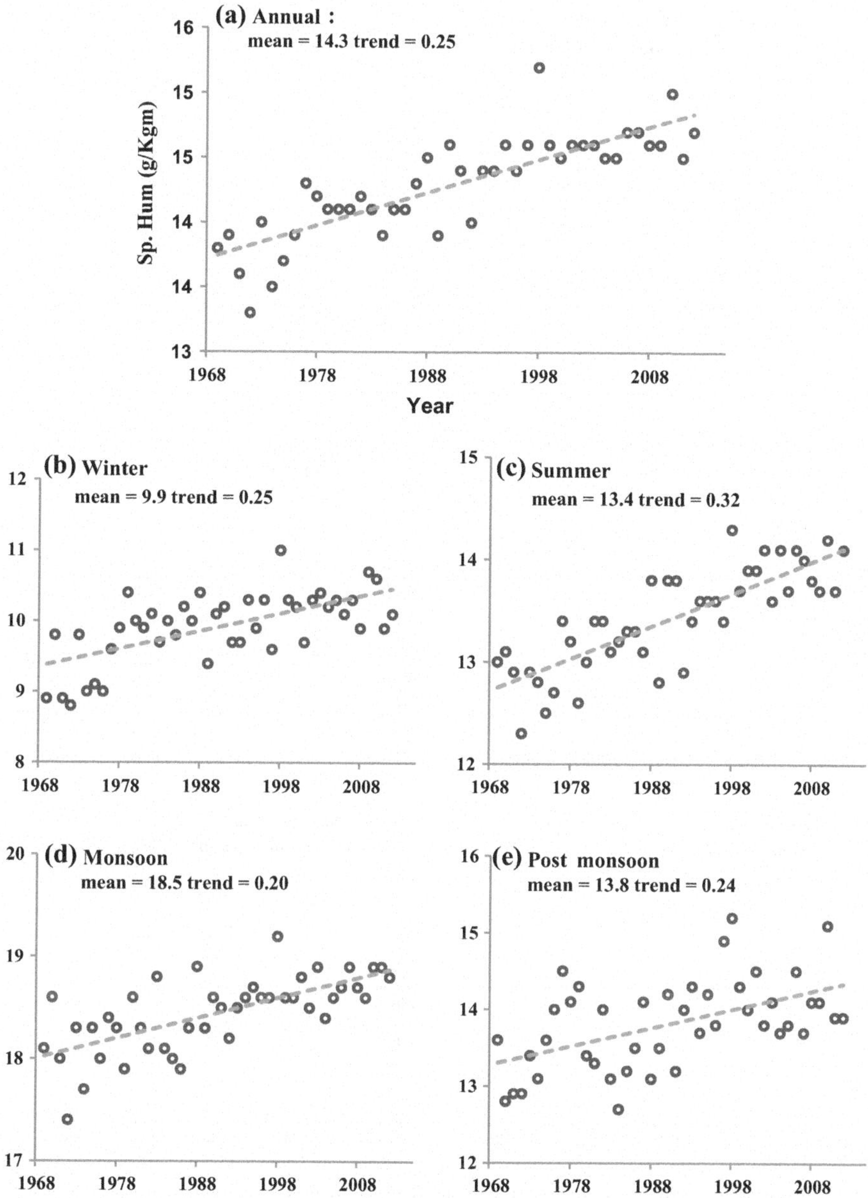

Fig. 2 Temporal variations in **a** annual, **b** winter, **c** summer, **d** monsoon and **e** post-monsoon surface specific humidity during 1969–2012. Data series are specific humidity (in g kg^{-1}) averaged over India based upon 244 stations under study. Specific humidity trends (in g kg^{-1} per decade) are statistically significant at 95 % level for the annual and the four seasons

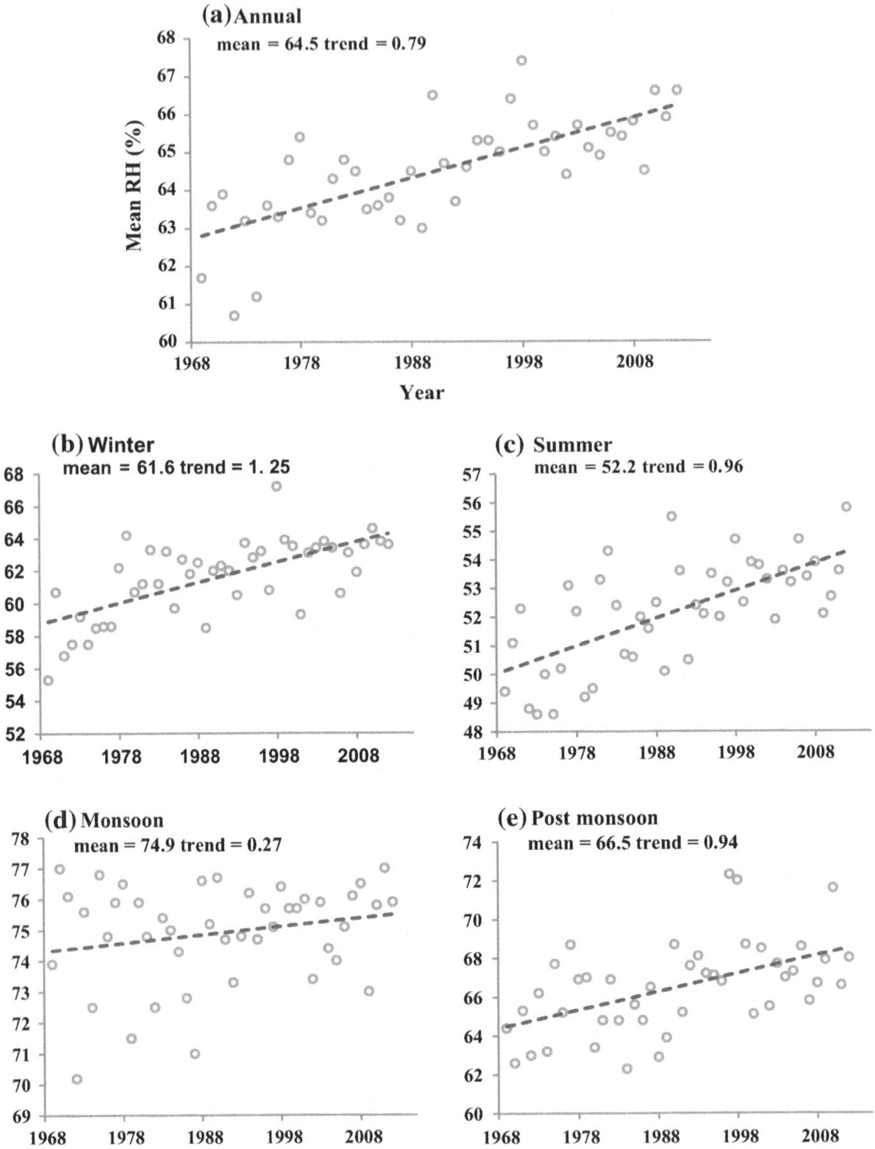

Fig. 3 Temporal variations in **a** annual, **b** winter, **c** summer, **d** monsoon and **e** post-monsoon surface relative humidity during the period of 1969–2012. Data series are relative humidity (in %) averaged over India based upon 244 stations under study. Relative humidity trends (in % per decade) are statistically significant at the 95 % significance level for all seasons except the monsoon season

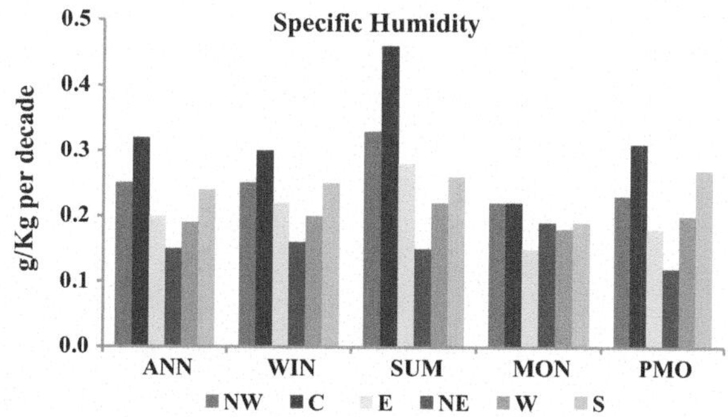

Fig. 4 Annual and seasonal rate of change in specific humidity in g kg^{-1} per decade in six regions of India during 1969–2012

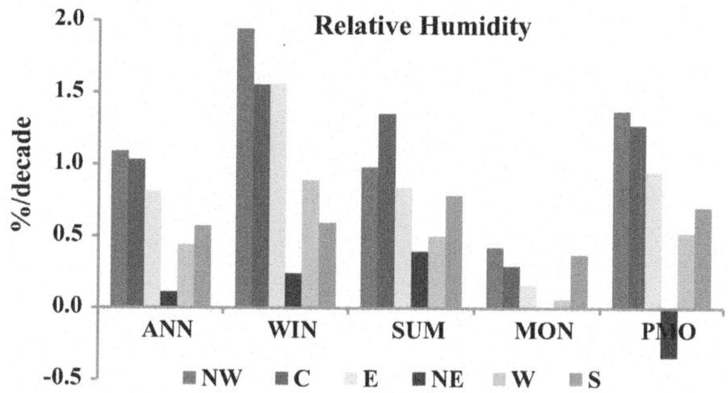

Fig. 5 Annual and seasonal rate of change in relative humidity in % per decade in six regions of India during 1969–2012

SPH and RH is prepared for 1969–2012. Temporal variations in all India averaged annual and seasonal mean SPH and RH are shown in Figs. 2 and 3, respectively. From regional time series, we have calculated rate of change in SPH and RH for annual and four seasons as shown in Figs. 4 and 5, respectively. Statistical significance of linear trends is tested at the 95 % confidence level using *t* test.

Soil moisture observations are taken once per week at several depths by gravimetric method. In this method, smaller soil samples are taken using coring devices at required depths and locations from each segment. The sample is weighed, oven-dried and weighed again. The difference in mass gives the total soil moisture in the sample, which is converted to volumetric units using the density of the soil. We have taken surface soil moisture data from 27 stations for the period of 1991 onwards. Location of these stations is given in Table 3. Spatial variations in annual mean surface soil moisture and linear trends are shown in Fig. 6, where trends statistically significant at 95 % level are marked by an outer circle. In evaluating the atmospheric moisture content, the specific humidity and relative humidity from ERA Interim (Dee et al. 2011), MERRA (Rienecker et al. 2011) reanalyses and from IMD stations observations have been analysed to evaluate the atmospheric moisture content over the Indian region (Singh et al. 2014). Further, to assess the variability of atmospheric water content, the precipitable water has been used. To link the atmospheric water content with the lower level, low-level atmospheric moisture content has been analysed.

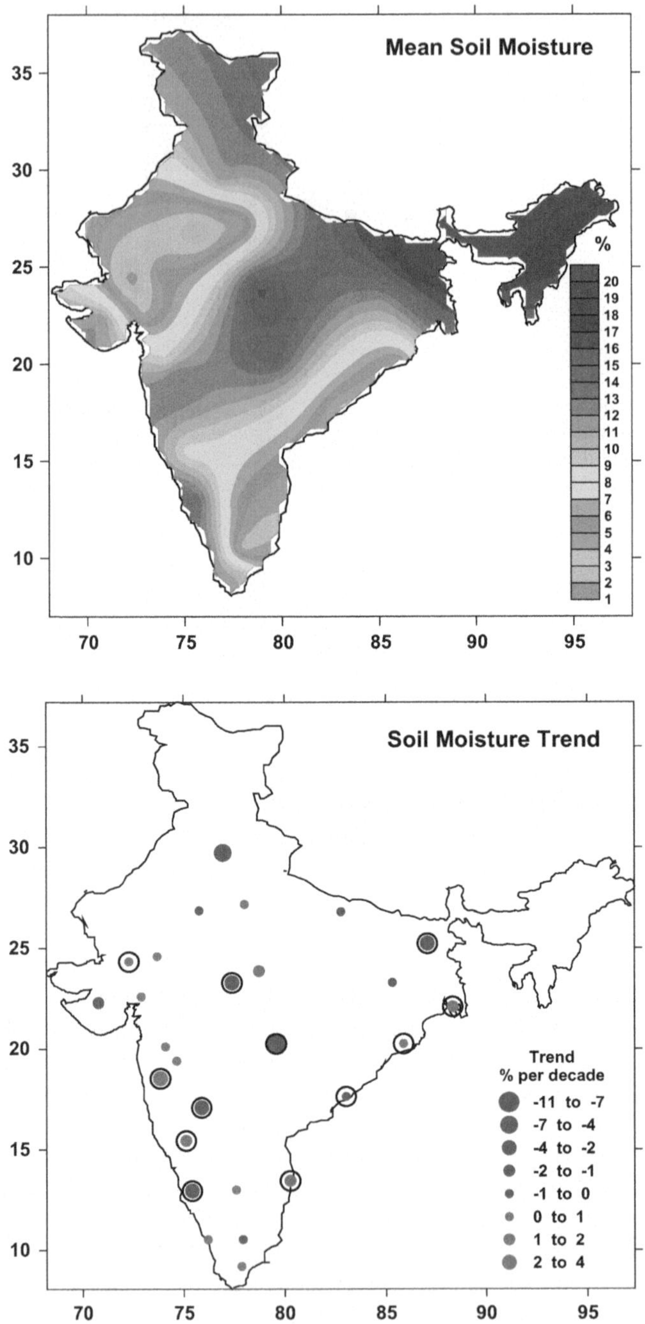

Fig. 6 Spatial distribution of annual surface soil moisture means and trends during the period of 1991–2013. Trends significant at the 95 % confidence level are circled

3 Results and Discussion

3.1 All India Trends

Specific Humidity

All India averaged annual mean SPH is 14.3 g/kg which is increasing significantly at the 95 % confidence level of confidence by +0.25 g/kg per decade during the period of 1969–2012 as shown in Fig. 2a. Country averaged seasonal means of SPH is the highest (18.5 g/kg) in monsoon season and the lowest (9.9 g/kg) in the winter season. On the seasonal scale, trends in SPH are significantly increasing for all the seasons, Fig. 2b–e. The magnitude of seasonal trends in SPH is the highest in summer, where it is increasing at the rate +0.32 g/kg per decade and the lowest in monsoon season where the rate of increase is +0.20 g/kg per decade. Temporal variations in annual mean SPH during the study period indicate the highest value 15.2 g/kg in 1998 and the lowest 13.3 g/kg in 1972. Similarly, seasonal mean values of SPH are also the highest in 1998 and the lowest in 1972 as shown in Fig. 2.

Relative Humidity

All India averaged annual mean RH is 64.5 % which is increasing significantly at the 95 % confidence level of confidence at the rate +0.79 % per decade during the period of 1969–2012 as shown in Fig. 3a. Country averaged seasonal means of RH is the highest (74.9 %) in monsoon season and the lowest (52.2 %) in the pre-monsoon season. On seasonal scale, trends in RH are significantly positive for all the seasons except for the monsoon season as shown in Fig. 3b–e. The magnitude of seasonal trends in RH is the highest in winter, where it is increasing at the rate +1.25 % per decade and the lowest in monsoon season where the rate of increase is +0.27 % per decade. Temporal variations in annual mean RH during the study period indicate the highest value of 67.4 % in 1972 and lowest 60.7 % in 1998 as shown in Fig. 3a. Winter season mean RH value is the highest (67.4 %) in 1998 and the lowest (55.3 %) in 1969. Summer season mean RH value is the highest (55.8 %) in 2012 and the lowest (48.6 %) in 1975. Monsoon season mean RH value is the highest (77.0 %) in 2011 and the lowest (70.2 %) in 1972. Post-monsoon season mean RH value is the highest (72.3 %) in 1997 and the lowest (62.3 %) in 1984.

3.2 Regional Trends

Specific Humidity

Annual and seasonal means and trends in region averaged SPH of six regions during the period of 1969–2012 are given in Table 1. Region averaged annual mean

Table 1 Mean and trends in surface specific humidity averaged over the six regions of India during the period of 1969–2012

Region	Annual		Winter		Summer		Monsoon		Post-monsoon	
	Mean	Trend	Mean	Trend	Mean	Trend	Mean	Trend	Mean	Trend
North-west	12.0	0.25[a]	6.9	0.25[a]	9.8	0.33[a]	18.1	0.22[a]	10.6	0.23[a]
Central	12.4	0.32[a]	8.0	0.30[a]	9.9	0.46[a]	18.0	0.22[a]	11.9	0.31[a]
East	15.5	0.20[a]	10.0	0.22[a]	15.1	0.28[a]	20.1	0.15[a]	15.1	0.18[a]
North-east	14.9	0.15[a]	9.4	0.16[a]	14.2	0.15[a]	19.5	0.19[a]	14.7	0.12[a]
West	14.5	0.19[a]	9.7	0.20[a]	14.2	0.22[a]	18.8	0.18[a]	13.8	0.20[a]
South	16.1	0.24[a]	13.4	0.25[a]	16.4	0.26[a]	17.6	0.19[a]	16.4	0.27[a]

While mean relative humidity values are in g/kg, trends are given in g/kg per decade. Trends statistically significant at the 95 % confidence level are marked 'a'

SPH is the highest (16.1 g/kg) in south and the lowest (12.0 g/kg) in north-west. Annual SPH trends are statistically significant at the 95 % confidence level for all regions. The rate of change in region averaged annual SPH is the highest (+0.32 g/kg per decade) in central and lowest (+0.15 g/kg per decade) in north-east as shown in Fig. 4. It is clear from Table 1 that seasonal mean SPH values averaged over the six regions are the highest in south (winter, summer and post-monsoon) and in east (monsoon). Similarly, the lowest seasonal mean SPH values are in north-west (winter, summer and post-monsoon) and in south (monsoon). Seasonal trends in region averaged SPH are significantly positive for all the seasons and across all the six regions of India indicating overall rise in water vapour content over the country. However, the rate of increase of SPH is the highest over central India for all the seasons as shown in Fig. 4.

Relative Humidity

Annual and seasonal means and trends in region averaged RH of the six regions during the period of 1969–2012 is given in Table 2. Region averaged annual mean RH is the highest (77.3 %) in north-east and the lowest (55.4 %) in central India. Annual RH trends are statistically significant at the 95 % confidence level for all the regions except over north-east. The rate of change in region averaged annual RH is the highest (+1.09 % per decade) in north-west and the lowest (+0.11 % per decade) in north-east as shown in Fig. 5. It is clear from Table 2 that seasonal mean RH values averaged over the six regions are the highest in north-east for all the seasons. Similarly, the lowest seasonal mean RH values are in central India (winter, summer and post-monsoon) and in north-west India (monsoon). Seasonal trends in region averaged RH are significantly positive for winter and summer in all regions except the north-east region. Monsoon season RH has shown significant increasing trend over the south region. Post-monsoon season RH is significantly increasing in north-west, central, east and south regions while it is significantly decreasing in the north-east region (Table 2).

Table 2 Mean and trends in surface relative humidity averaged over the six regions of India during the period of 1969–2012

Region	Annual		Winter		Summer		Monsoon		Post-monsoon	
	Mean	Trend	Mean	Trend	Mean	Trend	Mean	Trend	Mean	Trend
North-west	58.27	1.09[a]	62.02	1.94[a]	41.60	0.98[a]	67.87	0.42	58.42	1.37[a]
Central	55.40	1.03[a]	52.90	1.55[a]	33.15	1.35[a]	72.90	0.29	57.45	1.27[a]
East	70.30	0.81[a]	66.18	1.56[a]	59.06	0.84[a]	80.45	0.16	72.98	0.95[a]
North-east	77.09	0.11	72.83	0.24	71.02	0.39	83.91	0	78.85	−0.34[a]
West	63.73	0.44[a]	54.37	0.89[a]	54.85	0.50[a]	78.61	0.06	61.23	0.52
South	68.75	0.57[a]	65.44	0.59[a]	61.37	0.78[a]	74.06	0.37[a]	74.21	0.70[a]

While mean relative humidity values are in %, trends are given in % per decade. Trends statistically significant at the 95 % significance level are marked 'a'

3.3 Soil Moisture Trends

Figure 6a shows the patterns of spatial variation of annual mean surface soil moisture over India. Annual mean surface soil moisture is higher (>12 %) over north, central, west coast and north-east India. In contrast, western India and the east coast are having lower (<6 %) annual mean surface soil moisture. Trends in annual surface soil moisture are shown in Fig. 6b. Out of the 27 stations considered for the analysis, 15 stations are showing increasing trends. Stations showing significantly increasing trends are Dantiwada (0.75 % per decade), Kalyani (1.24 % per decade), Pune (3.98 % per decade), Bhubaneswar (0.97 % per decade), Dharwad (1.87 % per decade) and Tirupati (1.99 % per decade) while soil moisture is significantly decreasing at Nagpur (−10.65 % per decade), Bhopal (−2.73 % per decade), Sabour (−2.04 % per decade), Anakpalle (−0.78 % per decade), Solapur (−2.38 % per decade) and Vittal (−2.11 % per decade) as shown in Table 3.

3.4 Moisture Content in the Atmosphere

The relative humidity data at three representative levels, namely 850, 500 and 200 hPa, are analysed using the ECMWF Reanalyses (ERA) data. The data averaged over the central Indian region (18–28°N, 73–82°E) (Singh et al. 2014) are shown in Fig. 7a from 1980 to 2012. RH at all the levels (except 200 hPa) shows an increasing trend (significant at a confidence level of 99 % confidence level) during the period. This seems to be consistent with the fact of surface temperatures show increasing trend as reported by Jaswal and Koppar (2011). The increase in relative humidity is also consistent with the fact that the atmosphere shows increasing moist instability over the region for the period as mentioned by Goswami et al. (2006). To test the robustness of the result, another reanalyses data set MERRA relative humidity is analysed for the same period (Fig. 7b). The MERRA data also show an

Table 3 Annual means, standard deviation (Sdev) and trends of surface soil moisture at 27 stations

STATION	Period	Latitude (N)	Longitude (E)	Mean (%)	Std. dev. (%)	Trend (%/decade)
Karnal	1998–2008	29° 43′	76° 58′	11.76	3.32	−4.33
Agra	1991–2007	27° 10′	78° 02′	4.16	0.93	0.19
Durgapura	1993–2012	26° 51′	75° 47′	1.78	0.45	−0.19
Basti	1991–2008	26° 48′	82° 46′	15.11	1.34	−0.22
Sabour	1991–2008	25° 14′	87° 04′	18.84	2.21	−2.04[a]
Dantiwada	1996–2008	24° 10′	72° 29′	1.02	0.52	0.75[a]
Udaipur(RCA)	1993–2008	24° 35′	73° 42′	5.44	1.49	0.42
Bhopal	1991–2010	23° 16′	77° 25′	11.97	2.69	−2.73[a]
Sagar	1991–2008	23° 51′	78° 45′	16.31	2.79	1.10
Kanke	1998–2008	23° 17′	85° 19′	11.87	0.53	−0.41
Rajkot	1996–2008	22° 17′	70° 48′	11.83	1.70	−1.13
Anand	1996–2008	22° 35′	72° 55′	3.72	0.74	0.10
Kalyani	1993–2008	22° 05′	82° 20′	15.16	1.30	1.24[a]
Nagpur	1997–2008	20° 15′	79° 35′	14.51	4.87	−10.65[a]
Niphad	1991–2008	20° 06′	74° 06′	9.28	1.13	0.25
Bhubaneshwar	1994–2011	20° 15′	85° 52′	4.99	0.77	0.97[a]
Rahuri	1996–2013	19° 24′	74° 39′	13.10	0.75	0.42
Pune	1996–2012	18° 32′	73° 51′	12.94	3.04	3.98[a]
Anakapalle	1996–2013	17° 38′	83° 01′	3.51	0.66	−0.78[a]
Solapur	1996–2013	17° 04′	75° 54′	10.48	1.84	−2.38[a]
Dharwad (ARS)	1996–2008	15° 26′	75° 07′	8.32	1.05	1.87[a]
Tirupati	1991–2008	13° 27′	79° 76′	4.46	1.49	1.99[a]
Vittal	1991–2008	12° 57′	75° 25′	14.48	2.28	−2.11[a]
Hebbal	1996–2008	13° 00′	77° 37′	5.98	0.51	0.12
Vellanikara	1991–2007	10° 31′	76° 13′	12.68	1.55	0.11
Vedasandur	1992–2008	10° 32′	77° 57′	3.18	0.33	−0.20
Kovilpatti	1996–2011	09° 12′	77° 53′	10.39	0.15	0.05

Statistically significant trends are marked ‘a’

increasing trend (significant at the 99 % confidence level) for the three representative levels, e.g. 850, 500 and 200 hPa, over the central Indian region. Therefore, it indicates that the central Indian region which is considered to be the core monsoon zone shows an increasing trend in relative humidity in the middle and upper troposphere.

Along with the relative humidity, the specific humidity from ERA and MERRA analyses shows an increasing trend for the same period (Fig. 8a, b). The signal of enhanced humidity or moisture content in the atmosphere appears to be consistent with the fact as reported by Goswami et al. (2006) about increasing extreme events (Chap. 3). The precipitable water from ERA and MERRA data also shows an increasing trend (significant at the 99 % confidence level) during the period of

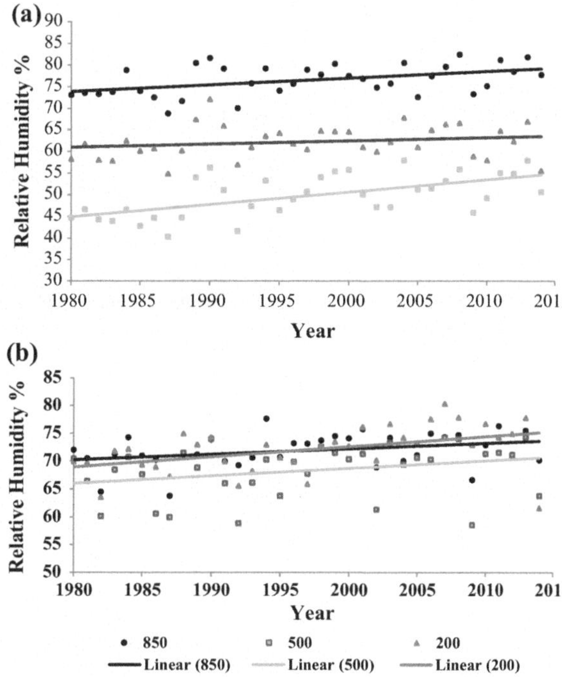

Fig. 7 Relative humidity at 850, 500 and 200 hPa from **a** ERA and **b** MERRA analysis averaged over the monsoon core region (18°–28°N, 73°–82°E)

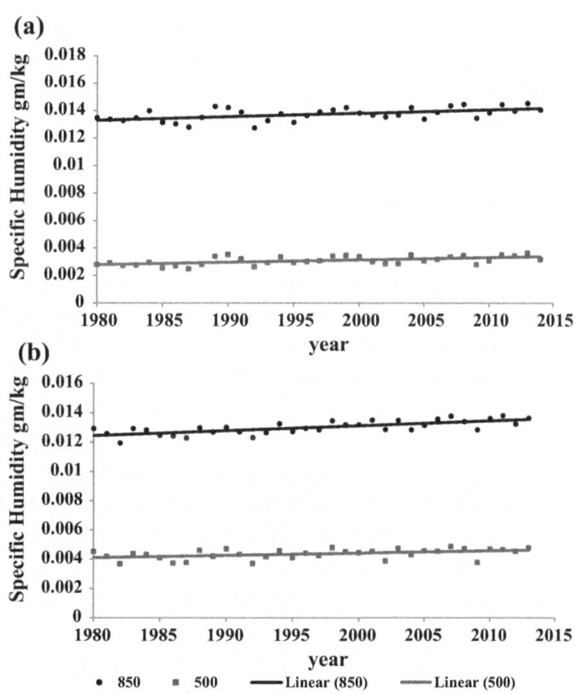

Fig. 8 Specific humidity at 850 and 500 hPa from **a** ERA and **b** MERRA analysis averaged over the monsoon core region (18°–28°N, 73°–82°E)

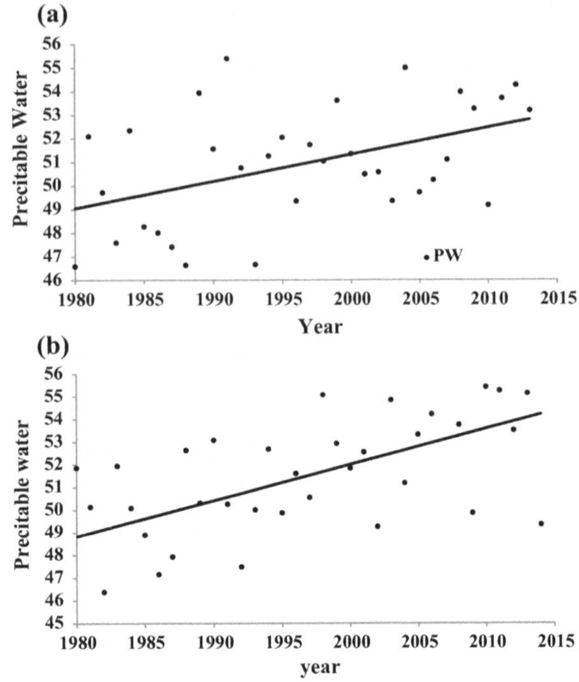

Fig. 9 Precipitable water (cm) from **a** ERA and **b** MERRA analysis averaged over monsoon core region (18°–28°N, 73°–82°E)

1980–2014 (Fig. 9a, b). The increasing precipitable water appears to be consistent with the analyses by Jaswal (Chap. 7) who showed that there is an increasing cloudiness over the core monsoonal region.

4 Conclusions

We have examined humidity and surface soil moisture data in this study using 1969–2012. The data analysis indicates increasing trends in both surface and upper air relative humidity and specific humidity. All India averaged specific humidity is significantly increasing for all the periods while relative humidity values are significantly increasing except for the monsoon season. The increase in surface humidity and moisture content in the atmosphere is consistent with the increase in surface and air temperatures over the region (Chap. 2). Variations in annual surface soil moisture in India show general agreement with the rainfall pattern. While the hilly regions, north-east and central India and west coast, have the highest soil moisture, the lowest soil moisture is found over the western region and along the east coast of India. The results obtained in this study provide evidence of significant increase in atmospheric moisture content over India.

References

Albergel, C., Dorigo, W., Reichle, R. H., Balsamo, G., de Rosnay, P., Muñoz-Sabater, J., Isaksen, L., de Jeu, R. and Wagner, W., 2013, "Skill and Global Trend Analysis of Soil Moisture from Reanalyses and Microwave Remote Sensing", J. Hydrometeor, **14**, 1259–1277.

Calla, O. P. N., Sharma, R., Gadri, K. L., Kalla, A., Agrahari, S. K. and Rathore, G., 2013, "Comparison of SMOS soil moisture/brightness temperatures for morning and evening passes", Ind. J. Radio and Space Phys., **42**, 420-424.

Dai, A, 2006, "Recent climatology, variability and trends in global surface humidity", J. Climate, **19**, 3589-3606.

De, U. S., Rao, G. S. P. and Jaswal, A. K., 2001, "Visibility over Indian airports during winter season", Mausam, **52**, 4, 717-726.

Dee, D. P., and Coauthors, 2011, "The ERA-Interim reanalysis: Configuration and performance of the data assimilation system", Quart. J. Roy. Meteor. Soc., **137**, 553–597, doi:10.1002/qj.828.

Dorigo, W., R. de Jeu, D. Chung, R. Parinussa, Y. Liu, W. Wagner, and D. Fernández-Prieto (2012), Evaluating global trends (1988–2010) in harmonized multi-satellite surface soil moisture, Geophys. Res. Lett., 39, L18405, doi:10.1029/2012GL052988.

Eswar, R., Sekhar, M. and Bhattacharya, B. K., 2013, "Simple model for spatial disaggregation of evaporative fraction: Comparative study with thermal sharpened land surface temperature data over India", J. Geophys. Res. Atmos., **118**, 12029–12044, *doi:*10.1002/2013JD020813.

Gaffen, D. J., and Ross, R. J., 1999, "Climatology and trends in U.S. surface humidity and temperature", J. Climate, **12,** 811–828.

Goswami B. N., Venugopal V., Sengupta, D., Madhusoodanan, M. S. and Prince, K. X., 2006, "Increasing Trend of Extreme Rain Events Over India in a Warming Environment", Science, **314**, 1442-1445.

Groisman, P. Ya., Knight, R. W., Karl, T. R., Easterling, D. R., Sun, B. M. and Lawrimore, J. H., 2004, "Contemporary changes of the hydrological cycle over the contiguous United States: Trends derived from in situ observations", J. Hydrometeor., **5,** 64–85.

Jaswal, A. K. and Koppar, A. L., 2011, "Recent climatology and trends in surface humidity over India for 1969-2007", Mausam, **62**, 145-162.

Jung, M., and Coauthors, 2010, "Recent decline in the global land evapotranspiration trend due to limited moisture supply", Nature, **467**, 951–954.

Koster, R. D. et al., 2004, "Regions of strong coupling between soil moisture and precipitation", Science, **305**, 1138–1140.

Murray, F. W., 1967, "On the computation of saturation vapor pressure", J. Appl. Meteor., **6**, 203-204.

New, M., Hulme, M., and Jones, P., 2000, "Representing twentieth century space-time climate variability. Part II: Development of 1901–96 monthly grids of terrestrial surface climate", J. Climate, **13,** 2217–2238.

Oort, A. H., 1983, "Global atmospheric circulation statistics, 1958–1973", NOAA Prof. Paper **14,** 180 pp.

Peixoto, J. P., and Oort, A. H., 1996, "The climatology of relative humidity in the atmosphere", J. Climate, **9,** 3443–3463.

Philipona, R., Durr, B., Marty, C., Ohmura, A. and Wild, M., 2004, "Radiative forcing- Measured at Earth's surface - Corroborate the increasing greenhouse effect", *Geophys. Res. Lett.,* **31,** L03202, doi:10.1029/2003GL018765.

Philipona, R., Durr, B., Ohmura, A., and Ruckstuhl, C., 2005, "Anthropogenic greenhouse forcing and strong water vapor feedback increase temperature in Europe", Geophys. Res. Lett., **32,** L19809, doi:10.1029/2005GL023624.

Rao, G. S. P., Jaswal, A. K. and Kumar, M. S., 2004, "Effects of urbanization on meteorological parameters", Mausam, **55**, 3, 429-440.

Rienecker, M. M., Suarez, M. J., Gelaro, R., Todling, R., Bacmeister, J., Liu, E., Bosilovich, M. G., Schubert, D., Takacs, L., Kim, G. K., Bloom, S., Chen, J., Collins, D., Conaty, A., Silva, A.

D., Gu, W., Joiner, J., Koster, R. D., Lucchesi, R., Molod, A., Owens, T., Pawson, S., Pegion, P., Redder, C. R., Reichle, R., Robertson, F. R., Ruddick, A. G., Sienkiewicz, M. and Woollen, J., 2011, "MERRA- NASA's modern-era retrospective analysis for research and applications", Journal of Climate, **24**, 14, 3624 – 3648

Robinson, P. J., 2000, "Temporal trends in United States dew point temperatures", Int. J. Climatol., **20,** 985–1002.

Robock, A., Vinnikov, K. Y., Srinivasan, G., Entin, J. K., Hollinger, S. E., Speranskaya, N.A., Liu, S., Namkhai, A., 2000, "The global soil moisture data bank", Bull. Am. Met. Soc., **81**, 1281–1300.

Robock, A., Mingquan, M., Konstantin, V., Iryna, V. T. and Tatyjana, I. A., 2005, "Forty five years of observed soil moisture in the Ukraine: No summer desiccation (yet)", Geophys. Res. Lett., **32**, L03401, doi:10.1029/2004GL021914.

Schönwiese, C. D., and Rapp, J., 1997, "Climate Trend Atlas of Europe Based on Observations 1891–1990", Kluwer Academic, 228 pp.

Sheffield, J. and Wood, E. F., 2008, "Projected changes in drought occurrence under future global warming from multi-model, multi-scenario, IPCC AR4 simulations", Climate Dynamics, **31**, 79–105.

Singh, P., Kumar, V. Thomas, T. and Arora, M., 2008, "Changes in rainfall and relative humidity in river basins in northwest and central India", Hydro. Proc, **22**, 16, 2982-2992.

Singh, D., Michael, T., Bala, R. and Noah, S.D., 2014, "Observed changes in extreme wet and dry spells during the South Asian summer monsoon season",.Nature, Climate Change, **4**, 456–461.

Sun, B., Groisman, P. Ya., Bradley, R. S. and Keimig, F. T., 2000, "Temporal changes in the observed relationship between cloud cover and surface air temperature", J. Climate, **13,** 4341–4357.

Tetens, O., 1930, "Über einige meteorologische Begriffe", Z. Geophys, **6,** 297–309.

Trenberth, K. E., Jones, P. D., Ambenje, P., Bojariu, R., Easterling, D., Klein Tank, A., Parker, D., Rahimzadeh, F., Renwick, J. A., Rusticucci, M., Soden, B. and Zhai, P., 2007, "Observations: surface and atmospheric climate change", In: Solomon S, Qin D, Manning M, Chen Z, Marquis M, Averyt KB, Tignor M, Miller HL (eds) Climate change 2007: the physical science basis. Contribution of working group I to the fourth assessment report of the intergovernmental panel on climate change. Cambridge University Press, Cambridge, UK/New York, NY, USA

Trenberth, K. E., 2011, "Changes in precipitation with climate change", Clim Res, **47**, 123–138.

Vincent, L. A., Van Wijngaarden, W. A. and Hopkinson, R., 2007, "Surface temperature and humidity trends in Canada for 1953–2005", J. Climate, **20**, 5100–5113.

Wang, J. X. L., and Gaffen, D. J., 2001, "Late-twentieth-century climatology and trends of surface humidity and temperature in China", *J. Climate,* **14,** 2833–2845.

Zhu C. M. and Lettenmaier, D. P., 2007, "Long-term climate and derived surface hydrology and energy flux data for Mexico, 1925-2004", Journal of Climate, **20**, 1936-1946.

Chapter 9
Trends in Radiative Fluxes Over the Indian Region

B. Padmakumari, Vijay Kumar Soni and M.N. Rajeevan

1 Introduction

The various fluxes of radiation to and from the Earth's surface are the important components of the Earth Radiation Budget. The incoming solar radiation is the primary source of energy for the Earth's environment that drives weather and climate. The evolution of Earth's climate is largely regulated by the global energy balance. The spatial and temporal variation in the global energy balance or Earth radiation budget affects various climatic parameters such as atmospheric and oceanic circulations, glaciers, the components of the hydrological cycle, plant productivity, and terrestrial carbon uptake (Ramanathan et al. 2001; Ohmura et al. 2007; Wild et al. 2008). Though Earth's radiation budget plays a major role in the climate system, there exists significant uncertainties in the quantification of its various components and depiction in climate models (Wild et al. 2012 and the references there in). Radiation budget of the Earth's system is perturbed through the modifications of the atmospheric composition of greenhouse gases and aerosols through natural as well as human activities and clouds.

Aerosols and clouds are two major controlling factors that affect the surface reaching solar radiation or radiative flux (Rs, also known as global irradiance), which in turn affect other radiative fluxes. Aerosols through their direct (by

B. Padmakumari (✉)
Indian Institute of Tropical Meteorology, Earth System Science Organization,
Pune 411008, India
e-mail: padma@tropmet.res.in

V.K. Soni
India Meteorological Department, Earth System Science Organization,
New Delhi 11003, India

M.N. Rajeevan
Earth System Science Organization, Ministry of Earth Sciences,
New Delhi 11003, India

M.N. Rajeevan and S. Nayak (eds.), *Observed Climate Variability and Change Over the Indian Region*, Springer Geology, DOI 10.1007/978-981-10-2531-0_9

scattering and absorption) and indirect (through changes in cloud microphysics) forcings have profound influence on solar radiation reaching the surface. Unlike the uniform distribution of greenhouse gases due to long lifetimes, aerosols show large spatial and temporal variability due to their short lifetimes of a week or less. As compared to global warming, global or solar dimming has received prominent attention only in the recent decades because of its profound climatic and environmental implications. The changes in the net solar radiation at the surface affect the surface energy budget and may affect the strength of the hydrological cycle and many other climatic parameters.

2 Solar Radiation Measurements

India has a long history of surface radiation measurements. The radiation measurements were made for the first time at Kolkata as early as November 1879, though the details of instruments were not clearly indicated. The radiation measurements were made at Mussoorie during 1883–1887 and Kodaikanal during 1895. More systematic measurements were started in 1930 at Pune using Ångström pyrgeometer for measuring the nocturnal radiation and direct solar irradiances with an Ångström pyrheliometer. Most of the solar radiation networks began to operate during the International Geophysical Year, 1957–58. India Meteorological Department (IMD) started four radiation stations at Pune, New Delhi, Kolkata, and Chennai in 1957. To begin with, measurements of global and diffuse solar irradiances (using MG pyranometers), direct solar irradiances (using Ångström pyrheliometer and three broad band-pass filters) and net terrestrial radiant energy two times in the night (using Ångström pyrgeometer) were started at Pune and New Delhi. By 1976, 31 stations were operating in the Indian network. The radiation data were measured with very high accuracy and much beyond the specifications recommended by World Meteorological Organization (WMO). Fourteen more stations were added to the radiation network during the years 1985 and 1986. The present strength of the IMD's network of radiation stations is 45 (Fig. 1). Recently, the UV-A and UV-B radiation measurements were also added to the network.

The Central Radiation Laboratory of IMD, Pune, has been designated by WMO as one of the two Regional Radiation Centers for Asia. The other regional radiation center, managed by Japan Meteorological Agency, is located at Tokyo. The Central Radiation Laboratory at IMD, Pune, maintains a number of standard equipments for calibration such as—primary, secondary, transfer, working and touring calibration standards. The radiation data are measured with reference to the World Radiometric Reference (WRR). To ensure the WRR traceability and stability of the primary standards, one instrument regularly participates in the WMO International Pyrheliometer Comparisons held once in five years at the World Radiation Centre, Davos, in Switzerland. The accuracy of the field instruments in solar radiation network of IMD is as follows.

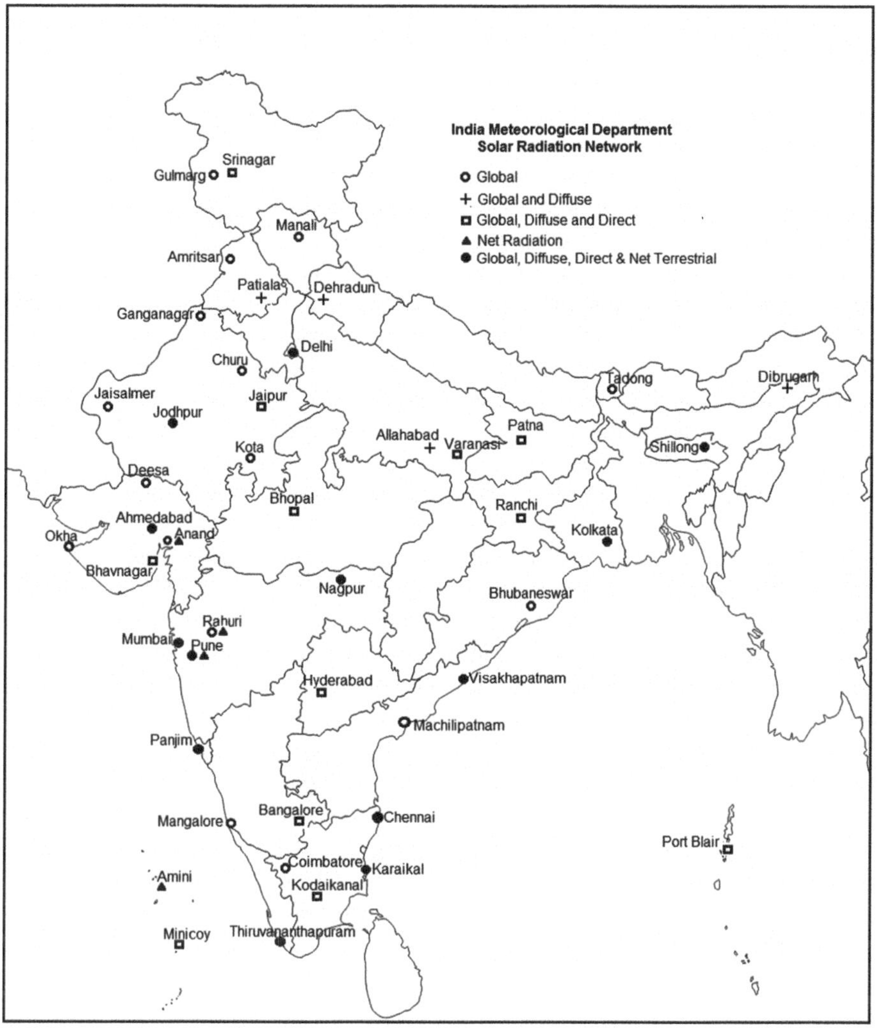

Fig. 1 India map showing the network of radiation measurement stations

Instrument	Accuracy recommended by WMO (%)	Accuracy achieved by IMD (%)
Ångström pyrheliometer	±0.5	±0.5
Thermoelectric pyrheliometer	±1.0	±1.0
Thermoelectric pyranometer	±3.0	±2.0
Bimetallic pyranograph	±5.0	±5.0
Net pyrradiometer	±7.0	±5.0
Ångström pyrgeometer	±1.0	

The Ministry of New and Renewable Energy (MNRE) started the Solar Radiation Resource Assessment (SRRA) project in 2011. SRRA has set up 115 automatic meteorological stations with solar radiation instruments including four advanced measurement stations. These stations measure direct, diffuse, and global irradiance along with meteorological parameters. The purpose of SRRA project is to quantify solar radiation data, modeling and making of solar atlas of the country. There are a number of other organizations in the country, who make solar radiation measurements at local level, though not necessarily as a routine.

The non-uniform distribution of observational sites and lack of observations from background and oceanic regions restricts the global analysis of solar radiation reaching the surface. Satellites provide an ideal way to obtain radiation data globally in view of temporal continuity and spatial homogeneity. The widely used satellite products of radiation include the Surface Radiation Budget project (SRB) and the International Satellite Cloud Climatology Project (ISCCP). Recently knowledge on the Earth's radiation budget has been improved with the advent of new satellite missions such as Clouds and the Earth's Radiant Energy System (CERES, Wielicki et al. 1996) and the Solar Radiation and Climate Experiment (SORCE, Anderson and Cahalan 2005). These satellites provide highly accurate radiative fluxes at the top of the atmosphere (TOA) (Loeb et al. 2012). However, much less is known about the energy distribution within the atmosphere and the Earth's surface. Unlike the TOA fluxes, the surface fluxes cannot be directly measured by the satellites. Overall, satellite data have an acceptable accuracy for radiation at global scale, but their accuracies on regional scale are largely unknown.

Modern Era Retrospective-analysis for Research and Applications (MERRA) is a NASA reanalysis for the satellite era using a major new version (V5) of the Goddard Earth Observing System (GEOS) Data Assimilation System (DAS). This data set is available from 1979 through present and provides net downward shortwave and long-wave fluxes and sensible, latent and ground heat fluxes. Detailed documentation on the data products and data access is found at http://gmao.gsfc.nasa.gov/MERRA.

3 Review of Earlier Studies in India

Over India, the important earlier investigations on solar radiation are reviewed in this section. For the first time in India, the results on measurements of the total radiation from the sun and sky made at Poona during 1935 were published in 1938. In the absence of observations from radiation network in India, several researchers earlier estimated surface solar radiation using sunshine duration and other meteorological parameters (Ramdas and Yengnanarayanan 1956; Mani et al. 1962a; Yadav 1965; Mani et al. 1967; Venkataraman and Krishnamurthy 1967; Ayyar and Krishnamurthy 1968; Ganesan 1970; Gangopadhyaya et al. 1970; Kumar and Manikiam 1988). One of the earliest and most comprehensive studies was made by Mani et al. (1962a) on distribution of sunshine duration and estimated surface solar radiation for 15 stations over India. They also found fair agreement between observed solar radiation and estimated values at Poona (now Pune), Delhi, Calcutta (now Kolkata), and Madras (now Chennai). Mani et al. (1962b) also brought out for the first time the observed seasonal and diurnal variation from sun and sky at the above four stations. Mooley and Raghavan (1963) assessed the percentage transmission of solar radiation through different types of clouds. Mani and Chacko (1963) presented the results of diffuse radiation received on a horizontal surface at Delhi and Poona during 1958 and 1959. Mani et al. (1967) prepared annual and monthly maps for four representative months of January, April, July, and October showing the distribution of global solar, net terrestrial and net radiation over the Indian Ocean by using the available observations, supplemented by calculations based on other meteorological measurements. It was observed that global irradiance shows minima over the equatorial and monsoon regions, while maxima over land in the high-pressure belts of the northern and southern hemispheres along the tropics of Cancer and Capricorn. Desikan et al. (1969) analyzed diffuse solar radiation measured at 10 stations in India during 1958–67 and found that under all-sky conditions the proportion of diffuse component in global radiation varies from 17 to 77 %. In highly polluted urban areas, the ratio of diffuse to global radiation was almost twice as that observed in less polluted regions. The diffuse radiation showed a marked increase in urban areas during this study period. Mani (1971) analyzed the albedo, outgoing long-wave radiation and the radiation balance over the globe using data from wide-field radiometer onboard satellite TIROS II, III, and IV. She also found that satellite measurements agree fairly well with ground-based and airborne measurements. Mani et al. (1977) studied direct solar radiation at different high-altitude stations in India. The low values of direct solar radiation over north and central Indian high-altitude stations were attributed to depletion of incoming solar radiation by the dense layer of dust in the pre-monsoon summer months which extends up to 6 km or more in the atmosphere. Mani and Rangrajan (1982) studied the radiation climate of India for the entire year, which gives the heat balance at the surface. The books "Handbook of Solar Radiation Data for India" by A. Mani published in 1980 and "Solar Radiation over India" by A. Mani and S. Rangarajan published in 1982 have become the reference material for researchers in radiative fluxes. Desikan et al. (1994a, b) studied the effect of atmospheric pollution on solar radiation and found that almost all stations showed a decreasing trend in the net outgoing terrestrial radiant energy. Systematic and detailed solar radiation studies on long-term variability of radiative fluxes have been undertaken mostly in recent years.

4 Annual and Seasonal Variability of Radiative Flux

Indian subcontinent receives significant amount of global solar radiant exposure in a year. On an annual average, India receives about 220 W m^{-2} (7000 MJ m^{-2} in a year) of global solar radiant exposure under all-sky conditions with annual bright sunshine of 2000–3200 h depending on the location. The major parts of peninsular India receive about 230 W m^{-2}, while the annual highest value of over 250 W m^{-2} (8000 MJ m^{-2} in a year) is observed over the Rann of Kutch. The observed long-term averages of the annual global irradiances ranged between 189 W m^{-2} at Kolkata to 229 W m^{-2} at Jodhpur. Significant seasonal variation in surface solar radiation is observed over India. The north Indian plains receive approximately 175 W m^{-2} and even less during January, which can be attributed to the low solar elevations and the shorter duration of the day. The weather conditions such as fog/smog also contribute to lower solar irradiance. The Kashmir valley receives less than 60 W m^{-2} during January, while the Deccan Plateau receives nearly 230 W m^{-2} of global solar irradiation. By April, the solar irradiance increases significantly to more than 230 W m^{-2} with Saurashtra and Rajasthan receiving more than 280 W m^{-2}. The similar solar radiation exposure continues in May also over the country except northeast India and the extreme south where the irradiation is less than 230 W m^{-2}. The northwest India and Kashmir, however, continue to receive more than 220 W m^{-2}. The Andaman and Nicobar Islands receive around 160 W m^{-2} only, while it is about 185 W m^{-2} over the Lakshadweep islands. Tamil Nadu and the adjoining region also continue to receive 220 W m^{-2} or more. October once again shows a generally uniform range 170–220 W m^{-2} except the Kashmir Valley where the value decreases to less than 160 W m^{-2} from a high value of more than 210 W m^{-2} in September. The NE monsoon mainly affects the south Andhra Pradesh and Tamil Nadu where the global irradiation is around 185 W m^{-2} in October and it is less than 175 W m^{-2} during November–December. The entire country receives more than 280 W m^{-2} of global irradiation on any cloudless day during the year, including the Northeast India and the Kashmir Valley.

The spatial patterns of annually averaged diffuse radiation are complex and inhomogeneous across India. All India annual diffuse radiation under all-sky conditions ranges from 80 to 100 W m^{-2} with an average value of approximately 90 W m^{-2}. In general the diffuse radiation is higher in coastal locations as compared to inland locations mainly due to more clouds. The relative portion of the diffuse radiation in the global solar radiation varies from 37 to 50 % in the annual mean. The coastal areas record more than 90 W m^{-2} of diffuse irradiance, mainly due to more clouding. The winter months of December and January, a period with minimum cloudiness, receive around 45–60 W m^{-2}. During the monsoon season, the diffuse radiation is about 115 W m^{-2} or more over most parts of the country. The post-monsoon sky conditions cause a steep fall in the diffuse component to an average of 70–80 W m^{-2} at most of the places. The annual mean diffuse irradiation values under cloud-free skies are in the range of 56–72 W m^{-2}. The diffuse irradiation is a measure of the extent of pollution over different parts of the country.

5 Long-Term Variability of Radiative Flux

Since 1950s, significant decadal changes in Rs were observed worldwide. Continuous observations of Rs at different locations around the world revealed reduction in surface solar radiation with time (Ohmura et al. 1989; Russak 1990;

Gilgen et al. 1998; Liepert 2002; Stanhill and Cohen 2001; Abakumova et al. 1996; Wild et al. 2005). This decreasing trend in surface solar radiation is known as global/solar dimming (Stanhill and Cohen 2001). After 1980s, a partial recovery has been observed at some stations around the globe (Wild et al. 2005; Pinker et al. 2005; Wild et al. 2007; Ruckstuhl et al. 2008; Philipona et al. 2009). Thereby, the term so-called brightening was coined. The updates beyond 2000 provide a less coherent picture as compared to the preceding dimming and brightening (Wild et al. 2009; Wild et al. 2012). After 2000, continuation of brightening was observed in Europe, USA, and some parts of Asia, leveling off at sites in Japan and Antarctica, and renewed dimming in China. These changes are mainly attributed to changes in aerosols, clouds, and their interactions (Wild et al. 2009). Anthropogenic aerosols are found to be important contributors to these changes (Streets et al. 2006). High-quality surface radiation measurements over seven different stations in USA for the period 1996–2011 showed increasing trend in shortwave radiation. This brightening was attributed to a decrease in cloud coverage; aerosols had only a minor effect (Augustin and Datton 2013). On the other hand, Tibetan Plateau, the cleanest region in the world, experienced a transition from brightening to dimming around the end of 1970s (Tang et al. 2011). Over the last three decades, solar radiation as well as total cloud cover over the Tibetan Plateau showed a decreasing trend. Also the extinction due to aerosol loading is one order lower in magnitude than the observed dimming. It is concluded that dimming over the Tibetan Plateau is mainly due to increase in the amount of water vapor and deep cloud cover but not due to aerosol loading (Yang et al. 2012).

Over India, aerosols and clouds are found to be the two important factors that lead to large variability in the solar radiation reaching at the surface. Aerosol concentrations over the Indian subcontinent show large seasonal variability and consistent increase. Long-term aerosol optical depth (AOD) variability observed using a network of multi-wavelength radiometers showed an increase of 2–9 % year^{-1} (Moorthy et al. 1993, 2001). Based on satellite data, AOD trends are increasing over all the major cities in India, with higher optical depths in the northern part of India as compared to southern part of India (Sarkar et al. 2006; Gautam et al. 2007). From surface observations, atmospheric turbidity is also found to be increasing significantly at different stations in India (Soni and Kannan 2003; Soni et al. 2012). There are many studies showing region specific short-term variations of radiative fluxes and aerosol radiative forcing. Simultaneous measurements of aerosol properties and radiative fluxes over an urban station, Pune, during the dry seasons of 2001 and 2002 showed a radiative forcing of −33, 0, and 33 W m^{-2} at the surface, top of the atmosphere, and the atmosphere, respectively (Pandithurai et al. 2004). Radiative forcing efficiency of −97 W m^{-2} was shown over Bengaluru (Babu et al. 2002). Over New Delhi, high aerosol radiative forcing during pre-monsoon is attributed to dust aerosols (Singh et al. 2005; Pandithurai et al. 2008), while during winter radiative forcing is influenced by haze and fog conditions (Ganguly et al. 2006). A recent study over Ranchi for the period February 2011–January 2012 showed an average reduction of 9 % in Rs due to aerosols (Latha et al. 2014). India is one of the few regions around the world, which showed decreasing trend in Rs (Ramanathan et al. 2005; Padmakumari et al. 2007;

Soni et al. 2012) under all-sky conditions, where aerosols and clouds together contributed to the annual trend. Over the Indian subcontinent, studies on long-term surface energy budget are lacking, due to lack of continuous long-term observational data.

Padmakumari et al. (2007) studied the long-term variability in Rs for the period 1981–2004 over 12 different stations in the Indian subcontinent. They found a decreasing trend in Rs over all the stations ranging from 0.17 to 1.44 W m^{-2} $year^{-1}$, and the average dimming is observed to be $\sim$0.86 W m^{-2} $year^{-1}$. The stations in the north showed strong decline as compared to the stations in the southern parts of India. Decadal variability showed a strong decline during the decade 1991–2000 as compared to 1981–1990. Soni et al. (2012) also studied the long-term variability in Rs as well as sunshine duration over 12 stations for the period 1971–2005. They observed decreasing trend in Rs over all the stations varying from 0.3 to 9.0 W m^{-2} per decade and sunshine duration also showed decreasing trend in correspondence with Rs.

Clouds display a wide variability on annual as well as on seasonal scales. Based on reanalysis data, earlier studies have shown increasing trend in cloud optical depths over the Indian subcontinent (Rajeevan et al. 2000; Badrinath et al. 2010). Cloud type information also helps to understand the variability in surface radiation budget. It was found that low and deep clouds produce more cooling effect in shortwave band, while warming effect in long-wave band, resulting in net cooling effect at the surface (Balachandran and Rajeevan 2007). They also observed that high clouds contribute more to shortwave radiative forcing over oceans, while both middle and high clouds over land.

To examine the possible reasons of decreasing trend in Rs, Padmakumari and Goswami (2010) examined the trends in Rs for clear-sky and cloudy-sky conditions. They studied the long-term Rs data from 1981 to 2006 over widely distributed 12 different stations in India. The study brought out new insights into the origin of solar dimming over India. The rate of dimming is found to be twice (12 W m^{-2} $decade^{-1}$) as large during cloudy conditions as compared to the clear-sky conditions (6 W m^{-2} $decade^{-1}$)(Fig. 2). Under cloudy conditions the changes in Rs is expected as clouds

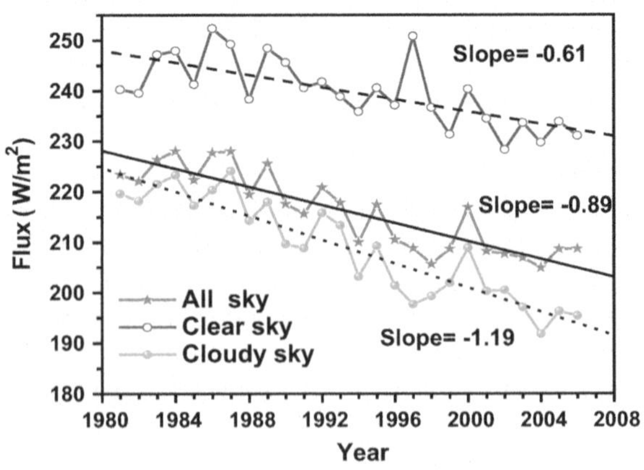

Fig. 2 Annual mean surface reaching radiative flux, averaged over 12 stations which are spatially distributed over the Indian subcontinent, under all-sky, clear-sky and cloudy-sky conditions. Straight lines represent linear trend lines. All the trends are statistically significant at the 99 % confidence level (Padmakumari and Goswami 2010)

are the strongest modulators of Rs. However,it is not very clear why it should show a strong decreasing trend. Figure 2 depicts that as compared to aerosols, clouds play a larger role in the observed solar dimming. To understand this fact, seasonal variation (MJJASO, NDJFMA) of Rs under clear and cloudy conditions is studied (Fig. 3).In both the seasons the clear-sky dimming is attributed to direct effect of aerosols (Fig. 3a). Under cloudy conditions, strong decreasing trend is observed during both the seasons (Fig. 3b). To understand the cloud effect on Rs, seasonal mean outgoing long-wave radiation (OLR) over the Indian region is considered (Fig. 3c). As OLR is linked to cloud top temperatures, it is used as a proxy to identify the intensity of cloud development. During the MJJASO, the decreasing trend of OLR represents increasing cloud amount, while during NDJFMA the increasing trend represents increasing shallower clouds. Also, during MJJASO a significant decreasing trend of OLR (<220 W/m^2) represents that deep clouds are increasing and the area covered by deep clouds is also found increasing (Fig. 3d) as compared to all other types of clouds. Thus, under cloudy conditions, solar dimming during the summer season may be due to increasing convective clouds covering a larger area, while during the winter season, it may be due to the aerosol indirect effect.

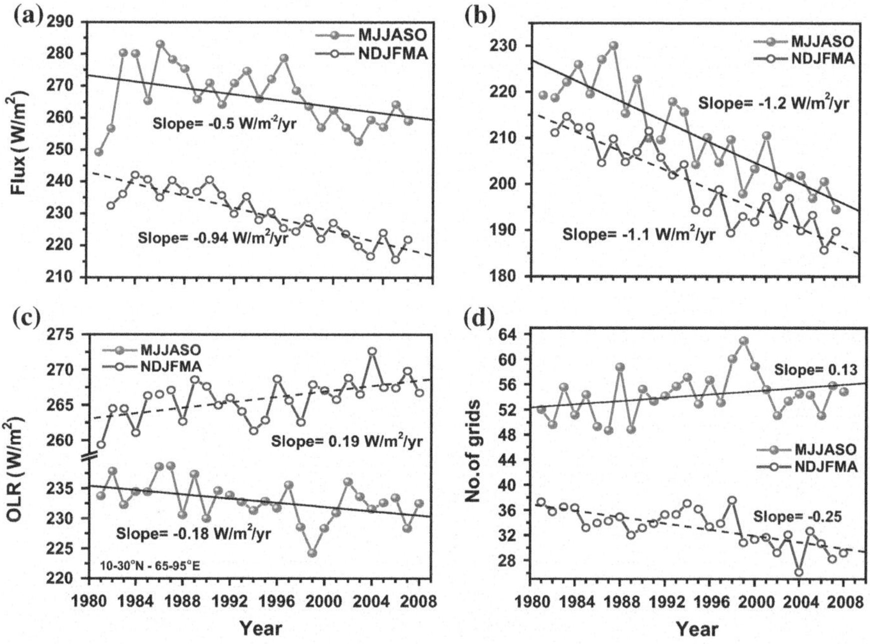

Fig. 3 Seasonal variation of surface reaching radiative flux (Rs) and OLR over India for two seasons MJJASO (May–October) and NDJFMA (November–April). **a** Rs under clear-sky conditions, **b** Rs under cloudy conditions, and **c** total OLR. **d** Seasonal variation of total number of grids covered by deep clouds (OLR < 220 W/m^2) over India (Padmakumari and Goswami 2010)

6 Radiative Flux Variability in the Recent Decade

To examine the solar radiation trend in the recent decade, Soni et al. (2016) further analyzed the data from 1971 to 2010 over 12 different stations in India (Fig. 4). The all-sky global irradiance showed a decreasing trend of 0.6 W m^{-2} $year^{-1}$ during 1971–2000 and 0.2 W m^{-2} $year^{-1}$ during 2001–2010. A reversal in all-sky global irradiance after 2001 at some stations also was noticed. Figure 4a shows that after 2001 there seems to be stagnation and slight upward trend. This indicates solar brightening over India, which may be ascertained further with long-term data sets.

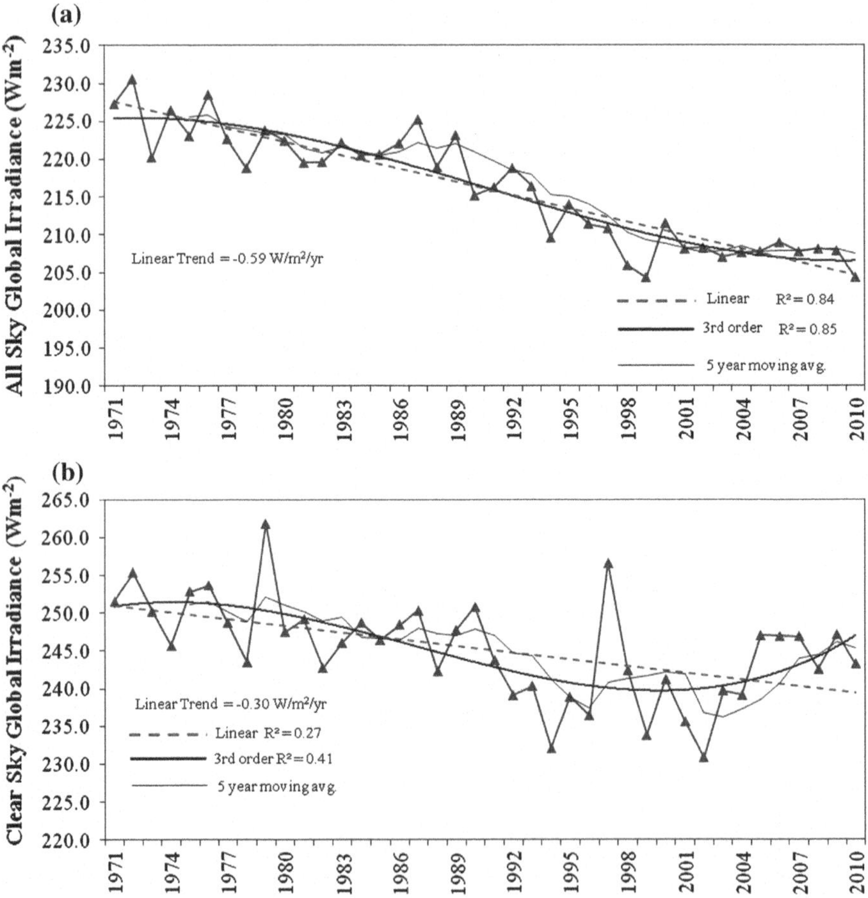

Fig. 4 Linear, third-order polynomial and 5-year moving average fits to annual time series of **a** all-sky global irradiance averaged over 12 solar radiation stations and **b** clear-sky global irradiance averaged over eight stations (Soni et al. 2016)

The clear-sky global irradiance showed a linear decreasing trend of 0.4 W m^{-2} $year^{-1}$ for the period 1971–2000 over India. While, interestingly, a positive and significant trend of magnitude 1.3 W m^{-2} $year^{-1}$ is observed during the latter period 2001–2010. The reversal in clear-sky solar trend is observed after 2001. However, station-wise analysis indicates reversal in trend at different years for different stations. It is noted that at Ahmedabad and Jodhpur the decreasing trend reversed from 2001, at Nagpur from 2003, at Pune from 2005, at Visakhapatnam and Shillong from 2004. During 2001–2010, the increasing linear trend in clear-sky irradiance was evident at stations Ahmedabad, Delhi, Jodhpur, Nagpur, and Pune. Thus, station-wise trends highlight the fact that dimming/brightening is region dependent due to regional sources and meteorology. The main factors that are responsible are changes in the atmospheric loading due to natural as well as various man-made activities, cloud cover and other properties of clouds. The data analysis of total cloud cover, based upon 172 surface meteorological stations (under the network of IMD) from 1961 to 2007, indicated a general decrease over most parts of India (Jaswal et al. 2010). The extension of their study to 195 stations from 1951 to 2010 showed that 74 stations reported decreasing trends, while 28 stations reported increasing trends; both the trends are statistically significant at the 95 % confidence level. The remaining stations showed either increasing or decreasing nonsignificant trends. For the country as a whole, annual mean total cloud cover showed statistically significant decreasing trend (0.44 % per decade), the trend being more after 2000. This seems to be consistent with the increasing trend in all-sky global irradiance in the late 2000.

Diffuse irradiance has shown a more complicated spatial and temporal variability. All-sky diffuse radiation, averaged over 12 stations, showed increasing trend from 1971 to 2010. The trend was 0.1 W m^{-2} $year^{-1}$ during the period 1971–2000 and 0.4 W m^{-2} $year^{-1}$ during 2001–2010 (Fig. 5). The diffuse irradiance showed high values in 1982 and 1983 followed by El-Chichon volcanic eruption in Chiapas (Mexico) in March 1982, and also during 1991–1992 peaking in 1992 followed by Mt. Pinatubo volcanic eruption in Philippines in June 1991. Seasonal variability of all-sky diffuse irradiance showed increasing trend for the period 1971–2000 and 2001–2010 except during the monsoon season. High trend values are observed during winter and post-monsoon seasons. These two seasons play an important role in the temporal variability of annual diffuse radiation. Under clear-sky conditions, the increasing trend in diffuse irradiance was much higher during the period 2001–2010 as compared to the period 1971–2000 over India. For changes in solar radiation under clear-sky conditions, aerosols are considered to be the main source of extinction in the atmosphere. Other possible causes could be the changes in frequency of occurrence of cloud-free days. However, further analysis is required to understand the changes in clear-sky global irradiance during 2001–2010.

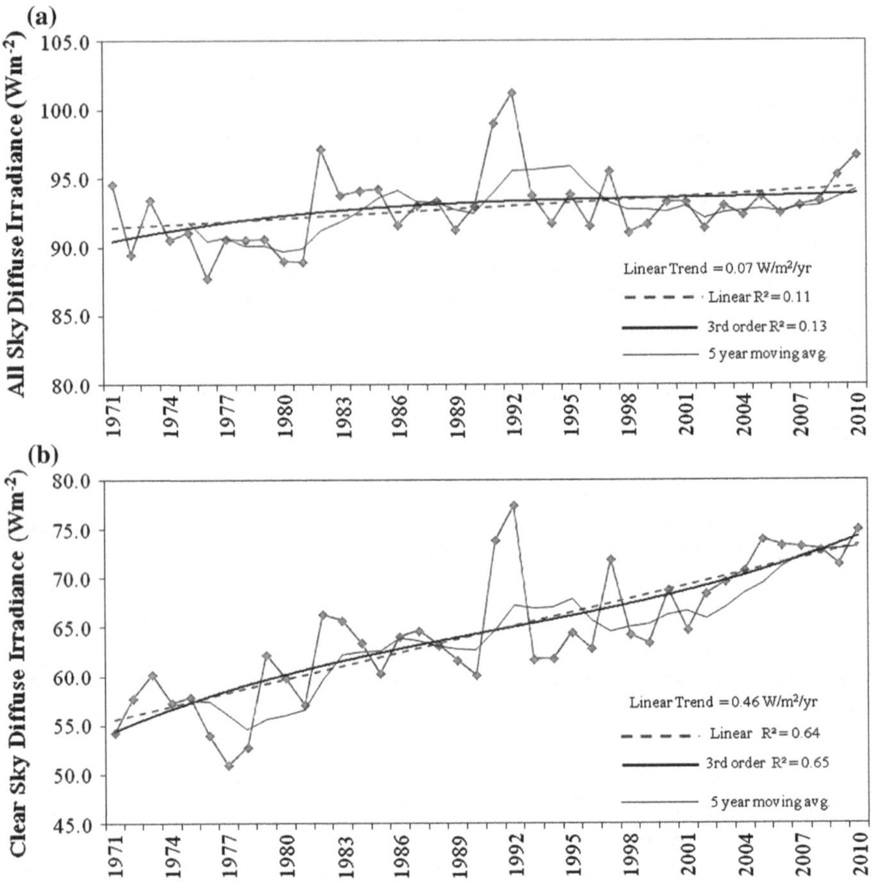

Fig. 5 Linear, third-order polynomial and 5-year moving average fits to annual time series of **a** all-sky diffuse irradiance averaged over 12 solar radiation stations and **b** clear-sky diffuse irradiance averaged over eight stations (Soni et al. 2016)

7 Radiative Flux from Other Data Sources

Decadal variability in Rs is studied from other data sources such as Modern Era Retrospective-analysis for Research and Applications (MERRA). Using long-term MERRA 2D data sets for the period 1979–2004, Kambezidis et al. (2014) studied the trends in net downward shortwave radiation (NDSWR) over South Asia covering continental India, Arabian Sea, Bay of Bengal, and tropical northern Indian Ocean. Over South Asia, they found a decreasing trend in shortwave radiation.

The NDSWR showed a pronounced decreasing trend of ~0.54 W m^{-2} $year^{-1}$ during 1979–2004. While under clear-sky conditions, much lower rate of −0.05 W m^{-2} $year^{-1}$ is observed. This depicts that aerosols could explain about one-tenth of the NDSWR trends over South Asia, with the clouds having the largest effect. Badrinath et al. (2010) studied the net incoming shortwave radiation over the urban location, Hyderabad, from 1979 to 2005 using MERRA data and showed a significant reduction of -0.69 W m^{-2} $year^{-1}$. The decline is more intense during 1993–2005 as compared to previous decade due to increase in high cloud and aerosol optical depths during the latter decade. Using the same data set Gayatri et al. (2012) showed greater positive trend in high clouds as compared to low clouds for the period 1979–2009 over different regions with different surface characteristics. They also found dimming over central and north-eastern part of India. Both net shortwave and net long-wave fluxes at the surface showed a decreasing trend for all the seasons. Area-averaged (68°–96°E; 8°–38°N) annual time series under the assumption of clear-sky (no clouds) for the period 1979–2014 shows a statistically significant (at the 99.9 % confidence level) negative trend with a reduction of 0.4 W m^{-2} per decade (Fig. 6). However, there remain some concerns over the accuracy of satellite-derived solar radiation data. MERRA generally overestimates incident solar radiation in the mid-latitudes but underestimates in the equatorial regions relative to the GEWEX-SRB (Yonghong et al. 2011).

CERES data (CERES_EBAF_Surface_Ed208) analyzed over the Indian subcontinent (68–96°E; 8–38°N) for the SW radiation at the surface under all-sky conditions for the period 2001–2013 is shown in Fig. 7. Under all-sky conditions a reduction of 0.2 W m^{-2} $year^{-1}$ is observed, which is not statistically significant. This trend is similar to that observed with surface observations for the period

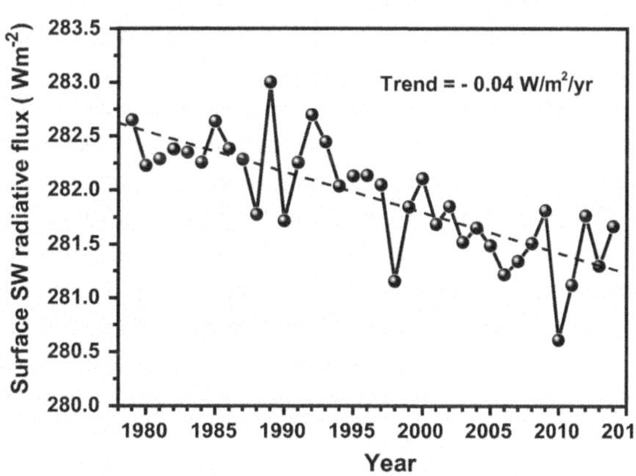

Fig. 6 Area-averaged (68°–96°E; 8°–35°E) annual time series of surface incident shortwave radiative flux under clear-sky conditions (no clouds), from 1979 to 2014 derived from MERRA reanalysis data. The trend is statistically significant at the 99.9 % confidence level

2001–2010 as discussed in the above section. The figure reveals that both aerosols and clouds play a major role in reducing the surface reaching solar radiation. The CERES data products use cloud and aerosol properties derived from MODIS

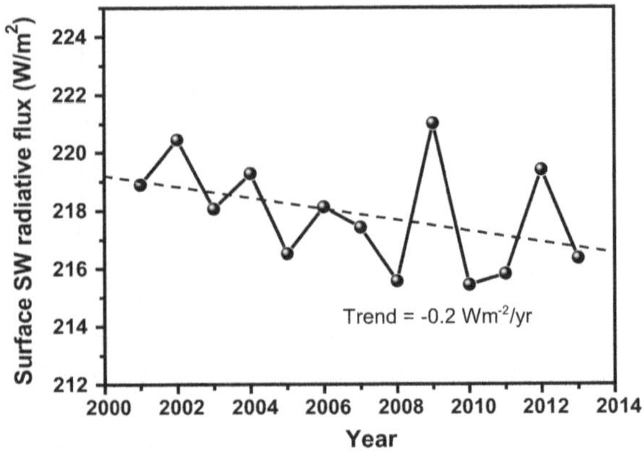

Fig. 7 Area-averaged (68°–96°E; 8°–35°E) mean annual time series of surface incident shortwave radiative flux under all-sky conditions from 2001 to 2014 derived from CERES_EBAF_Surface_Ed2.8 data. The trend is not statistically significant

radiances, meteorological assimilation data from GEOS version 4 and 5 models, and aerosol assimilation from the Model for Atmospheric Transport and Chemistry (MATCH).

8 Implications for Climate Change

Significant changes in surface radiative fluxes on decadal timescales have the potential to affect various aspects of climate change. Surface reaching radiative flux is the key determinant of surface air temperature. Despite solar dimming, the mean annual air temperature has been increasing due to increase in greenhouse gases. Earlier studies have shown that strong greenhouse forcing outweighs decreasing solar radiation (Philipona and Durr 2004 over Central Europe, Padmakumari et al. 2007 over the Indian subcontinent).

Several of earlier studies reported that radiative forcing is more effective in altering the strength of hydrological cycle than thermal forcing due to changes in the greenhouse gases. Pan evaporation (*E*pan) showed significant decreasing trend over different stations in India (Jaswal et al. 2008; Padmakumari et al. 2013). For the four decades considered, *E*pan showed high correlation with Rs (Fig. 8) as compared to surface air temperature and relative humidity (Padmakumari et al. 2013). The interpretation of variability in pan evaporation trends has been attributed to various reasons; most important among them is the changes in Rs. Vegetation growth, cover, and crop production are also affected by the changes in Rs.

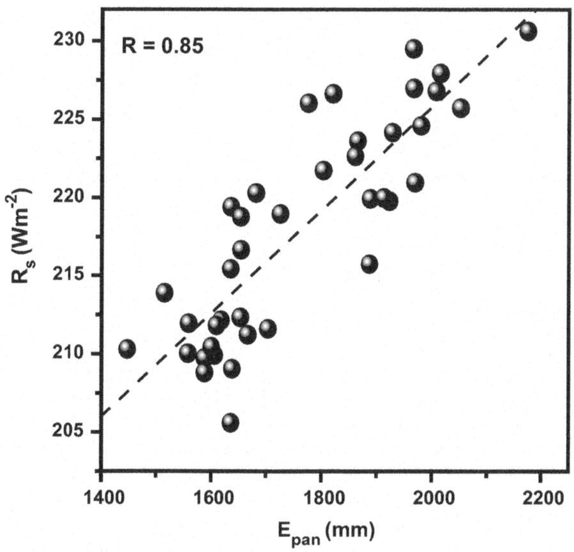

Fig. 8 Comparison of observed surface reaching solar radiative flux (Rs) and pan evaporation (*E*pan) averaged for 11 stations for the period of 40 years (1971–2010)

9 Summary

Long-term observations of surface radiative fluxes from ground-based instrumentation are more accurate to understand their spatial and temporal variability in the climate change scenario. Satellite measurements are accurate for the radiative fluxes at the TOA. In the present scenario, inter-comparison and validation of satellite and reanalysis data sets with the ground-based measurements is highly required to understand the Earth Radiation Budget and therefore the global energy balance.

Over the Indian subcontinent, Rs showed significant seasonal, annual, and decadal variability depending upon the variability in aerosol loading and cloud cover. Dust aerosols, from the Thar desert and West Asian countries through long-range transport, anthropogenic aerosols during winter, and large variability in cloud cover and cloud types during the monsoon season play crucial role in radiative flux variability over India.

Global irradiance data from 1981 to 2006 represented a significant decreasing trend of 0.89 W m^{-2} year^{-1} under all-sky conditions, called solar/global dimming. It is also reported that the rate of dimming under cloudy conditions is twice as large as that during clear-sky conditions. The clear-sky dimming is mainly due to increasing aerosol loading. While cloudy-sky dimming during the summer season is attributed to increasing deeper clouds covering increasingly larger area and winter dimming in cloudy conditions is attributed to aerosol indirect effects.

Long-term all-sky global irradiance data from 1971 to 2010 illustrated a decreasing trend of 0.6 W m^{-2} year^{-1} during 1971–2000 and 0.2 W m^{-2} year^{-1} during 2001–2010. The lower trend value in the latter part is due to the reversal trend observed at some stations after 2001. This is also consistent with the decreasing trend in total cloud cover at some stations. This indicates that there seems to be an indication of brightening after 2010. Although dimming/brightening is region dependent, on a large-scale dimming still persists. In the present scenario of increasing aerosol loading, diffused radiation is increasing but the lower trend value in global irradiance might be due to the changes in aerosol optical properties, which needs to be further investigated. Indication of brightening late 2000 may be further ascertained with long-term data sets.

References

Abakumova, G.M., Feigelson, E.M., Russak, V., Stadnik, V.V. (1996), Evaluation of Long-Term Changes in Radiation, Cloudiness, and Surface Temperature on the Territory of the Former Soviet Union. J. Clim 9, 1319-1327.

Anderson, D.E. and Cahalan, R.F. (2005), The solar radiation and climate experiment (SORCE) mission for the NASA earth observing system (eos). Sol Phys 230(1-2):3-6. doi:10.1007/S11207-005-1592-6.

Augustin, J A., and Dutton, E.G. (2013), Variability of the surface radiation budget over the United States from 1996 through 2011 from high-quality measurements. J. Geophys. Res., 118. doi:10.1029/2012JD018551, 43-53.

Ayyar, P.S.H. and Krishnamurthy, V. (1968), Net Radiation Climate of India. IJMG 19, 2,203-208.

Babu, S.S., Satheesh, S.K. and Moorthy, K.K. (2002), Aerosol radiative forcing due to enhanced black carbon at an urban site in India. Geophys. Res. Lett. 29(18), 1889, doi:10.1029/2002GL015826.

Badrinath, K.V.S., Sharma, A.R., Kaskaoutis, D.G., Kharol, S.K. and Kambezidis, H.D. (2010), Solar dimming over the tropical urban region of Hyderabad, India: Effect of increased cloudiness and increased anthropogenic aerosols. J. Geophys. Res. 115, D21208, doi:10.1029/2009JD013694.

Balachnadran, S. and Rajeevan, M. (2007), Sensitivity of surface radiation budget to clouds over the Asian monsoon region. J. Earth Syst. Sci. 116, No. 2, 159-169.

Desikan, V., Iyer N.V. and Rahalkar, C.G. (1969), Diffuse solar (sky) radiation measurements over India. Ind. J. Met. Geophys 20, 4, 389-394.

Desikan, V., V. R. Chivate and V. V. Abhyankar (1994a), Reaction of radiometric parameters to atmospheric pollution: Part I -Variation over time. Mausam 45, 1, 79-86

Desikan, V., Chivate, V.R. and Abhyankar, V. V. (1994b), Reaction of radiometric parameters to atmospheric pollution: Part II -Variation over time. Mausam 45, 2, 139-148.

Ganesan, H.R. (1970), Estimates of solar radiation over India. Mausam 21, 4, 629-636.

Gangopadhyaya, M., Datar, S.V. and George, C.J. (1970), On the global solar radiation climate and evapotranspiration estimates in India. Mausam 21, 1, 23-30.

Ganguly, D., Jayaraman, A., Rajesh, T. A., Gadhavi, H. (2006), Winter aerosol properties during foggy and nonfoggy days over urban centre Delhi and their implication for shortwave radiative forcing. J. Geophys. Res. 111, D15217, doi:10.1029/2005JD007029.

Gautam, R., Christina Hsu, N., Kafatos, M., and Tsay, S. C. (2007), Influences of winter haze on fog/low cloud over the Indo-Gangetic plains. J. Geophys. Res. 112. D05207, doi:10.1029/2005JD007036.

Gayatri, U., Prabha, T.V., Pandithurai, G., Pallavi, Achuthavarier, D.and Goswami, B.N. (2012), Aerosol and cloud feedbacks on surface energy balance over selected regions of the Indian subcontinent. J. Geophys. Res. 117, D04210, doi:10.1029/2011JD016363.

Gilgen, H., Wild, M. and Ohmura, A. (1998), Means and trends of shortwave irradiance at the surface estimated from global energy balance archive data. J. Clim 11(8): 2042-2061.

Jaswal A.K, Prakasa Rao, G.S. and De, U.S. (2008), Spatial and temporal characteristics of evaporation trends over India during 1971–2000. Mausam 59 (2): 149–158.

Jaswal, A.K. (2010), Changes in total cloud cover over India based upon 1961-2007 surface observations. Mausam 61, 4, 455-468.

Kambezidis, H.D., Kaskaoutis, D.G., Kharol, S.K., Moorthy, K.K., Satheesh, S.K., Kalapureddy, M.C.R., Badarinath, K.V.S., Sharma, A.R., Wild, M. (2014), Multidecadal variation of the net downward shortwave radiation over south Asia: the solar dimming effect. Atmos. Environ 50, 360–372. doi:10.1016/j.atmosenv.2011.11.008.

Kumar, G.V. and B. Manikiam (1988), Global radiation derived from sunshine data. Mausam 39, 1,87-90

Latha, R., Murthy, B.S., Lipi, K., Manoj Kumar, Das, S.K., and Mahanti, N.C. (2014), Temporal variation of solar dimming induced by composite and carbonaceous aerosols: Observations from mineral-rich eastern Indian region. J. Geophys. Res. 119, 6246-6259, doi:10.1002/2014JD021483.

Loeb, N.G., Lyman, J.M., Johnson, G. C., Allan, R.P., Doelling, D.R., Wong, T., Soden, B.J., Stephens, G.L. (2012), Observed changes in top-of-the-atmosphere radiation and upper-ocean heating consistent within uncertainty. Nat Geosci 5(2): 110-113, doi:10.1038/Ngeo1375.

Liepert B.G. (2002), Observed reductions of surface solar radiation at sites in the United States and worldwide from 1961 to 1990. Geophys. Res. Lett. 29, 1421 doi:10.1029/2002GL014910.

Mani, A., Swaminathan, M.S. and Venkiteshwaram, S.P. (1962a), Distribution of Sunshine and Solar Radiation over the Indian Peninsula. Ind. J. Met. Geophys 13, 2, 195-212.

Mani, A., Chacko, O. and Venkateshwaram, S. P. (1962b), Measurements of total Radiation from Sun and Sky in India during IGY. Ind. J. Met. Geophys 13, 3, 337-366.

Mani, A. and Chacko, O. (1963), Measurements of Diffuse Solar Radiation at Delhi and Poona, IJMG 14, 4, 416-432.

Mani, A., Chacko, O., Krishnamurthy, V. and Desikan, V. (1967), Distribution of global and net radiation over the Indian Ocean and its environments. Theoretical and Applied Climatology 15, 82-98.

Mani A. (1971), Meteorological measurements from satellites -A review. Mausam, 22, 3, 369-375.

Mani, A., Chacko, O. and Desikan, V. (1977), Solar radiation measurements and studies of atmospheric transmission at high altitude stations in India. Mausam 28, 1, 51-62.

Mani, A. and Rangarajan, S. (1982), Solar radiation over India. Allied Publishers, New Delhi.

Mooley, D.A. and Raghavan, S. (1963), Total radiation from sun and sky on a horizontal surface in relation to cloud type at Madras. Ind. J. Met. Geophys 14, 4, 482-485.

Moorthy, K.K., Nair, P.R., Prasad, B.S.N., Muralikrishnan, N., Gayathri, H.B., Narasimha Moorthy, B. (1993), Results from the MWR Network of IMAP. Indian Journal of Radio and Space Physics 22: 243-258.

Moorthy, K.K., Saha, A., Prasad, B.S.N., Niranjan, K., Jhurry, D., Pillai, P.S. (2001), Aerosol optical depths over the peninsular India and adjoining ocean during the INDOEX campaigns: spatial, temporal and spectral characteristics. J. Geophys. Res 106 (28): 539-554.

Ohmura, A., Bauder, A., Mueller, H., and Kappenberger, G. (2007), Long-term change of mass balance and the role of radiation. Ann. Glaciol 46, 367-374, doi:10.3189/172756407782871297.

Ohmura, A., Gilgen, H., and Wild, M. (1989), Global energy balance archieve GEBA, World Climate Program – Water Project A7, Zuercher Geografsche Schriften, 34, 62 pp., Verlag der Fachvereine, Zurich.

Padmakumari, B., Londhe, A.L., Daniel, S., Jadhav, D.B. (2007), Observational evidence of solar dimming: Offsetting surface warming over India. Geophys Res Lett 34: L21810. doi:10.1029/2007GL031133.

Padmakumari, B. and Goswami, B.N. (2010), Seminal role of clouds on solar dimming over the Indian monsoon region. Geophys Res Lett 37: L06703. doi:10.1029/2009GL042133

Padmakumari B., Jaswal, A.K. and Goswami, B.N. (2013), Decrease in evaporation over the Indian monsoon region: implication on regional hydrological cycle. Clim Change, doi:10.1007/s10584-013-0957-3.

Pandithurai, G., Pinker, R.T., Takamura, T. and Devara, P.C.S. (2004), Aerosol radiative forcing over a tropical urban site in India. Geophys. Res. Lett. 31, L12107, doi:10.1029/2004GL019702.

Pandithurai, G., Dipu, S., Dani, K.K., Tiwari, S., Bisht, D.S., Devara, P.C.S., and Pinker, R.T. (2008), Aerosol radiative forcing during dust events over New Delhi, India. J. Geophys. Res. 113, D13209, doi:10.1029/2008JD009804.

Pinker, R.T., Zhnag, B., Dutton, E.G. (2005), Do satellites detect trends in surface solar radiation? Science 308: 850-854.

Philipona, R., and Durr, B. (2004), Greenhouse forcing outweighs decreasing solar radiation driving rapid temperature rise over land. Geophys. Res. Lett., 31, L2208, doi:10.1029/2004GL020937.

Philipona, R., K. Behrens and C. Ruckstuhl (2009), How declining aerosols and rising greenhouse gases forced rapid warming in Europe since the 1980s. Geophys. Res. Lett. *36,* L02806. doi:10.1029/2008GL036350.

Rajeevan, M., De, U.S. and Prasad, R.K. (2000), Decadal variation of sea surface temperatures, cloudiness and monsoon depressions in the north Indian ocean. Curr. Sci. 79(3), 283-285.

Ramanathan, V., Crutzen, P.J., Kiehl, J.T. and Rosenfield, D. (2001), Aerosol, climate and the hydrological cycle. Science 294, 2119-2124. doi:10.1126/science.1064034.

Ramanathan, V., Chung, C., Kim, D., Bettge, T., Buja, L., Kiehl, J.T., Washington, W.M., Fu, Q., Sikka D.R., Wild, M. (2005), Atmospheric brown clouds: Impacts on south Asian climate and hydrological cycle. PNAS 102, 5326-5333.

Ramdas, L.A. and Yegnanarayanan S. (1956), Solar Energy in India. Proc. UNESCO Symp. Wind and Solar Energy, New Delhi pp 188-197.

Ruckstuhl, C. et al. (2008), Aerosol and cloud effects on solar brightening and the recent rapid warming. Geophys. Res. Lett., 35, L12708, doi:10.1029/2008GL034228.

Russak, V. (1990), Trends of solar radiation, cloudiness and atmospheric transparency during recent decades in Estonia. Tellus 42B, 206-210.

Sarkar, S., Chokngamwong, R., Cervone, G., Singh, R.P. and Kafatos, M. (2006), Variability of aerosol optical depth and aerosol forcing over India. Adv. Space Res. 37(12), 2153-2159.

Soni, V.K., Pandithurai, G., and Pai, D.S. (2012), Evaluation of long term changes of solar radiation in India. Int. J. Climatol doi:10.1002/joc.2294.

Soni, V.K., Pandithurai, G. and Pai, D.S. (2016), Is there a transition of solar radiation from dimming to brightening over India? Atmos Res. 169, 209–224, doi:10.1016/j.atmosres.2015.10.010

Soni, V.K. and Kannan, P.S. (2003), Temporal variation and the effect of volcanic eruptions on atmospheric turbidity over India. Mausam 54(4): 881-890.

Singh, S., Nath, S., Kohli, R., and Singh, R. (2005), Aerosols over Delhi during pre-monsoon months: Characteristics and effects on surface radiation forcing. Geophys. Res. Lett. 32, L13808. doi:10.1029/2005GL023062.

Stanhill G. and Cohen, S. (2001), Global dimming: a review of the evidence for a wide spread and significant reduction in global radiation with discussion of its probable causes and possible agricultural consequence. Agricultural and Forest Meteorology 107: 255-278.

Streets, D. G., Wu, Y., and Chin, M. (2006), Two decadal aerosol trends as a likely explanation of the global dimming/brightening transition. Geophys. Res. Lett. 33, L15806, doi:10.1029/2006GL026471.

Tang, W., Yang, K., Qin, J., Cheng C.C.K. and He J. (2011), Solar radiation trend across China in recent decades: A revisit with quality-controlled data. Atmos. Chem. Phys., 11, 393-406, doi:10.5.194/acp-11-393-2011.

Venkataraman, S. and Krishnamurthy, V. (1967), The radiation climate over India. IJMG, 18, 1, 39-44.

Wielicki, B.A., Barkstrom, B.R, Harrison, E.F, Lee, R. B, Smith, G. L. and Cooper, J. E (1996), Clouds and Earth's radiant energy system (CERES): an earth observing system experiment. Bull Am Meteorol Soc 77(5): 853-868.
Wild M., et al. (2005), From dimming to brightening: Decadal changes in solar radiation at Earth's surface. Science 308: 847-850.
Wild, M., Ohmura, A. and Makowski, K. (2007), Impact of global dimming and brightening on global warming. Geophys. Res. Lett. 34, L04702, doi:10.1029/2006GL028031.
Wild M., Grieser, J. and Schar, C. (2008), Combined surface solar brightening and increasing greenhouse effect support recent intensification of the global land-based hydrological cycle. Geophys. Res. Lett. 35, L17706, doi:10.1029/2008GL034842.
Wild M., et al. (2009), Global dimming and brightening: An update beyond 2000. J. Geophys. Res. 114, D00D13, doi:10.1029/2008JD011382.
Wild M., et al., (2012), The global energy balance from a surface perspective. Clim Dyn. doi:10.1007/s00382-012-1569-8.
Yadav, B. R. (1965), Total Solar Radiation in relation to duration of sunshine. IJMG 16, 2, 261-266.
Yang, K., et al. (2012), Can aerosol loading explain the solar dimming over the Tibetan Plateau. Geophy. Res. Lett. 39, L20710, doi:10.1029/2012GL053733.
Yonghong Yi, Kimball, J.S., Jones L.A., Reichle, R.H. and McDonald, K.C. (2011), Evaluation of MERRA Land Surface Estimates in Preparation for the Soil Moisture Active Passive Mission. J. Clim 24, 3797–3816. doi:10.1175/2011JCLI4034.1

Chapter 10
Variability and Trends of Sea Surface Temperature and Circulation in the Indian Ocean

C. Gnanaseelan, M.K. Roxy and Aditi Deshpande

1 Observations in the Indian Ocean

The history of the Indian Ocean observations stretches back approximately to the 1870s (Deser et al. 2010), and a large share of these measurements came from ships of opportunity, mostly linking Mediterranean through the Suez Canal to the maritime continent. Physical oceanographic expeditions were rare in the Indian Ocean until 1950s, with usually more than a couple of decades between two observations in the same region (Behrman 1981). It was only during 1959–1965 that a systematic recording of the Indian Ocean characteristics started with the International Indian Ocean Expedition (IIOE). This was later appended with large-scale sampling through regional programs such as MONEX, BOBMEX, JASMINE and ARMEX (Murakami 1979; Bhat et al. 2001; Webster et al. 2002; Shenoi et al. 2005) along with the use of Argo array profiling floats. Since the advent of satellite era, high-resolution satellite data at frequent intervals became available, helping scientists to decipher the spatiotemporal variability of the Indian Ocean even at finer scales (Bhat et al. 2004). Another major advancement in comprehending the Indian Ocean dynamics took place with the altimeter mission Topex/Poseidon. Understanding of the Indian Ocean sea surface temperature (SST) variability entered a new phase with the advent of microwave radiometer-based satellite observations from December 1997, especially during the cloudy period. Currently, the northern Indian Ocean has a reasonable observational coverage dating back to approximately 1870s (Deser et al. 2010) which supports a reliable analysis of Indian Ocean SST variability on both short and long timescales. The next section discusses the SST trends and variability over the tropical

C. Gnanaseelan (✉) · M.K. Roxy · A. Deshpande
Indian Institute of Tropical Meteorology, Earth System Science Organization, Pune, India
e-mail: seelan@tropmet.res.in

A. Deshpande
Savitribai Phule Pune University, Pune, India

M.N. Rajeevan and S. Nayak (eds.), *Observed Climate Variability and Change Over the Indian Region*, Springer Geology, DOI 10.1007/978-981-10-2531-0_10

Indian Ocean, at annual, interannual and intraseasonal timescales and their impact on regional climate. Section 3 discusses the variability and trends in the Indian Ocean circulation.

2 Sea Surface Temperature (SST) Variability in the Indian Ocean

Among the tropical oceans, the Indian Ocean is the smallest and the warmest. The tropical Indian Ocean forms the major part of the largest warm pool (SST > 28 °C) among the global oceans. Tropical Indian Ocean SST plays a significant role in shaping climate as well as its variability on both regional (Chowdary et al. 2015) and global scales (Schott et al. 2009), and it is hence important to know the characteristics of Indian Ocean SST on spatial and temporal scales. Indian Ocean exhibits climate variability in several timescales, ranging from diurnal to interannual, and is strongly coupled to the seasonal cycle (Schott et al. 2009). The Indian Ocean also exhibits long-term trends in temperatures at both the surface (Roxy et al. 2014) and the subsurface (Lee et al. 2015), with an impact on the local monsoon Hadley circulation (Roxy et al. 2015), interannual variability (Chakravorty et al. 2014a) and the intraseasonal variability (Sabeerali et al. 2014).

The tropical ocean–atmosphere system exhibits marked variability in the intraseasonal timescales (Lau and Waliser 2012). Central to the intraseasonal variability is the Madden–Julian Oscillation (MJO), which was first discovered as an atmospheric phenomenon (Madden and Julian 1971, 1972, 1994) over the equatorial oceans. Subsequent experiments like the MONEX (see Murakami 1979) and BOBMEX revealed strong intraseasonal signals in the Bay of Bengal as well (Krishnamurti et al. 1988; Bhat et al. 2001). This intraseasonal variability in SST is closely associated with the atmospheric variability at similar scales (10–20 days and 30–90 days) and acts as a coupled phenomenon (Sengupta et al. 2001; Vecchi and Harrison 2002; Roxy and Tanimoto 2007; Vialard et al. 2012; Roxy et al. 2012). On the other hand, the intraseasonal SST variability in the Arabian Sea is forced by the oceanic processes (Vialard et al. 2012). A recent field campaign called the Dynamics of MJO/Cooperative Indian Ocean Experiment on Intraseasonal Variability in Year 2011 (DYNAMO/CINDY2011) traces the initiation of the MJO events to local ocean atmospheric processes over the Indian Ocean (Li et al. 2015). Several model studies addressed the intraseasonal SST variability associated with MJO and the associated processes (e.g. Jayakumar et al. 2011).

On interannual and decadal timescales, the Indian Ocean SSTs are influenced by two prominent modes of variability, namely El Niño Southern Oscillation (Rasmusson and Carpenter 1982; ENSO) and the Indian Ocean Dipole (IOD; Saji et al. 1999; Webster et al. 1999; Murtugudde et al. 2000).

2.1 *Response of Indian Ocean to El Niño Southern Oscillation (ENSO)*

ENSO is characterized by quasi-periodical events of warm and cool SST anomalies in the east equatorial Pacific. The large-scale shift in convection during an ENSO event alters the atmospheric circulation over both the tropics and extra-tropics. Tropical Indian Ocean (TIO) climate is strongly influenced by El Niño through an atmospheric bridge (e.g. Alexander et al. 2002) and ocean bridge such as pathways of Indonesian throughflow (ITF, e.g. Vaid et al. 2007; Sprintall et al. 2014). The basin-wide warming or cooling is found to be the first leading mode of the interannual SST variability in the Indian Ocean (Klein et al. 1999; Alexander et al. 2002; Chowdary and Gnanaseelan 2007; Yang et al. 2007; Du et al. 2009; Schott et al. 2009; Chowdary et al. 2015) and is induced by El Niño (La Niña) in boreal winter (Klein et al. 1999; Alexander et al. 2002; Chowdary and Gnanaseelan 2007). Ekman divergence/convergence-induced Rossby waves and El Niño-related subsidence-induced variations in heat flux play crucial roles in inducing TIO basin-wide warming (e.g. Chowdary and Gnanaseelan 2007; Du et al. 2009; Chowdary et al. 2015). This TIO warming persists until the following boreal summer (Xie et al. 2009; Chakravorty et al. 2014b), especially during the recent years (e.g. Chakravorty et al. 2014a), whereas El Niño-related warm SST anomalies in the eastern Pacific weakens or terminates by the following spring (e.g. Xie et al. 2010).

2.2 *Indian Ocean Dipole (IOD)*

IOD events are characterized by cool SST anomalies in the south-eastern equatorial Indian Ocean and warm SST anomalies in the western equatorial Indian Ocean (Saji et al. 1999; Webster et al. 1999; Vinayachandran et al. 1999; Murtugudde et al. 2000). The SST dipole is accompanied by a similar dipole in the precipitation anomalies, with suppressed precipitation in the east and enhanced precipitation in the western Indian Ocean. They are also closely associated with the equatorial Indian Ocean wind anomalies (e.g. Gadgil et al. 2004). Dipole events are generally initiated during June–August, but the events peak during September–November. The magnitude of any given event is represented by a normalized time series of the difference between west and south-east equatorial Indian Ocean anomalies known as the Dipole Mode Index (DMI). Positive IOD events often coincide with El Niño or El Niño-like events (Drbohlav et al. 2007; Roxy et al. 2014). Good Asian summer monsoon may also trigger IOD events (Ashok et al. 2003; Krishnan and Swapna 2009; Cai et al. 2013). Sayantani et al. (2014) showed the evolution of premonsoon Arabian Sea warming as a trigger for pure IOD events. A few studies indicate that the decadal variability in the Indian Ocean and the IOD is associated with the Pacific Decadal Oscillation (PDO) of the North Pacific SST (Crueger et al. 2008; Krishnamurthy and Krishnamurthy 2016). IOD has significant impact on Indian Ocean circulation, and interbasin mass and salt transport (Thompson et al. 2006; Jensen 2007) as well as the climate of the adjoining region. The east African fall rainfall variability is strongly affected by the developing phase of IOD, and significant impact on the Indian summer monsoon is observed only during strong IOD years (Deshpande et al. 2014). In addition to the surface dipole variability, Rao et al. (2002) and Sayantani and Gnanaseelan (2015) reported, respectively, the east–west and north–south dipole variability in the subsurface temperature (thermocline).

2.3 Indian Ocean Warming

The global oceans account for approximately 93 % of the warming of the earth system that has occurred since 1955. Indian Ocean exhibits enhanced warming in the surface and subsurface. Observations indicate a substantial surface warming of the Indian Ocean during the past half-century (Alory et al. 2007; Ihara et al. 2008; Alory and Meyers 2009; Swapna et al. 2014; Roxy et al. 2015). These studies indicate a basin-wide surface warming in the Indian Ocean (Du and Xie 2008; Hoerling et al. 2004), with an expansion of the warm pool region (Fig. 1a). The peak surface warming is displayed over the central equatorial region. This is favourable for weakening the monsoon Hadley cell. Ihara et al. (2008), using observed data, indicated that the western and the eastern regions of the Indian Ocean display significant warming trends since 1950. Over the past 60 years, the Indian Ocean warmed two to three times faster than the tropical Pacific (Williams and Funk 2011). Radiative forcing due to increased greenhouse gases has been given as an obvious cause of the warming (Du and Xie 2008), but that does not explain the enhanced warming in the Indian Ocean relative to the other tropical regions, especially with the decreasing trends in the net heat flux (Rahul and Gnanaseelan 2013). A few of these studies

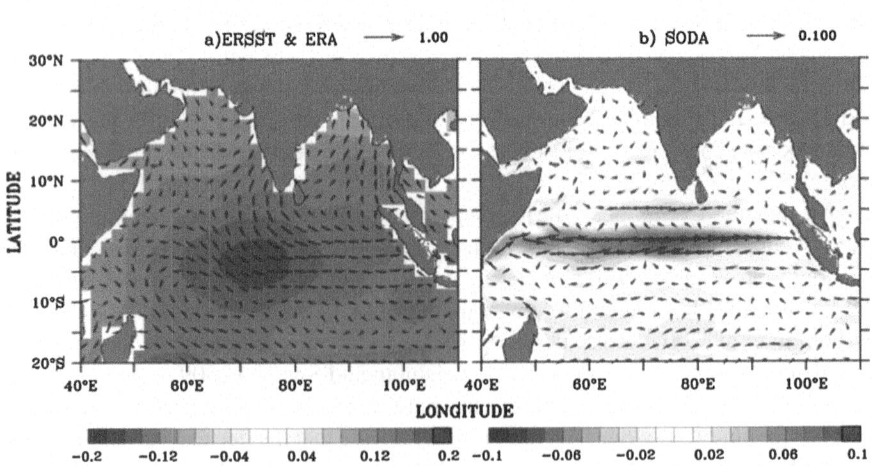

Fig. 1 **a** Trend in annual SST (*shaded*, per decade) superimposed with trend in ERA 10 m annual winds (*vectors*, per decade) for the period of 1958–2015, **b** trend in annual zonal current (per decade) superimposed with linear trend in current (vectors, per decade) for the period of 1958 to 2008

suggest that warm SSTs in this region trigger a local air–sea coupled interaction (Lau and Nath 2000; Du et al. 2009), which raises the warm pool temperature further. Meanwhile, other studies (Swapna et al. 2014) suggest that the weakening monsoon winds are responsible for the increasing surface warming over the Indian Ocean during the monsoon season. Roxy et al. (2014) point out that the western Indian Ocean has been warming consistently in summer during the past century and links the warming to the asymmetry and skewness in the ENSO forcing. They suggest that the increase in the frequency and magnitude of El Niño events in the recent years has resulted in a warming of the western Indian Ocean during summer, which persists for several months due to local air–sea interactions (Du et al. 2009) and ocean dynamics (Chowdary and Gnanaseelan 2007).

Though the century-long warming indicates large warming trend in the western Indian Ocean, SST trends during 1958–2015 display a basin-scale warming with peaks in the central equatorial Indian Ocean (Fig. 1a). The basin-scale warming is the leading mode of tropical Indian Ocean SST variability on interannual timescales and hence known as the Indian Ocean Basin (IOB) mode (e.g. Chowdary and Gnanaseelan 2007). It is also the prominent feature of the interdecadal SST trend in the recent decades. The IOB mode and Nino indices are highly correlated (Klein et al. 1999; Saji et al. 2006) though part of the SST variability is attributed to ocean dynamic processes and local air–sea interaction in the TIO (Lau and Nath 2000; Chowdary and Gnanaseelan 2007; Du et al. 2009).

The trend in the heat absorbed by the oceans contributes significantly to the variability at different depths (Levitus et al. 2012). About 60 % of the variability in the upper Pacific Ocean (top 1000 m) and Atlantic Ocean (top 1500 m) is contributed by linear trends (Levitus et al. 2012). The scenario is so complex in the Indian Ocean with the significant variance contributed by the linear trend observed in the 0–100 m, 200–500 m and 1100–1500 m layers (Levitus et al. 2012). Previous studies indicate the prominent subsurface warming of the Indian Ocean. Increase in the heat content of the upper 700 m of the southern Indian Ocean is reported by Levitus et al. (2012). According to Pierce et al. (2006), most of the tropical Indian Ocean warming is trapped in the top 125 m. Lee et al. (2015) reported an increase in the heat transport from the Pacific to the Indian Ocean during the recent decade through the Indonesian pathways. They pointed out that this heat gain in the past decade is more than 70 % of the global ocean heat gain in the upper 700 m. These studies indicate the growing importance of Indian Ocean as a heat reservoir in a warming world.

Figure 2a, b shows the trends in the upper ocean heat content, respectively, from SODA (Carton and Giese 2008) during 1958–2008 and Hadley centre EN4.0.2 (Good et al. 2013) during 1958–2015. Both products show warming trends in the north Indian Ocean, especially over the eastern equatorial Indian Ocean, Bay of Bengal and Arabian Sea. The warming pattern is similar for both the products in the upper 100 m, whereas they display some differences below 100 m, especially over the western Bay of Bengal. Figure 2c shows the increasing trend in the north Indian Ocean heat content of the upper 700 m from SODA and EN4. Both the products show similar trends though they display some differences in the interannual variability. Figure 2d shows the evolution of the annual mean over the north Indian Ocean and its linear

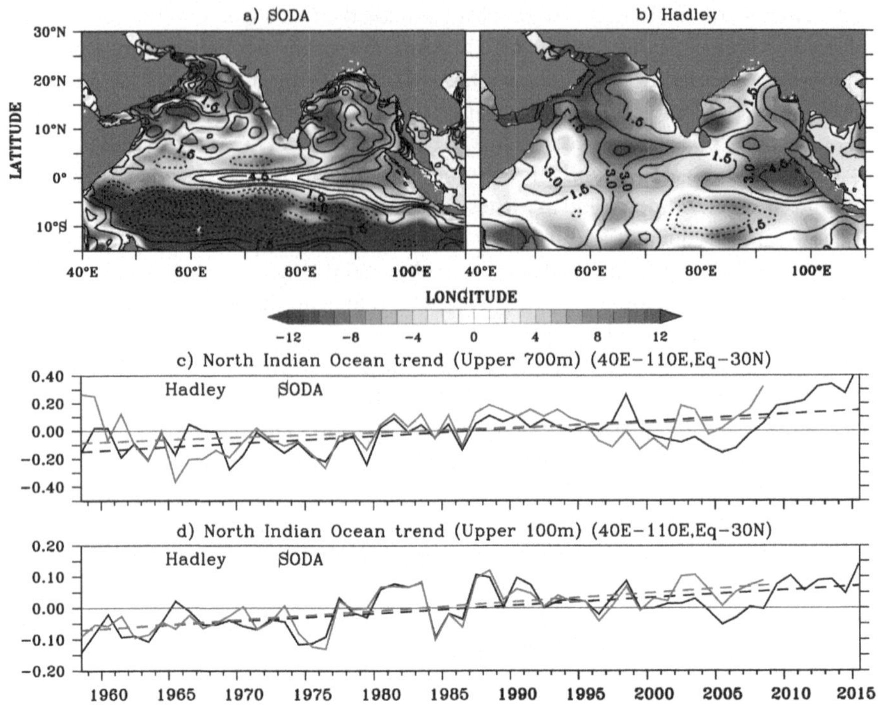

Fig. 2 *Upper* 700 m ocean heat content trend per decade (*shaded*, $\times 10^{-7}$) superimposed with upper 100 m heat content trend per decade (contour, $\times 10^{-7}$) from **a** SODA (for the period of 1958–2008), **b** Hadley centre EN4 (for the period of 1958–2015), **c** annual upper 700 m heat content anomaly over north Indian Ocean basin (40°E–110°E, Eq-30°N) (*solid*) superimposed with its trend (*dash*) from SODA (*red*) (for the period of 1958–2008) and Hadley centre EN4 (*black*) (for the period of 1958–2015), **d** is same as **c** but the upper 100 m heat content anomaly and its trend (color figure online)

trend. The correspondence between SODA and EN4 is better for 100 m heat content compared to that of 700 m. Both the products show steady warming of north Indian Ocean.

2.4 Local and Remote Impacts of Indian Ocean Warming

The Indian Ocean warming has local and remote impacts on regional climate change. Observations and model simulations indicate that the enhanced warming in the Indian Ocean increases rainfall over the Indian Ocean, but results in subdued convection over the Indian subcontinent via a weakened local Hadley circulation (Mishra et al. 2012; Kulkarni 2012; Saha et al. 2014; Roxy et al. 2015). Associated with the warming, the weak south-westerly winds during summer (Fig. 1a) and the coupled ocean–atmosphere interaction keep the surface ocean warm. Sabeerali et al. (2014) indicate that the SST warming in the Indian Ocean during the recent decades has changed the space–time characteristics of the northward-propagating monsoon intraseasonal oscillations. They found that the excess warming during the recent years triggers stationary convection over the equatorial Indian Ocean, stalling the northward propagation. Additionally, the warming trend has the potential to influence the Australian precipitation (Ashok et al. 2003) and the East Asian monsoon circulation (Li et al. 2010).

Model simulations indicate that the Indian Ocean warming plays a significant role on the African and Sahel droughts (Bader and Latif 2003; Giannini et al. 2003; Hoerling et al. 2006). According to Hoerling and Kumar (2003), widespread drying over the mid-latitudes are due to warm SST anomalies over the tropical Indian and the western Pacific oceans.

Several studies suggest that a warmer Indian Ocean may modulate the tropical Pacific variability. Terray et al. (2015), using model sensitivity experiments, showed that positive SST anomalies in the Indian Ocean dampen the magnitude of ENSO and also shorten its life cycle. This involves modulations of wind-induced eastward-propagating oceanic Kelvin waves by altering the thermocline. Luo et al. (2012) indicate that enhanced tropical Indian Ocean warming is favourable for the strengthening of trade winds in the western Pacific through atmospheric processes, contributing to La Niña-like conditions in the Pacific, as observed in the recent decades. Their analysis based on historical and future projections suggests that the Indian Ocean warming is modulating the Pacific climate in the twentieth and twenty-first centuries.

The rapid ocean warming has also resulted in increased surface stratification over the Indian Ocean. Increasingly stratified ocean waters suppress the upwelling of nutrients from the subsurface waters and reduce the marine primary production (Roxy et al. 2016). Future climate projections suggest the possibility of further warming of the Indian Ocean, which may affect the marine productivity further, which, combined with the fishing pressure, can deprive this region from its enhanced biological productivity.

Another contributing factor in the modulation of SST is the heat and mass transport carried by currents. Indian Ocean currents are dominated by the strong seasonal cycle of winds and also exhibit intraseasonal and interannual variability. The details of Indian Ocean circulation and its variability and trends are discussed in the following sections.

3 Indian Ocean Surface Circulation

Ocean currents transport mass and energy from one region to another around the world. The large movement of heat and salt associated with these currents makes the ocean current one of the primary drivers of global climate. The ocean circulation stabilizes the global atmospheric circulation and regulates the local weather and temperature extrema. Unlike the other tropical oceans, the atmospheric circulation over the tropical Indian Ocean is characterized by the seasonally reversing cross-equatorial flow. The circulation of the northern Indian Ocean is therefore strongly influenced by the reversing monsoon winds. The most prominent current systems in the northern Indian Ocean are the seasonally reversing Somali current, the west and east India coastal currents and semi-annual equatorial jets or Wyrtki jets. In addition to the seasonality in these current systems, they undergo large interannual and intraseasonal variability. Several studies in the recent years have addressed the Indian Ocean currents to a large extent, but more focused long-term observations are needed for better understanding of the Indian Ocean circulation (Schott and McCreary 2001; Shankar et al. 2002). Satellite-derived surface current products and the observations based on moored buoy array (e.g. RAMA) provide some relief despite their limitations. Rahul and Gnanaseelan (2016) reported decadal changes in the large-scale circulation over Indian Ocean contributing to surface temperature trends and variability.

3.1 Seasonal Mean Circulation in the Indian Ocean

The Somali current, monsoon currents (coastal currents), Wyrtki jets and ITF are the major current systems affecting the north Indian Ocean. Somali current is closely associated with the Findlater jet and the resultant coastal upwelling. The strong upwelling brings down the seasonal mean SST to below 26 °C off Somalia and the Arabian Peninsula. This further undergoes significant intraseasonal (Roxy and Tanimoto, 2007; Vialard et al. 2012) and interannual (Schott et al. 2009) variability, making it an important process in this part of the Indian Ocean. The monsoon currents are the pathways for the interbasin mass transport between Arabian Sea and Bay of Bengal. This plays an important role in the salt balance in the Bay of Bengal and Arabian Sea. This interbasin mass and salt transport displays strong interannual variability through changes in the circulation patterns of Bay of Bengal (e.g. Thompson et al. 2006; Jensen 2007). The equatorial Indian Ocean is characterized by strong eastward surface currents (Wyrtki jets, Wyrtki 1973) during boreal spring and fall, driven by the prevailing westerlies. The Wyrtki jets are closely related to the heat budget of the tropical Indian Ocean. They deepen the thermocline in the east and raise it in the west (Wyrtki 1973), inducing an east–west thermocline (and temperature) gradient (Gnanaseelan et al. 2012). The ITF links Pacific and Indian Oceans by providing a pathway and modifies the stratification within each of these oceans. In addition to the heat and freshwater balance in the Indian Ocean, ITF plays a significant role in the global circulation (Godfrey and Golding 1981). ITF also displays intraseasonal to decadal variability (Valsala et al. 2010).

3.2 Intraseasonal Variability in the Indian Ocean Circulation

Intraseasonal variability in SST over the Indian Ocean is very important in the air–sea interaction associated with MJO and monsoon intraseasonal oscillation (Lau and Waliser 2012). Intraseasonal variability in the ocean currents has a major role in the evolution of intraseasonal SST over Indian Ocean, especially over the Arabian Sea (Vialard et al. 2012). So we provide a brief review of intraseasonal oscillations in current over the Indian Ocean. Reppin et al. (1999) reported small-scale variability in currents near Sri Lanka. Sengupta et al. (2007) showed that both spring and fall Wyrtki jets are modulated on intraseasonal timescale. They showed that the spring Wyrtki jet is modulated by a single intraseasonal event, whereas the fall Wyrtki jet is modulated by two to three intraseasonal events. In addition to the intraseasonal variability in the spring and fall Wyrtki jets, Senan et al. (2003) reported intraseasonal jets during the summer monsoon season. Vialard et al. (2009) reported intraseasonal variability in the west India coastal current from ADCP observations off Goa coast. Mukherjee et al. (2014) reported intraseasonal variability in the east India coastal current. Shenoi (2010) provided a review on the evidences of intraseasonal variability of the coastal currents around India.

3.3 Interannual Variability in the Indian Ocean Circulation

Interannual variability over the equatorial Indian Ocean is dominated by Wyrtki jet variability. Interannual wind variability in winds associated with IOD and El Niño is primarily responsible for the Wyrtki jets variability. Anomalous westward currents in response to the easterly wind anomalies during the 1997 IOD event are reported by Grodsky et al. (2001). Thompson et al. (2006) also analysed the anomalous circulation associated with IOD. Reppin et al. (1999) from ship drift observations speculated that the El Niño is primarily responsible for the interannual variability of the Wyrtki jets. Recent studies have attributed this interannual variability mainly to IOD forcing (Chowdary and Gnanaseelan 2007; Nagura and McPhaden 2010; Gnanaseelan et al. 2012). The coastal region of the western Indian Ocean is a relatively unexplored area but displays strong variability in currents. Modelling studies on the formation and interannual variability of Great Whirl are few, but have shown that the Great Whirl variability is primarily dominated by its internal processes rather than the wind field (Schott and McCreary 2001). The Somali current as well as the coastal currents in both Arabian Sea and Bay of Bengal are studied by Schott and McCreary (2001) extensively using observations as well as models. The mean clockwise Bay of Bengal circulation is intensified during El Niño and IOD years (Thompson et al. 2006; Jensen 2007) resulting in excess transport of high saline Arabian Sea water to Bay of Bengal. On the other hand, the transport of Arabian Sea water to Bay of Bengal reduces during the La Niña years (Jensen 2007). Tropical Indian Ocean climate is strongly influenced by ITF (e.g. Sprintall et al. 2014). ITF displays strong interannual and intraseasonal variability and is dominated by El Niño. There is a large reduction in ITF transport during El Niño years (e.g. Valsala et al. 2010). Lee et al. (2015) suggest an increase in the heat transport from the Pacific Ocean to the Indian Ocean by ITF during the recent decade.

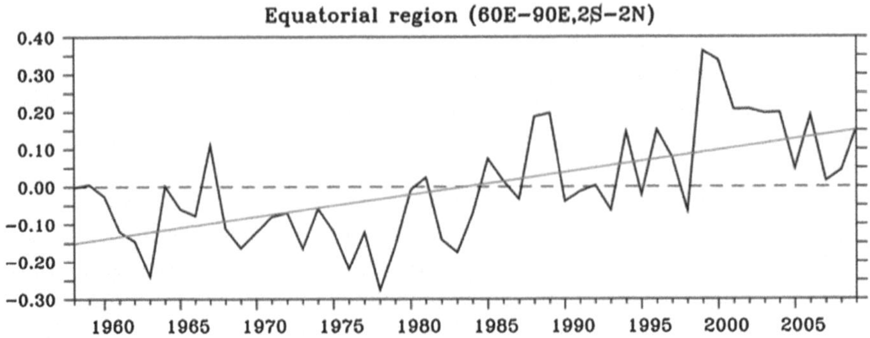

Fig. 3 Annual zonal current anomaly averaged over the region 60–90°E, 2°S–2°N (*black*) superimposed with its linear trend during 1958–2008 (*red*) (color figure online)

3.4 Trends in Indian Ocean Circulation

The Indian Ocean warming and the associated wind changes have strong impact on the Indian Ocean currents. Rahul and Gnanaseelan (2016) showed the existence of coupled feedback between Indian Ocean warming trends and circulation changes. It is important to note that there is a strengthening of mean westerly winds over the equatorial Indian Ocean (Fig. 1a) in the recent years (1958–2015). These westerlies force eastward surface currents along the equator. The trends in Indian Ocean surface circulation are shown in Fig. 1b. The trends in major currents over the Indian Ocean are evident. The most significant features are strengthening of the mean eastward currents along the equator and westward currents along 5°N. There is a significant increasing trend in the equatorial Indian Ocean currents (Fig. 3) and weak (and insignificant) reduction in the Somali current (figure not shown) during 1958–2008. The significant increasing trend in the equatorial currents or Wyrtki jets in turn increases eastward heat transport. These have tremendous consequences on the heat content of Bay of Bengal and the eastern Indian Ocean.

Acknowledgments The authors acknowledge the support of Director, IITM and MoES. We thank Hadley centre, ECMWF, for different data sets used in this study.

References

Alexander MA, Blade I, Newman M, Lanzante JR, Lau NC, Scott JD (2002) The atmospheric bridge: the influence of ENSO teleconnections on air-sea interaction over the global oceans. Journal of Climate 15(16):2205–2231. doi:10.1175/1520-0442(2002)015<2205:tabtio>2.0.co;2

Alory G, Meyers G (2009) Warming of the Upper Equatorial Indian Ocean and Changes in the Heat Budget (1960–99). Journal of Climate 22 (1)

Alory G, Wijffels S, Meyers G (2007) Observed temperature trends in the Indian Ocean over 1960–1999 and associated mechanisms. Geophysical Research Letters 34 (2)

Ashok K, Guan Z, Yamagata T (2003) A look at the relationship between the ENSO and the Indian Ocean Dipole. Journal of the Meteorological Society of Japan 81 (1):41-56
Bader J, Latif M (2003) The impact of decadal scale Indian Ocean sea surface temperature anomalies on Sahelian rainfall and the North Atlantic Oscillation. Geophysical Research Letters 30 (22).
Behrman D (1981) Assault on the largest unknown; the International Indian Ocean Expedition 1959-65.
Bhat GS, Gadgil S, Hareesh Kumar P, Kalsi S, Madhusoodanan P, Murty V, Prasada Rao C, Ramesh Babu V, Rao L, Rao R (2001) BOBMEX: The Bay of Bengal monsoon experiment. Bulletin of the American Meteorological Society 82 (10):2217-2243
Bhat GS, Vecchi GA, Gadgil S (2004) Sea surface temperature of the Bay of Bengal derived from the TRMM Microwave Imager. Journal of Atmospheric and Oceanic Technology 21 (8):1283-1290
Cai W, Zheng X-T, Weller E, Collins M, Cowan T, Lengaigne M, Yu W, Yamagata T (2013) Projected response of the Indian Ocean Dipole to greenhouse warming. Nature geoscience 6 (12):999-1007
Carton JA, Giese BS (2008) A Reanalysis of Ocean Climate Using Simple Ocean Data Assimilation (SODA). Monthly Weather Review, 136, 2999-3017.
Chakravorty S, Chowdary JS, Gnanaseelan C (2014a) Epochal changes in the seasonal evolution of tropical Indian Ocean warming associated with El Niño. Climate Dynamics, 42, 805-822, DOI:10.1007/s00382-013-1666-3.
Chakravorty S, Gnanaseelan C, Chowdary JS, Luo J-J (2014b) Relative role of El Niño and IOD forcing on the southern tropical Indian Ocean Rossby waves. Journal of Geophysical Research, 119, 5105-5122, DOI:10.1002/2013JC009713.
Chowdary JS, Gnanaseelan C (2007) Basin-wide warming of the Indian Ocean during El Niño and Indian Ocean dipole years. International Journal of Climatology 27 (11):1421-1438.
Chowdary JS, Bandgar AB, Gnanaseelan C, Luo J-J (2015) Role of tropical Indian Ocean air-sea interactions in modulating Indian summer monsoon in a coupled model. Atmospheric Science Letters, 16, 170-176, DOI:10.1002/asl2.561.
Crueger T, Zinke J, Pfeiffer M (2008) Patterns of Pacific decadal variability recorded by Indian Ocean corals. Int J Earth Sci 98:41–52.
Deser C, Alexander MA, Xie SP, Phillips AS (2010) Sea surface temperature variability: Patterns and mechanisms. Annual review of marine science 2:115-143
Deshpande A, Chowdary JS, Gnanaseelan C (2014) Role of thermocline–SST coupling in the evolution of IOD events and their regional impacts. Climate Dynamics, 43 (1:2): 163–174, DOI:10.1007/s00382-013-1879-5.
Drbohlav HKL, Gualdi S, Navarra A (2007) A diagnostic study of the Indian Ocean dipole mode in El Niño and non-El Niño years. Journal of Climate 20 (13):2961-2977. doi:10.1175/Jcli4153.1
Du Y, Xie S-P, Huang G, Hu K (2009) Role of Air-Sea Interaction in the Long Persistence of El Niño-Induced North Indian Ocean Warming. Journal of Climate 22 (8):2023-2038
Du Y, Xie S-P (2008) Role of atmospheric adjustments in the tropical Indian Ocean warming during the 20th century in climate models. Geophysical Research Letters 35 (8)
Gadgil S, Vinayachandran PN, Francis PA, Gadgil S (2004) Extremes of the Indian summer monsoon rainfall, ENSO and equatorial Indian Ocean oscillation. Geophysical Research Letters 31:L12213. doi:10. 1029/2004GL019733.
Giannini A, Saravanan R, Chang P (2003) Oceanic forcing of Sahel rainfall on interannual to interdecadal time scales. Science 302 (5647):1027-1030.
Gnanaseelan C, Deshpande A, McPhaden MJ (2012) Impact of Indian Ocean Dipole and El Niño/Southern Oscillation wind-forcing on the Wyrtki jets. Journal of Geophysical Research, 117, 1-11, DOI:10.1029/2012JC007918, C08005.
Godfrey J.S. and Golding TJ (1981) The Sverdrup relationship in the Indian Ocean, and the effect of Pacific–Indian Ocean throughflow on Indian Ocean circulation and on the east Australian current. Journal of Physical Oceanography, 11, 771–779.

Good, S. A., M. J. Martin and N. A. Rayner (2013). EN4: quality controlled ocean temperature and salinity profiles and monthly objective analyses with uncertainty estimates, Journal of Geophysical Research: Oceans, 118, 6704-6716, doi:10.1002/2013JC009067.
Grodsky SA, Carton JA and Murtugudde M (2001) Anomalous surface currents in the tropical Indian Ocean. Geophysical Research Letters, 10.1029/2001GL013592.
Hoerling M, Hurrell J, Eischeid J, Phillips A (2006) Detection and attribution of twentieth-century northern and southern African rainfall change. Journal of climate 19 (16):3989-4008
Hoerling M, Kumar A (2003) The perfect ocean for drought. Science 299 (5607):691-694
Hoerling MP, Hurrell JW, Xu T, Bates GT, Phillips A (2004) Twentieth century North Atlantic climate change. Part II: Understanding the effect of Indian Ocean warming. Climate Dynamics 23 (3-4), 391-405
Ihara C, Kushnir Y, Cane MA (2008) Warming trend of the Indian Ocean SST and Indian Ocean dipole from 1880 to 2004. Journal of Climate 21, 2035–2046.
Jayakumar A, Vialard J, Lengaigne M, Gnanaseelan C, McCreary JP, Praveen Kumar B (2011) Processes controlling the surface temperature signature of the Madden-Julian Oscillation in the thermocline ridge of the Indian Ocean. Climate Dynamics, 37, 2217-2234, DOI 10.1007/s00382-010-0953-5.
Jensen T (2007) Wind-Driven Response of the Northern Indian Ocean to Climate Extremes. Journal of Climate, 20, 2978-2993.
Klein SA, Soden BJ, Lau N-C (1999) Remote sea surface temperature variations during ENSO: Evidence for a tropical atmospheric bridge. Journal of Climate 12 (4):917-932.
Krishnamurthy L, and Krishnamurthy V (2016) Decadal and interannual variability of the Indian Ocean SST. Climate Dynamics 46, 57-70.
Krishnamurti TN, Oosterhof D, Mehta A (1988) Air-sea interaction on the time scale of 30 to 50 days. Journal of the atmospheric sciences 45 (8):1304-1322
Krishnan R, Swapna P (2009) Significant Influence of the Boreal Summer Monsoon Flow on the Indian Ocean Response during Dipole Events. Journal of Climate, 22, 5611-5634.
Kulkarni A (2012) Weakening of Indian summer monsoon rainfall in warming environment. Theoretical and Applied Climatology 109 (3-4):447-459
Lau NC, Nath MJ (2000) Impact of ENSO on the variability of the Asian-Australian monsoons as simulated in GCM experiments. Journal of Climate 13 (24):4287-4309
Lau WKM, Waliser DE (2012) Intraseasonal Variability in the Atmosphere-Ocean Climate System. Springer Berlin Heidelberg, Berlin, Heidelberg
Lee, S.-K. et al. (2015) Pacific origin of the abrupt increase in Indian Ocean heat content during the warming hiatus. Nature Geoscience 8, 445-449, doi:10.1038/ngeo2438.
Levitus, Sydney, J. Antonov, and T. Boyer. "Warming of the world ocean, 1955–2003." Geophysical Research Letters 32.2 (2005).
Levitus, Sydney, et al. "World ocean heat content and thermosteric sea level change (0–2000 m), 1955–2010." Geophysical Research Letters 39.10 (2012).
Li H, Dai A, Zhou T, Lu J (2010) Responses of East Asian summer monsoon to historical SST and atmospheric forcing during 1950–2000. Climate Dynamics 34 (4):501-514
Li T, Zhao C, Hsu P-c, Nasuno T (2015) MJO Initiation Processes over the Tropical Indian Ocean during DYNAMO/CINDY2011. Journal of Climate (2015).
Luo, Jing-Jia, Wataru Sasaki, and Yukio Masumoto. (2012): "Indian Ocean warming modulates Pacific climate change." Proceedings of the National Academy of Sciences 109.46 18701-18706.
Madden RA, Julian PR (1971) Detection of a 40-50 day oscillation in the zonal wind in the tropical Pacific. Journal of the Atmospheric Sciences 28 (5):702-708
Madden RA, Julian PR (1972) Description of global-scale circulation cells in the tropics with a 40-50 day period. Journal of the Atmospheric Sciences 29 (6):1109-1123
Madden RA, Julian PR (1994) Observations of the 40-50-Day Tropical Oscillation - a Review. Monthly Weather Review 122 (5):814-837.

Mishra V, Smoliak BV, Lettenmaier DP, Wallace JM (2012) A prominent pattern of year-to-year variability in Indian Summer Monsoon Rainfall. Proceedings of the National Academy of Sciences 109 (19):7213-7217

Mukherjee A, Shankar D, Fernando V, Vernekar SP (2014) Observed seasonal and intraseasonal variability of the East India Coastal Current on the continental slope. Journal of Earth System Science 123(6):1197-1232.

Murakami T (1979) Scientific objectives of the Monsoon Experiment (MONEX). GeoJournal 3 (2):117-136

Murtugudde R, McCreary JP, Busalacchi AJ (2000) Oceanic processes associated with anomalous events in the Indian Ocean with relevance to 1997–1998. Journal of Geophysical Research, 105:3295–3306.

Nagura, M., and M. J. McPhaden (2010), Dynamics of zonal current variations associated with the Indian Ocean dipole, *J. Geophys. Res., 115*, C11026, doi:10.1029/2010JC006423.

Pierce DW, Barnett TP, AchutaRao KM, Gleckler PJ, Gregory JM, Washington WM (2006) Anthropogenic warming of the oceans: observations and model results. Journal of Climate 19 (10):1873-1900

Rahul S, Gnanaseelan C (2013) Net heat flux over the Indian Ocean: Trends, driving mechanisms, and uncertainties. IEEE Geoscience and Remote Sensing Letters, 10, 776-780, DOI:10.1109/LGRS.2012.2223194.

Rahul S, Gnanaseelan C (2016) Can large scale surface circulation changes modulate the sea surface warming pattern in the Tropical Indian Ocean?. Climate Dynamics, 46: 3617-3632, DOI:10.1007/s00382-015-2790-z.

Rao SA, Behera SK, Masumoto Y, Yamagata T (2002) Interannual subsurface variability in the tropical Indian Ocean with a special emphasis on the Indian Ocean Dipole. Deep Sea Research, Part II, 49, 1549–1572, doi:10.1016/S0967-0645(01)00158-8.

Rasmusson EM, Carpenter TH (1982) Variations in tropical sea surface temperature and surface wind fields associated with the Southern Oscillation/El Niño. Mon Wea Rev 110 (5):354-384

Reppin, J., F. A. Schott, J. Fischer, and D. Quadfasel (1999), Equatorial currents and transports in the upper central Indian Ocean: Annual cycle and interannual variability, *J. Geophys. Res., 104* (C7), 15,495–15,514, doi:10.1029/1999JC900093.

Roxy M, Tanimoto Y (2007) Role of SST over the Indian Ocean in Influencing the Intraseasonal Variability of the Indian Summer Monsoon. Journal of the Meteorological Society of Japan 85 (3):349-358. doi:10.2151/jmsj.85.349

Roxy M, Tanimoto Y, Preethi B, Pascal T, Krishnan R (2012) Intraseasonal SST-precipitation relationship and its spatial variability over the tropical summer monsoon region. Climate Dynamics 41 (1):45-61. doi:10.1007/s00382-012-1547-1

Roxy MK, Ritika K, Terray P, Masson S (2014) The curious case of Indian Ocean warming. Journal of Climate 27 (22):8501-8509. doi:10.1175/JCLI-D-14-00471.1

Roxy MK, Ritika K, Terray P, Murtugudde R, Ashok K, Goswami BN (2015) Drying of Indian subcontinent by rapid Indian Ocean warming and a weakening land-sea thermal gradient. Nature Communications 6:7423. doi:10.1038/ncomms8423.

Roxy M. K., A. Modi, R. Murutugudde, V. Valsala, S. Panickal, S. Prasanna Kumar, M. Ravichandran, M. Vichi and M. Levy, 2016: A reduction in marine primary productivity driven by rapid warming over the tropical Indian Ocean. Geophys. Res. Lett., 43, 826-833, doi:10.1002/2015gl066979.

Sabeerali C, Rao SA, George G, Rao DN, Mahapatra S, Kulkarni A, Murtugudde R (2014) Modulation of monsoon intraseasonal oscillations in the recent warming period. Journal of Geophysical Research: Atmospheres 119 (9):5185-5203.

Saha A, Ghosh S, Sahana A, Rao E (2014) Failure of CMIP5 climate models in simulating post-1950 decreasing trend of Indian monsoon. Geophysical Research Letters 41 (20):7323–7330.

Saji NH, Goswami BN, Vinayachandran PN, Yamagata T (1999) A dipole mode in the tropical Indian Ocean. Nature 401 (6751):360-363.

Saji NJ and Yamagata T (2003) Possible impacts of Indian Ocean Dipole mode events on global climate. Climate Research, 25, 151–169.

Saji NJ, S.-P. Xie, and T. Yamagata, (2006) Tropical Indian Ocean variability in the IPCC twentieth-century climate simulations. J. Climate, 19, 4397–4417.

Sayantani O, Gnanaseelan C (2015) Tropical Indian Ocean subsurface temperature variability and the forcing mechanisms. Climate Dynamics, 44, 2447–2462, DOI:10.1007/s00382-014-2379-y

Sayantani O., C. Gnanaseelan, J.S. Chowdary (2014) Role of Arabian Sea in the evolution of Indian Ocean Dipole, International Journal of Climatology, 34: 1845-1859, DOI:10.1002/joc.3805

Schott FA, Xie SP, McCreary JP (2009) Indian Ocean Circulation and Climate Variability. Reviews of Geophysics 47. doi:10.1029/2007rg000245

Schott, F., and J. P. McCreary (2001), The monsoon circulation of the Indian Ocean, *Progress in Oceanography, 51*, 1–123.

Senan, R., D. Sengupta, and B. N. Goswami (2003) Intraseasonal "monsoon jets" in the equatorial Indian Ocean. Geophysical Research Letters, 30, 1750, doi:10.1029/2003GL017583.

Sengupta D, Goswami BN, Senan R (2001) Coherent intraseasonal oscillations of ocean and atmosphere during the Asian summer monsoon. Geophysical Research Letters 28 (21), 4127-4130.

Sengupta D, Senan R, Goswami BN, Vialard J (2007) Intraseasonal Variability of Equatorial Indian Ocean Zonal Currents. Journal of Climate, 20, 3036-3055.

Shankar D, Vinayachandran PN, Unnikrishnan A (2002) The monsoon currents in the north Indian Ocean. Progress in Oceanography, 52(1), 63-120. DOI: 10.1016/S0079-6611(02)00024-1.

Shenoi SSC (2010) Intra-seasonal variability of the coastal currents around India: A review of the evidences from new observations. Indian Journal of Marine Sciences, 39 (4), 489-496.

Shenoi S, Shankar D, Michael G, Kurian J, Varma K, Ramesh Kumar M, Almeida A, Unnikrishnan A, Fernandes W, Barreto N, Gnanaseelan C, Mathew R, Praju KV, Mahale V (2005) Hydrography and water masses in the southeastern Arabian Sea during March-June 2003. Journal of Earth System Science 114 (5):475-491.

Sprintall, J., A. L. Gordon, T. Lee, J. T. Potemra, K. Pujiana, *and* S. E. Wijffels *(*2014*)*, The Indonesian seas and their role in the couple ocean-climate system, Nature Geoscience, 7, 487–492, doi:10.1038/ngeo2188.

Swapna P, Krishnan R, Wallace J (2014) Indian Ocean and monsoon coupled interactions in a warming environment. Climate Dynamics 42 (9-10):2439-2454.

Terray P., S. Masson, C. Prodhomme, M. K. Roxy, K. P. Sooraj, 2015: Impacts of Indian and Atlantic oceans on ENSO in a comprehensive modeling framework. Climate Dynamics, doi:10.1007/s00382-015-2715-x

Thompson, B., C. Gnanaseelan, and P. S. Salvekar (2006), Variability in the Indian Ocean circulation and salinity and its impact on SST anomalies during dipole events, *Journal of Marine Research, 64*, 853–880, doi:10.1357/002224006779698350

Vaid BH, Gnanaseelan C, Polito PS, Salvekar PS (2007) Influence of Pacific on Southern Indian Ocean Rossby Waves. Pure and Applied Geophysics, 164, 1765-1785.

Valsala V, Maksyutov S and Murtugudde R (2010) Possible Interannual to Interdecadal variabilities of the Indonesian Throughflow (ITF) water pathways in the Indian Ocean, Journal of Geophysical Research, doi:10.1029/2009JC005735.

Vecchi GA, Harrison DE (2002) Monsoon breaks and subseasonal sea surface temperature variability in the Bay of Bengal. Journal of Climate 15 (12), 1485-1493.

Vialard J, Jayakumar A, Gnanaseelan C, Lengaigne M, Sengupta D, Goswami B (2012) Processes of 30–90 days sea surface temperature variability in the northern Indian Ocean during boreal summer. Climate Dynamics 38 (9-10):1901-1916, DOI:10.1007/s00382-011-1015-3.

Vialard J, S.S.C Shenoi, J.P. McCreary, D. Shankar, F. Durand, V. Fernando and S.R. Shetye (2009) Intraseasonal response of Northern Indian Ocean coastal waveguide to the Madden-Julian Oscillation. Geophysical Research Letters, 36(14), 2009, doi:10.1029/2009GL038450.

Vinayachandran PN, Saji NH, Yamagata T (1999) Response of the equatorial Indian Ocean to an unusual wind event during 1994. Geophys. Res. Lett., 26(11), 1613–1616, doi:10.1029/1999GL900179.

Webster P, Bradley EF, Fairall C, Godfrey J, Hacker P, Houze Jr R, Lukas R, Serra Y, Hummon J, Lawrence T (2002) The JASMINE pilot study. Bulletin of the American Meteorological Society 83 (11):1603-1630

Webster PJ, Moore AM, Loschnigg JP, Leben RR (1999) Coupled ocean-atmosphere dynamics in the Indian Ocean during 1997-98. Nature 401 (6751):356-360

Williams AP, Funk C (2011) A westward extension of the warm pool leads to a westward extension of the Walker circulation, drying eastern Africa. Climate Dynamics 37 (11-12):2417-2435

Wyrtki, K. (1973), An equatorial jet in the Indian Ocean, *Science, 181*, 262–264.

Xie S-P, Hu K, Hafner J, Tokinaga H, Du Y, Huang G, Sampe T (2009) Indian Ocean capacitor effect on Indo-Western Pacific climate during the summer following El Niño. Journal of Climate 22:730–747

Xie S-P, Du Y, Huang G, Zheng X-T, Tokinaga H, Hu K, Liu Q (2010) Decadal shift in El Niño influences on Indo-western Pacific and East Asian climate in the 1970 s. Journal of Climate 23:3352–3368.

Yang J, Liu Q, Xie S-P, Liu Z, Wu L (2007) Impact of the Indian Ocean SST basin mode on the Asian summer monsoon. Geophysical Research Letters 34:L02708. doi:10.1029/2006GL028571.

Chapter 11
Sea Level Variability and Trends in the North Indian Ocean

Anant Parekh, C. Gnanaseelan, J.S. Deepa, Ananya Karmakar and J.S. Chowdary

1 Introduction

Sea level is a crucial oceanic parameter which shows variability on different timescales including the long-term trends. The reported increasing trend in sea level has very strong socioeconomic consequences as it inundates to the surrounding coastal area (e.g., Nicholls et al. 2007). Sea level rise depends on the following processes: (1) thermal expansion due to temperature rise (which is known as thermosteric effect), (2) salinity change (halosteric effect) (temperature- and salinity-induced ocean volume change results in the sea level change and is known as "steric change"), (3) melting of sea ice, (4) terrestrial water storage (different components of water reservoir on the Earth such as rivers, lakes, ice sheets, glaciers, and precipitation) exchanges water mass with oceans, they modify the volume of the ocean water time to time, which is known as a "eustatic" and (5) sediment deposition at the coast. The role of above processes to the sea level change is highly uncertain due to the lack of information on contributions by the natural and anthropogenic activities. Church et al. (2001) using global coastal tide gauge observations reported that sea level rise during the twentieth century was 1.5 ± 0.5 mm/day. In the past, several attempts were made to estimate sea level changes by incorporating different processes. The resultant estimate of total sea level change was 0.7 ± 1.5 mm/day, suggesting more uncertainty in such computations (Fig. 1). Thus, it is easy to observe the sea level change from observations, but it is difficult to quantify the role of individual processes to the resultant sea level change, probably due to the existence of some unknown processes. Several studies reported a sea level rise of 3.2 ± 0.4 mm/year during 1993–2009 (e.g., Church and White 2011) using satellite observations, much of which is due to steric effect (Cabanes et al. 2001b). This recent trend is much higher than the

A. Parekh · C. Gnanaseelan (✉) · J.S. Deepa · A. Karmakar · J.S. Chowdary
Indian Institute of Tropical Meteorology, Pune 411008, India
e-mail: seelan@tropmet.res.in

M.N. Rajeevan and S. Nayak (eds.), *Observed Climate Variability and Change Over the Indian Region*, Springer Geology, DOI 10.1007/978-981-10-2531-0_11

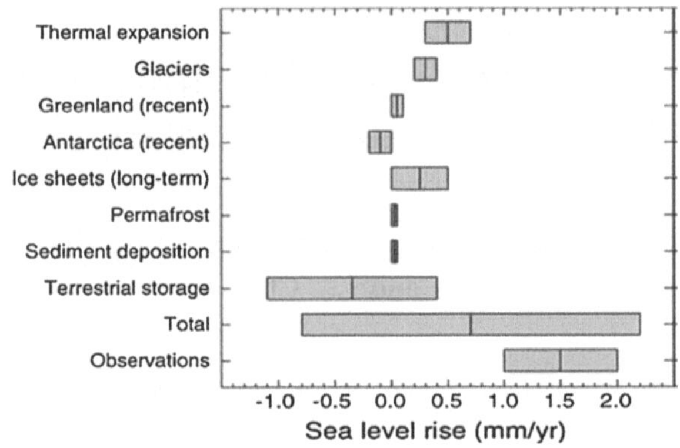

Fig. 1 Contribution of different component to sea level rise and level of uncertainty for individual component estimation (adopted from Cazenave and Nerem 2004)

twentieth-century trend. Southern subtropical Indian Ocean displayed stronger sea level rise compared to the northern counterpart (Cabanes et al. 2001a). It is important to note that oceans absorb about 80 % of the heat supplied to the climate system, leading to the upper ocean (up to 3000 m depth) temperature rise. The global warming induced melting of glacier and sea ice caps also contribute to the sea level change. Satellite data confirms that the ice cover of the Greenland and polar region displays decreasing trend, suggesting its possible role on sea level rise.

Sea level does influence the ecosystems and humans in many ways. The inundation of coastal area due to the sea level change is closely related to the slope of the beach as well. It is estimated that sea level rise to beach inundation ratio is about hundred, and hence, sea level rise of 1 cm can contribute to the inundation by 1 m. A sea level rise also affects freshwater availability to the coastal and island population as the freshwater sources may get contaminated. Also, the increase in sea level and the associated vertical rise in water column result in waterlogging, thereby destroying the coastal ecology such as mangroves and dependent biota. Sea level rise accelerates the impact of flooding and cyclone-driven storm surges and giant waves. Sea level rise is dramatically changing the physical character of the low-lying river deltas, for example, the Ganges, Mississippi, Nile, and numerous others. Sea level rise generally creates fears of possible submerge of islands and low-lying coastal cities, which invites the international attention. The present chapter provides the current status and knowledge of Indian Ocean sea level change and variability. Further, it includes a brief description on future projections of sea level changes over the Indian Ocean.

2 Sea Level Observations

Sea level is usually measured by “tide gauges” and “altimeter on board of satellite”. Tide gauge observations are mostly from coastal and island stations, and some of the global stations have provided more than hundred years of data. One such Indian station is Mumbai, which is located in the west coast of India. In addition to the above two conventional measurement strategies, temperature and salinity observations over a long period can also provide sea level rise estimates accounting the steric effect. Melting of ice sheets and mountain glaciers add freshwater to the oceans, which may also increase the water mass of the ocean. The continuous increase in water mass of the ocean can modify the Earth's rotation rate as well as gravitational oblateness. Measurements of these possible changes due to water mass increase can in turn help to estimate sea level change due to melting of sea ice and glacier (Cazenave and Nerem 2004).

2.1 Tide Gauges

Tide gauge observations are the main source for understanding the long-term sea level change (Douglas 2001; Woodworth and Player 2003), and it measures the sea level in mm. One of the limitations of tide gauge observations is that it records sea level only at the coast (it cannot provide observations of open ocean sea level); secondly, they are mounted to land which may be experiencing vertical movements (Cazenave et al. 1999; Nerem and Mitchum 2002). The second limitation can be rectified by geodetic leveling to some extent. These limitations can introduce error in the long recorded sea level information, which indeed mislead the rate of sea level estimation (Cabanes et al. 2001b). Tide gauge in general measures the sea level at every hourly interval. To extract sea level trend information, monthly or yearly averaged tide gauge data are used. However, the factors such as data length and impact of short-term climate variability in a specific region add uncertainty in the estimate of sea level trend from these data. Permanent Service for Mean Sea Level (PSMSL), is the nodal agency for the global tide gauge data.

2.2 Satellite Altimetry

Altimeter is an active microwave sensor to retrieve the range between the ocean and satellite. Many missions (Geodynamics Experimental Ocean Satellite-3, Seasat, Geosat, and European Remote Sensing satellite-1) prior to TOPEX/Poseidon (T/P) were carried out to retrieve accurately the sea level modulation over the global ocean. However, T/P was the first most accurate and successful altimeter, mainly due to technical and orbital specifications (Tapley et al. 1994; Christensen et al. 1994; Mitchum 1994). Chelton et al. (2001) estimated that the accuracy of T/P is about 2–3 cm with concurrent collocated in situ observations. Jason-1 was the first follow-on to the highly successful T/P, which provided decade long observations of sea level from the satellite. Altimeter mainly provides the range between the sensor on board of satellite and the ocean surface at different frequencies. Atmospheric corrections due to dry as well as moist constitute of atmosphere, ionosphere corrections, oceanic tide corrections, atmospheric column pressure corrections, etc., are necessary for deriving the final sea level product. Advanced methods are developed to include all these corrections in the retrieval algorithm for accurate sea level change estimate (e.g., Chelton et al. 2001). In contrast to the tide gauge, altimeter observations provide spatial distribution of global ocean sea level. Hence, altimeter played significant role in understanding the present status of sea level globally as well as regionally. Recent missions such as Satellite with ARgos and ALtiKa claim the sea level accuracy of 8 mm. Since last few years, multiple altimeter missions and advanced technology helped us to improve the sampling rate of sea level observations leading to explore sea level variability at high frequency using merged product of these satellites. These merged products of seal level data are available from AVISO (http://www.aviso.altimetry.fr/en/data.html).

3 Long-Term Trend in Indian Ocean Sea Level

Indian Ocean sea level change displays large diverging nature as compared to the rest of the global ocean, and its future projection is also equally ambiguous (Church et al. 2006). Han et al. (2010) using satellite and in situ observations along with climate model simulations showed decreasing trend of sea level in the south tropical Indian Ocean and positive trend in the northern Indian Ocean. They attributed the sea level trends to changes in the atmospheric general circulation (Hadley and Walker cell of circulation) associated with climate change. Rahul and Gnanaseelan (2015) hypothesized a coupled ocean atmospheric feedback mechanism to develop such sea level trends especially in the southern tropical Indian Ocean through changed surface circulation.

Sea level rise of 1.4 mm/year is reported by Church et al. (2006) based on tide gauge observations at different ports along the Indian coast during 1950–2001. However, with different corrections, the figure rose to 2 mm/year, a value comparable with the global trends. The long-term trends from tide gauge data for selected station Charchanga (Bengladesh), Chennai, and Mumbai are shown in Fig. 2. Longest time series observations are available for Mumbai, while for rest of the stations, at least 20 years data are available. From the linear trend analysis, it is found that the Charchanga station data for the period of 1980–2000 displayed highest sea level rise (8.8 mm/year, Fig. 2a), whereas sea level rise at Mumbai is 0.68 mm/year (Fig. 2c) and at Chennai is 0.42 mm/year. Parekh (2010) reported sea level change of 4.84 mm/year at the Diamond Harbour, 1.38 mm/year over Kochi, and 0.74 mm/year over Visakhapatnam. Thus, the highest sea level rise is reported in the north Bay of Bengal and the lowest among these stations are at Chennai and Mumbai. These are in agreement with the findings of Unnikrishnan and Shankar (2007). Any such sea level rise along the east coast has serious impacts as the east coast of India is very sensitive to sea level rise (Shetye et al. 1990).

Figure 3a shows the spatial distribution of sea level trend from the Simple Ocean Data Analysis (SODA, 1959–2008). Sea level change is positive in the Bay

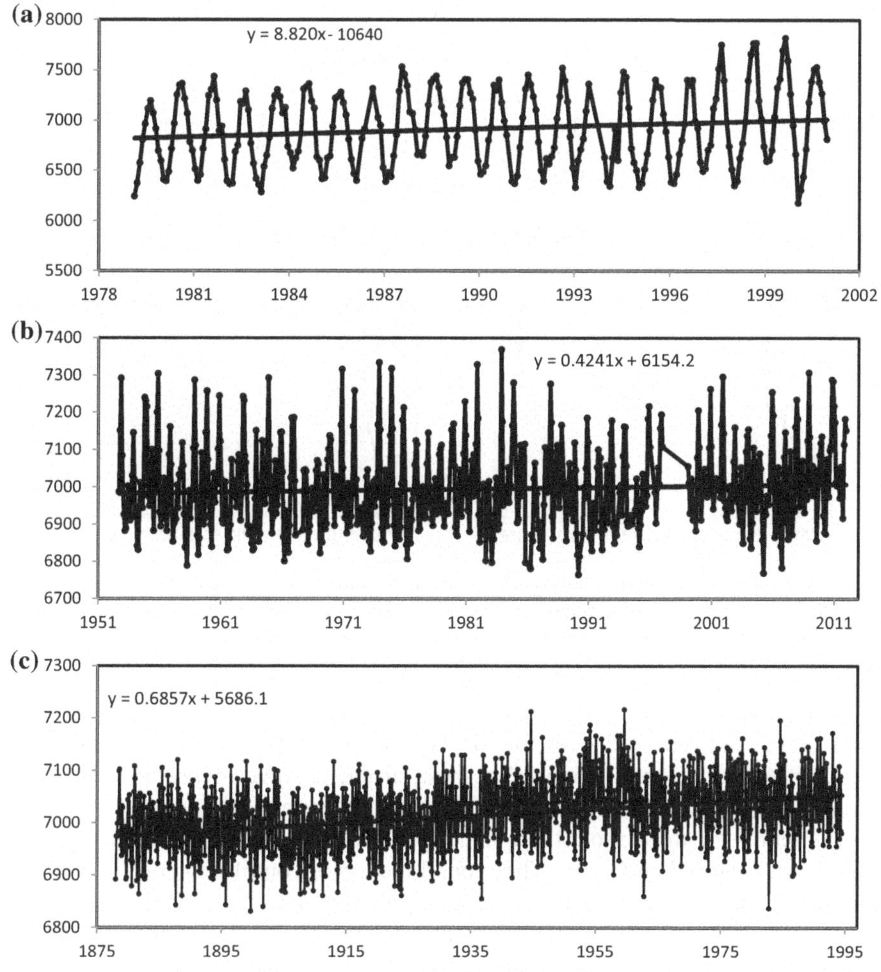

Fig. 2 Tide gauge sea level (mm) plot from **a** Charchanga (Bangladesh), **b** Chennai, and **c** Mumbai. Linear trends are shown by the *thick straight line*

of Bengal and southern subtropical Indian Ocean; however, negative trends are reported in the Arabian Sea and southern tropical Indian Ocean. The maximum trends are observed over the coastal region of Bay of Bengal, indicating the trends in the coastally trapped waves. Figure 3b shows the sea level trend over the north Indian Ocean from the altimeter observations of 1993–2015. It is important to note the large differences in the reanalysis and satellite observations, partly contributed by the differences in the time period. The rising trends of sea level are higher in the Bay of Bengal than in the Arabian Sea. This is consistent in both reanalysis and satellite data. The sea level rise over the Arabian Sea and Bay of Bengal from the T/P altimeter monthly mean observations depict that the rate over the Arabian Sea is

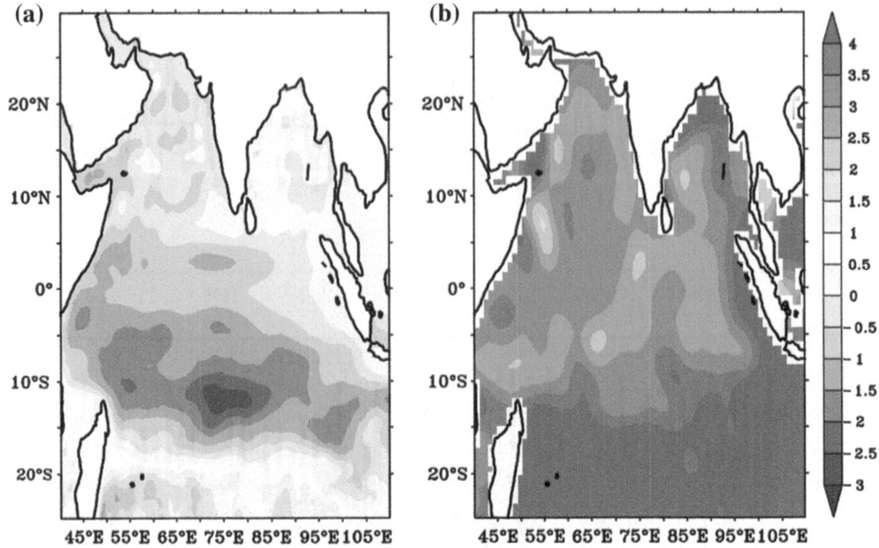

Fig. 3 Sea level anomaly trend (mm/year) over the north Indian Ocean from **a** SODA analysis for 1959–2008 and **b** altimeter observations for 1993–2015 (GIA (glacial isostatic adjustment) and IB (inverse barometer) corrected merged (T/P, Jason-1 and Jason-2) data from http://www.cmar.csiro.au/sealevel/sl_data_cmar.html)

about 0.5–3 mm/year and over the Bay of Bengal is 0.75 to about 6 mm/year (Fig. 3b). Altimeter observations also show strong sea level trend (4–7 mm/year) south of 10°S in the Indian Ocean. Palanisamy et al. (2014) used sea level reconstruction for the period 1950–2009 to understand the sea level change and variability in the Indian Ocean. They found the major contribution of the total sea level rise (about 1.5 mm/year) is of steric origin. They also revealed that variability such as Indian Ocean Dipole (IOD) do influence the sea level change in the Indian Ocean. Overall, there is a consensus that sea level rise of Indian Ocean is robust. Recent developments in satellite oceanography such as Global Positioning System (GPS) and Doppler Orbitography and Radiopositioning Integrated by Satellite (DORIS) help to bifurcate estimated sea level change into the components due to climate change and due to vertical motion of land. This method of separation is vital for estimating the climate change induced by sea level change. Unnikrishnan et al. (2015) reported very high sea level change (>5 mm/year) in the northern and eastern coast of Bay of Bengal and more uniform trend (~ 3.2 mm/day) in the rest of the Indian Ocean. Kusche et al. (2016) separated the mass and steric contributions to sea level variability by applying inverse approach (Rietbroek et al. 2012) to the Jason-1/2 radar altimetry and Gravity Recovery and Climate Experiment (GRACE) data and revealed that steric origin sea level change in the Bay of Bengal dominates by a factor of two over the mass-driven sea level change.

4 Inter-annual to Decadal Variability of Sea Level in the Indian Ocean

The different scale of variability (inter-annual to decadal) in the sea level can influence its long-term trend (Unnikrishnan et al. 2015). Understanding of the sea level variability in the north Indian Ocean at different temporal and spatial scales is therefore very important. Recently, many studies explored the impact of dominant climate modes such as IOD and El Niño/La Niña on the variability and trend of sea level in the north Indian Ocean (Sreenivas et al. 2012; Aparna et al. 2012). They showed from both observations and reanalysis products that the sea level in the Bay of Bengal and east coast of India displays negative anomalies during positive IOD years and El Niño years. Most of the variability associated with the IOD and El Niño are observed at the eastern coast, whereas the western coast did not display any coherent response to the large-scale variability. Figure 4 displays the first and second modes of Empirical Orthogonal Function (EOF) of satellite-derived sea level anomaly. The first mode (Fig. 4a) displays a north–south mode as reported in Sayantani and Gnanaseelan (2015), and the second mode (Fig. 4b) represents a dominant zonal mode. The respective principal components are shown in Fig. 4c, d for the period of 1993–2012. During El Niño, negative sea level anomalies are observed twice (April–December and November–July), with a relaxation between the two peaks. However the signatures during negative IOD and La Niña events are much weaker.

Nidheesh et al. (2013) found that decadal scale variability strength was higher than the long-term trend of sea level, but it was half of the inter-annual variability. Decadal variability is also observed over the eastern equatorial Indian Ocean and in the Bay of Bengal but is not related to the Pacific decadal oscillations. Shankar et al. (2010) reported from wavelet analysis of altimeter sea level data that the Indian Ocean regions have variability at all timescales. Further, they reported that direct wind forcing and reflected Rossby waves interfere destructively (constructively) to produce inter-annual periods of minima (maxima). Figure 5 is a wavelet analysis (similar to Torrence and Compo 1998) of sea level from the tide gauges for the selected stations (Charchanga, Chennai, and Mumbai). It clearly indicates that the sea level varies from place to place and has different scales of variabilities, which are consistent with Shankar et al. (2010).

5 CMIP Projections on Future Sea Level Rise

The future projections of sea level at global as well as regional scales due to heat absorbed by the ocean and changes in wind forcing were studied in the past using Coupled Model Intercomparison Project Phase 5 (CMIP5) projections. The fifth assessment report of Intergovernmental Panel on Climate Change (IPCC) noted that sea level variability associated with the El Niño or La Niña and equatorial current systems in future can be up to 8 cm in the tropical Indo-Pacific region. Another way

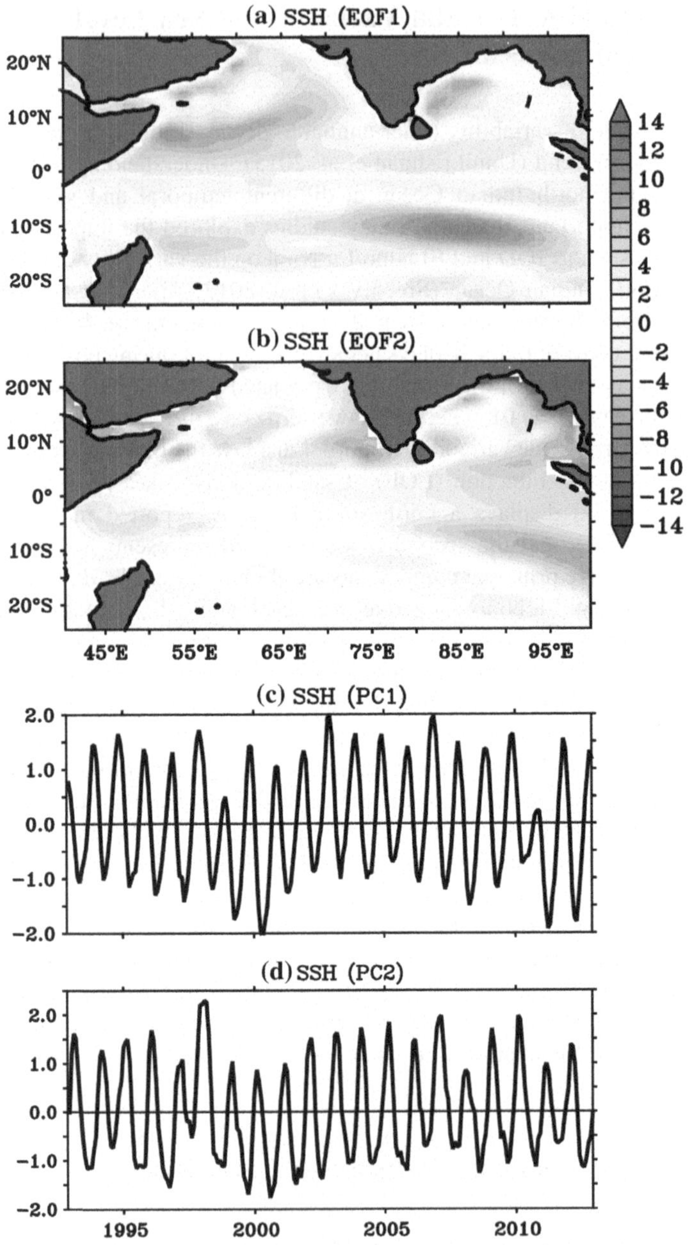

Fig. 4 Leading modes of variability in altimeter SLA using the EOF analysis **a** first mode, **b** second mode, **c** PC-1, and **d** PC-2

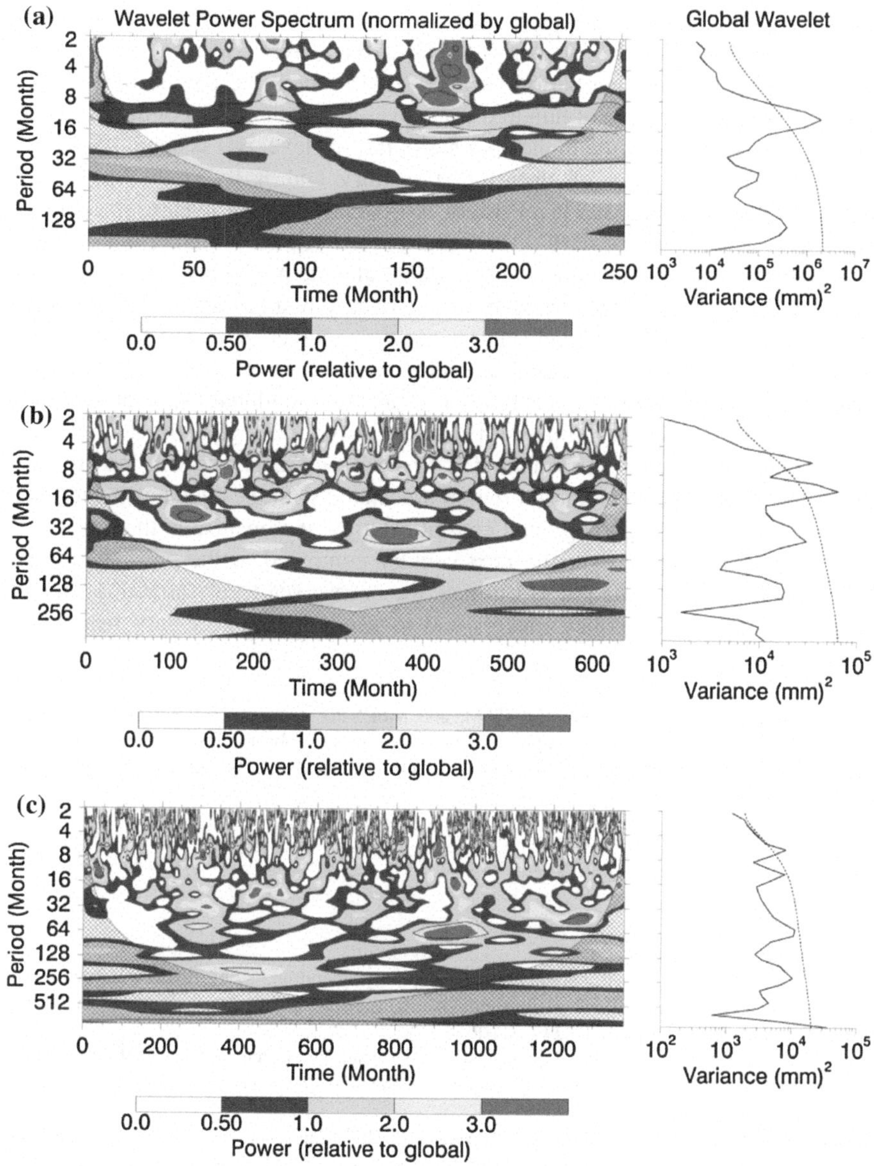

Fig. 5 Wavelet analysis of tide gauge sea level (mm) from **a** Charchanga (Bangladesh), **b** Chennai, and **c** Mumbai

to explore future projection of sea level is to use high-resolution regional climate models (RCMs) under different prescribed emissions to simulate future changes in the forcing to the ocean. Such simulated changes can be used to force storm surge model for estimating sea level projection and extremes (Unnikrishnan et al. 2006, 2011). Projections for A2 scenario (2071–2100) gave sea level increase of 40–67 cm with the baseline period of 1921–1990, where the contribution of tide was 6–12 cm for the period of hundred years over the northern part of the east coast of India. However, under the highest scenario, the sea level rise is about 45–82 cm for the last twenty years of twenty-first century and it is 52–98 cm for the twenty-first century; thus, sea level rise of last twenty years of twenty-first century will be very critical. These rates of sea level rise are ten times higher than what is reported for the twentieth century. Using global climate model predictions, Church et al. (2001) found that future sea level for the twenty-first century will be 9–88 cm with respect to the 1990 sea level state, witnessing an average rise of 4.4 mm/year. The IPCC reported that the global sea level rise will be 18–59 cm due to thermosteric effect associated with the 1.1–6.4 °C increase in global temperature due to the global warming. These estimates of sea level may further go up by 20 cm, since melting of polar ice caps is not included. Overall, sea level rise will lead to very severe storm surge over the coastal region associated with the cyclones, which will affect the beach geomorphology and coastal ecology. In conclusion, the rate of sea level rise will be higher over the next century compared to the present century. Efforts should therefore be put to estimate the correct sea level variability, trends, and projections for the policy makers and to avoid any unnecessary fear in the people's mind.

Acknowledgments We thank Director, IITM, and MoES for support. PSMSL and AVISO are acknowledged for providing tide gauges and altimeter-merged observations on their Web site. Some of the figures are prepared using Ferret. PAOS-Colorado is acknowledged for providing wavelet software online.

References

Aparna, S. G., McCreary, J. P., Shankar, D. and Vinayachandran, P. N. (2012), Signatures of Indian Ocean Dipole and El Nino–Southern Oscillation events in sea level variations in the Bay of Bengal. *J. Geophys. Res. C*, **117**, C10012, 13.

Cabanes, C., A. Cazenave, and C. Le Provost (2001a), Sea level change from TOPEXPoseidon altimetry for 1993 –1999 and possible warming of the southern oceans, Geophys. Res. Lett., 28 (1), 9–12.

Cabanes, C., A. Cazenave, and C. Le Provost (2001b), Sea level rise during past 40 years determined from satellite and in situ observations, Science, 294, 840–842. Cazenave, A., K. Dominh, M. C. Gennero, and B. Ferret (1998), Global mean sea level changes observed by TOPEX-Poseidon and ERS-1, Phys. Chem. Earth, 23(9–10), 1069– 1075.

Cazenave, A., and R. S. Nerem (2004), Present-day sea level change: Observations and causes, Rev. Geophys., 42, RG3001, doi:10.1029/2003RG000139.

Cazenave, A., K. Dominh, F. Ponchaut, L. Soudarin, J.-F.Cretaux, and C. Le Provost (1999), Sea level changes from TOPEXPoseidon altimetry and tide gauge, and vertical crustal motions from DORIS, Geophys. Res. Lett., 26(14), 2077–2080.

Chelton, D. B., J. C. Ries, B. J. Haines, L.-L.Fu, and P. S. Callahan (2001), Satellite altimetry, in Satellite Altimetry and Earth Sciences, edited by L.-L. Fu and A. Cazenave, pp. 1– 131, Academic, San Diego, Calif.

Christensen, E. J., et al. (1994), Calibration of TOPEX/Poseidon at Platform Harvest, J. Geophys. Res., 99(C12), 24,465–24,486.

Church, J. A., Gregory, J. M., Huybrechts, Kuhn, M., Lambeck, K., Nhuan, M. T., Qin, D. and Woodworth, P. L.(2001), Changes in sea level. In *Climate Change 2001: The Scientific Basis. Contribution of Working Group I to the Third Assessment Report of the Intergovernmental Panel on Climate Change*, Cambridge University, Press, Cambridge, UK, pp. 639–693.

Church J A, and White N J (2011), Sea-Level Rise from the Late 19th to the Early 21st Century, Surv Geophys (2011) 32:585–602

Church J A, White N J and Hunter J R (2006), Sea-level rise at tropical Pacific and Indian Ocean islands; Global and Planetary Change, 53 155–168.

Douglas, B. C. (2001), Sea level change in the era of the recording tide gauge, in Sea Level Rise: History and Consequences, Int. Geophys. Ser., vol. 75, edited by B. C. Douglas, M. S. Kearney, and S. P. Leatherman, chap. 3, pp. 37 – 64, Elsevier, New York.

Han, W. *et al.*(2010), Patterns of Indian Ocean sea-level change in a warming climate. *Nature Geoscience*, **3**, 546–550; DOI:10.1338/NGEO 901.

Kusche, J., B. Uebbing, R. Rietbroek, C. K. Shum, and Z. H. Khan (2016), Sea level budget in the Bay of Bengal (2002–2014) from GRACE and altimetry, J. Geophys. Res. Oceans, 121, doi:10.1002/2015JC011471.

Mitchum, G. T. (1994), Comparison of TOPEX sea surface heights and tide gauge sea levels, J. Geophys. Res., 99(C12), 24,541– 24,554.

Nerem, R. S., and G. T. Mitchum (2002). "Estimates of vertical crustal motion derived from differences of TOPEX/POSEIDON and tide gauge sea level measurements." *Geophysical Research Letters* 29: 40-1

Nicholls, R.J., P.P. Wong, V.R. Burkett, J.O. Codignotto, J.E. Hay, R.F. McLean, S. Ragoonaden and C.D. Woodroffe, (2007) Coastal systems and low-lying areas. Climate Change 2007: Impacts, Adaptation and Vulnerability. Contribution of Working Group II to the Fourth Assessment Report of the Intergovernmental Panel on Climate Change, M.L. Parry, O.F. Canziani, J.P. Palutikof, P.J. van der Linden and C.E. Hanson, Eds., Cambridge University Press, Cambridge, UK, 315-356.

Nidheesh, A. G., Lengaigne, M., Vialard, J., Unnikrishnan, A. S. and Dayan, H.(2013), Decadal and long-term sea level variability in the tropical Indo-Pacific Ocean. *Climate Dyn.*, **41**(2), 381–402.

Parekh A.(2010), A Note on Sea Level Rise, Int. J. Earth Sciences and Engineering,,03, CAS Ref. No.: 172238.

Palanisamy, H., Cazenave, A., Meyssignac, B., Soudarin, L., Wöppelmann, G. and Becker, M. (2014) Regional sea level variability, total relative sea level rise and its impacts on islands and coastal zones of Indian Ocean over the last sixty years. *Global Planet. Change*, doi:10.1016/j.gloplacha.2014.02.001.

Rahul S, C. Gnanaseelan (2015) Can large scale surface circulation changes modulate the sea surface warming pattern in the Tropical Indian Ocean?. Clim Dyn 10.1007/s00382-015-2790-z

Rietbroek R, Brunnabend SE, Kusche J, Schröter J (2012) Resolving sea level contributions by identifying fingerprints in time-variable gravity and altimetry. J Geodyn 59:72–81.

Sayantani Ojha, C. Gnanaseelan (2015) Tropical Indian Ocean subsurface temperature variability and the forcing mechanisms, Clim Dyn 44:2447–2462.

Shankar, D. *et al. (*2010), Minima of interannual sea-level variability in the Indian Ocean. *Prog. Oceanogr.*, 84(3–4), 225–241.

Shetye, S.R.,Gouveia, A.D, Pathak, M.C.(1990) Vulnerability of the Indian coastal region to damage from sea level rise. *Current Science*, Vol.59; 152-156

Sreenivas, P., Gnanaseelan, C., and Prasad, K.V.S.R., (2012), Influence of El Niño and Indian Ocean Dipole on sea level variability in the Bay of Bengal, *Global and Planetary Change*, 80-81, 215-225, doi:10.1016/j.gloplacha.2011.11.

Tapley, B. D. (1994), et al., Precision orbit determination for TOPEX/POSEIDON, J. Geophys. Res., 99(C12), 24,383 – 24,404.

Torrence, C., G. Compo (1998), A practical guide to wavelet analysis, *Bull. Am. Meteorol. Soc.*, **79**,61–78.

Unnikrishnan, A. S., K. Rupa Kumar, Sharon E. Fernandes, G. S. Michael and S. K. Patwardhan (2006), Sea level changes along the Indian coast: Observations and projections, Curr. Sci.,, Vol. 90, 362-368.

Unnikrishnan, A. S. and Shankar, D. (2007), Are sea-level-rise trends along the north Indian Ocean coasts consistent with global estimates? *Global Planet. Change*, **57**, 301–307.

Unnikrishnan A. S., A. G. Nidheesh and M. Lengaigne (2015), Sea-level-rise trends off the Indian coasts during the last two decades,, Curr.Sci., Vol. 108, 966-971.

Unnikrishnan, A.S.; RameshKumar, M.R.; Sindhu, B. (2011) Tropical cyclones in the Bay of Bengal and extreme sea-level projections along the east coast of India in a future climate scenario, Curr. Sci.: 101(3); 327-331.

Woodworth, P. L., and R. Player (2003), The Permanent Service for Mean Sea Level: An update to the 21st century, J. Coastal Res., 19, 287–295.

Chapter 12
Variability of Glaciers and Snow Cover

Anil V. Kulkarni, Shailesh Nayak and S. Pratibha

1 Introduction

The Himalayan region has one of the largest concentrations of glaciers. Major rivers such as the Indus, the Ganga, the Brahmaputra and their numerous tributaries originate from this glacier bound region. The snow and glacier melt are important to sustain water requirement in the North India. The contribution of glacier melt in annual stream run-off is the highest in the Indus, followed by the Brahmaputra and the Ganga basins. (Immerzeel et al. 2010; Singh and Jain 2009). The Indus basin produces almost 96 and 27 % of the food production in Pakistan and India, respectively, due to well-developed canal network (RBI report 2011; Khan et al. 2010). Therefore, changes in glacier melt run-off pattern can significantly influence the water and food security. However, distribution of the Himalayan glaciers, seasonal snow cover and possible changes in its dimension due to regional and global climate change is difficult to obtain due to rugged terrain and extreme weather conditions. Therefore, remote sensing techniques have been extensively used to monitor numerous parameters of snow and glaciers. These include areal extent of snow cover and glaciers, albedo of seasonal snow cover, glacier mass balance, formation and development of moraine-dammed lakes and debris cover on the glaciers.

Significant improvement in satellite techniques has made it possible to measure and model these parameters using combination of satellite and field measurements. The characteristics of satellite sensor such as temporal coverage, multiple spatial resolutions, and reflectance and emission characteristics in different wavelengths

A.V. Kulkarni (✉) · S. Pratibha
Divecha Centre for Climate Change, Indian Institute of Science, Bengaluru, India
e-mail: anilkulkarni@caos.iisc.ernet.in

S. Nayak
Earth System Science Organization, Ministry of Earth Sciences, New Delhi, India

M.N. Rajeevan and S. Nayak (eds.), *Observed Climate Variability and Change Over the Indian Region*, Springer Geology, DOI 10.1007/978-981-10-2531-0_12

have been extensively used in glacial investigations. For example, medium-resolution satellite sensors with high repetitivity such as Moderate Resolution Imaging Spectroradiometer (MODIS) and Advanced Wide Field Sensor (AWiFS) have been extensively used for snow cover and albedo monitoring. Data from Landsat TM Thematic Mapper and Linear Imaging Self Scanning Sensor (LISS) II and III of Indian Remote Sensing (IRS) have been extensively used for glacier inventory, estimation of glacier velocity and modelling depth of glacial ice. High-resolution stereo data have been used to generate digital elevation model (DEM) and to estimate glacier retreat. DEM can also be generated using microwave data and to estimate glacial mass balance. In addition, since recently, gravity anomaly is also measured using satellite such as Gravity Recovery and Climate Experiment (GRACE) to estimate changes in glacial mass for the entire Himalayan region. These techniques have aided in measuring various glacial parameters of individual glaciers and on basin and mountain scale, providing invaluable information about Himalayan glaciers, which were not available earlier. Studies carried out by Indian and global scientific community using remote sensing and field investigations are reviewed here.

2 Glacier Extent and Volume of Glacier-Stored Water in the Indian Himalaya

2.1 Glacial Extent

In the Himalaya, glacier inventory has been published by numerous research workers (Kulkarni and Buch 1991; Sangewar and Shukla 2009; Bajracharya and Shrestha 2011; Arendt et al. 2012). These studies suggest that the glacial extent in the Indian Himalaya is between 20,785 and 27,915 km^2, indicating large discrepancies in the estimates (Fig. 1). These discrepancies are possibly due to the difference in the time period, scale and methodology of mapping. One of the earliest inventories was carried out in the year 1991 by the Space Application Centre (SAC-ISRO) (Kulkarni and Buch 1991) using satellite data of year 1987, estimating an areal extent of the glaciers in the Indian Himalaya as 23,314 km^2. Subsequently, the Geological Survey of India (GSI) also published an inventory using topographic maps and aerial photographs taken in the year 1962–1963 (Sangewar and Shukla 2009). It estimates an areal extent of the Indian Himalaya as 26,775 km^2. The difference in SAC and GSI estimates could be because the glacierets were not mapped in the SAC inventory and also, possibly due to the difference in methodology. In the due course, the inventories by the International Centre for Integrated Mountain Development (ICIMOD) and Randolph Glacier Inventory (RGI) were also published, suggesting an areal extent of glaciers in Indian Himalaya as 20,785 and 27,915 km^2 respectively, (Bajracharya and Shrestha 2011; Arendt et al. 2012).

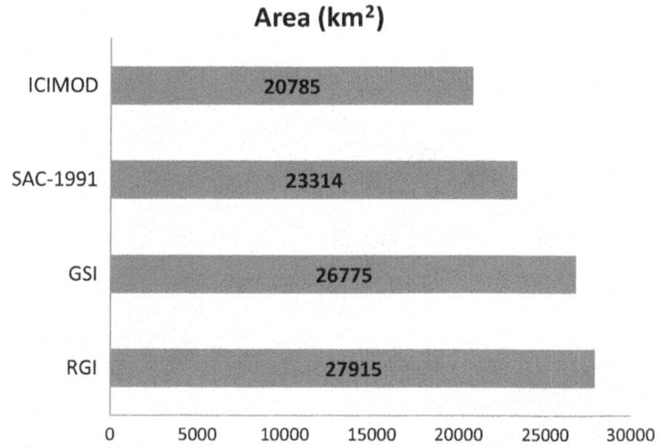

Fig. 1 Comparison of areal extent of glaciers in the Indian Himalaya mapped by the International Centre for Integrated Mountain Development (ICIMOD), Space Application Centre (SAC), Geological Survey of India (GSI) and Randolph Glacier Inventory (RGI) (Bajracharya and Shrestha 2011; Kulkarni and Buch 1991; Sangewar and Shukla 2009; Arendt et al. 2012)

However, as the ICIMOD and RGI inventories are available in digital form, they were further analysed to assess their accuracies in the Baspa basin, Himachal Pradesh. The glacier outlines were manually interpreted using Landsat imagery of 2013 and 2014. The glacierised area was estimated at 194 ± 14 km^2. The glacierised extent of Baspa basin recorded by the inventory of ICIMOD and RGI 4.0 suggests an area of 163 ± 12 and 193 ± 16 km^2, respectively. As RGI glacier outlines are closer to our interpretation, RGI outlines were further used to assess glacier volume in Indian Himalaya (Fig. 2). RGI is a compilation of glacier boundaries from different agencies and covers almost all glaciers around the globe.

2.2 *Glacial Volume*

The spatial distribution of glacier depth and assessment of glacial-stored water is essential information to manage water resources in North India. Glacier ice thickness can be estimated using ground penetrating radar (GPR), and this technique has been used to estimate depth of few glaciers such as the Dokriani Bamak, Patseo and Chotta Shigri (Gergan et al. 1999; Singh et al. 2010, 2012). An attempt was also made to measure the ice thickness of Samudra Tapu but failed due to the limitations of the instrument (Singh et al. 2010). The ice thickness of the Dokriani and the Chotta Shigri glaciers measured by GPR ranged from 15 to 120 m and 110 to 150 m, respectively. The depth of the Patseo glacier was estimated at 40 m. But these in situ measurements are clearly limited in number, and moreover, these records are point measurements and do not cover the entire glacier.

Alternatively, glacier volume is estimated through glacier parameters such as area, length and slope. Various scaling methods estimate volume by developing empirical relationships between volume and glacier parameters (Chen and Omura 1990; Bahr 1997; Liu and Sharma 1998; Arendt et al. 2006; Cuffey and Paterson 2010). However, glacier-stored water in the Indian Himalayas estimated by

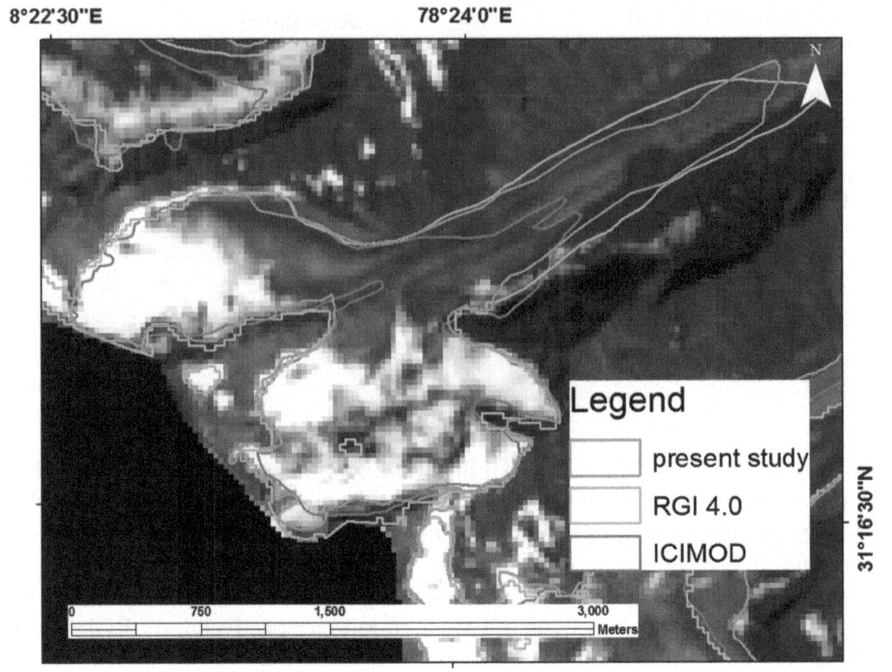

Fig. 2 Glacier outline from the inventory of the International Centre for Integrated Mountain Development (ICIMOD), Randolph Glacier Inventory (RGI) and the interpretation of glacier extents from the present study overlaid on Landsat imagery of 2014 of Naradu Garang in the Baspa basin, Himachal Pradesh

different scaling methods shows large variability (Fig. 3). Also, these methods have their own limitations. Few volume-scaling methods are developed for glaciers of different geomorphological settings than those of glaciers in the Indian Himalayas and hence, are inherently not free of uncertainties. Error in delineating glacier boundary can magnify the error in volume estimates that are deduced from volume area scaling method. Moreover, scaling methods estimate only the amount of glacier-stored water and not the spatial distribution of thickness.

To address this limitation, a method based upon surface velocity and flow law of ice was developed to estimate the ice thickness distribution (Gantayat et al. 2014). In this approach, surface velocity fields were estimated using remote sensing data, and then ice thickness is determined using flow law of ice (Eqs. 1 and 2). The methodology is applied to the Gangotri Glacier (Figs. 4 and 5).

$$U_s = U_b + \frac{2A}{n+1}\tau^n H \quad (1)$$

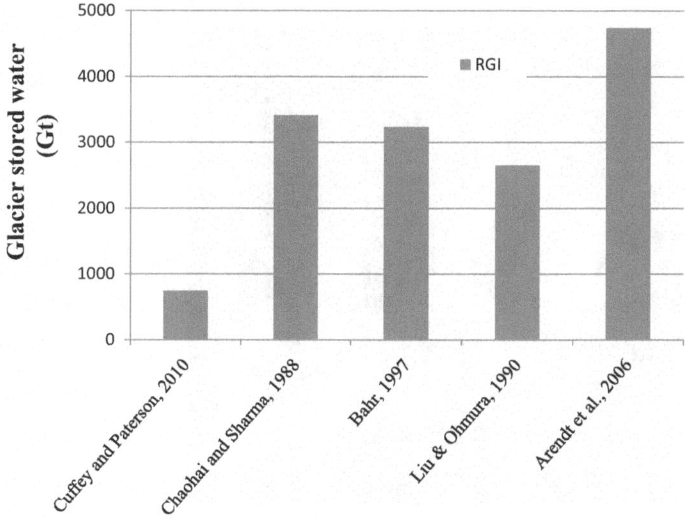

Fig. 3 Estimation of glacier-stored volume in the Indian Himalaya using Randolph Glacier Inventory (RGI), and various scaling methods show large variability

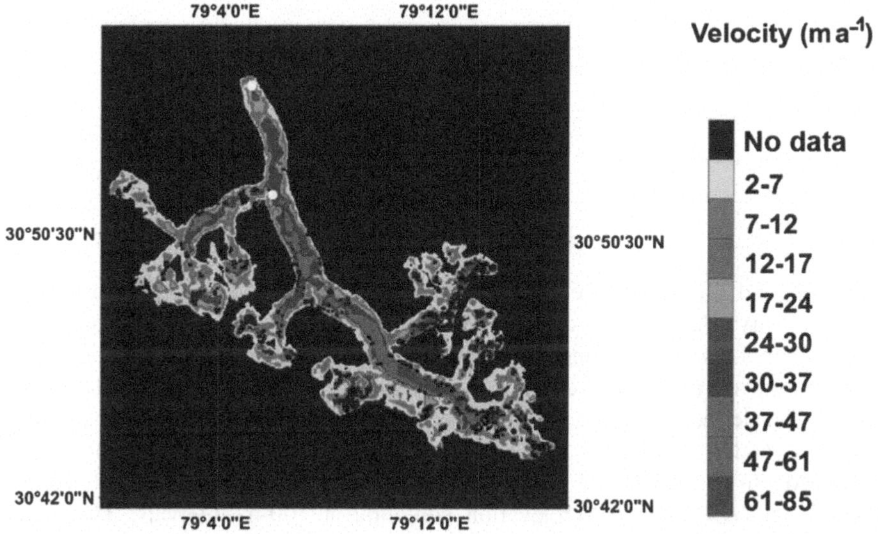

Fig. 4 Surface velocity field of the Gangotri Glacier, varying from 85 to 5 ma^{-1}. Maximum velocity at higher reaches and minimum velocities along the glacier boundary were observed. The two large *white dots* represent the sites where surface velocity was validated using field measurements. *Source* Gantayat et al. (2014)

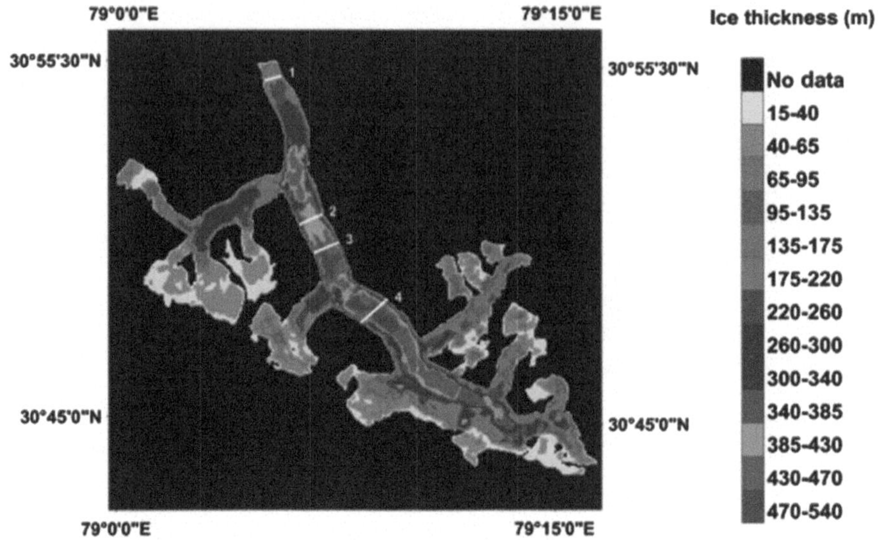

Fig. 5 The model thickness of ice was varying from 40 to 540 m, and maximum thickness was observed in the central part of the main trunk, whereas the thickness at the snout is estimated to be in the range of 40–65 m. *Source* Gantayat et al. (2014)

where,

$$\tau = f\rho g H \sin\alpha$$

$$H = \sqrt[4]{\frac{1.5U_s}{A(f\rho g \sin\alpha)^3}} \tag{2}$$

where U_s and U_b are surface and basal velocities, respectively. In the absence of basal velocity, U_b *is* assumed to be 25 % of the surface velocity (Swaroop et al. 2003). Glen's flow law exponent, n, is assumed to be 3, H is ice thickness and A is a creep parameter. τ is basal stress. ρ is the ice density, g is acceleration due to gravity and f is a scale factor, i.e. the ratio between the driving stress and basal stress along a glacier, and f is assumed to be 0.8 (Haeberli and Hoelzle 1995). Slope α is estimated from Advanced Spaceborne Thermal Emission and Reflection Radiometer (ASTER) DEM elevation contours, at 100-m intervals. This gives depth for each area between successive 100-m contours which are all plotted together to provide an ice-thickness distribution for the entire glacier. The total ice volume was estimated to be around 1.8 ± 0.5 km^3 (Gantayat et al. 2014). This compares well with the estimates of Gabbi et al. (2012) and Farinotti et al. (2009), as 1.36 ± 0.07 and 1.96 km^3, respectively.

3 Retreat and Area Loss

There are reports of changes in the glacial extent and length in the Himalayas. The reports include changes for individual glaciers and many times for the group of glaciers. An individual glacier is normally studied using field and satellite data. The changes are generally estimated using remote sensing data with limited field studies. The data on long-term retreat are available for 83 glaciers in Table 1. The amount and rates of retreat are given in Table 2. The location of glaciers and amount of retreat that occurred during the period 1960–2000 are given in Fig. 6. The average loss of glacial length during this period is approximately 550 ± 419 m. The large standard deviation suggests variation in glacier retreat. In

Table 1 Retreat of Glaciers in the Himalaya

Glacier ID	Glacier	Period	Retreat rate (m/decade)	References
1	Zanskar glacier 1	1999–2004	80	Kamp et al. (2011)
2	Parkachik	1990–2003	−110	Kamp et al. (2011)
3	Zanskar glacier 3	1990–2003	220	Kamp et al. (2011)
4	Zanskar glacier 4	1975–2003	270	Kamp et al. (2011)
5	Zanskar glacier 5	1990–2003	40	Kamp et al. (2011)
6	Zanskar glacier 6	1990–2003	290	Kamp et al. (2011)
7	Zanskar glacier 8	1975–2003	10	Kamp et al. (2011)
8	Zanskar glacier 7	1975–2003	80	Kamp et al. (2011)
9	Zanskar glacier 9	1975–2006	320	Kamp et al. (2011)
10	Zanskar glacier 10	1975–2006	610	Kamp et al. (2011)
11	Drang Drung	1975–2008	90	Kamp et al. (2011)
12	Zanskar glacier 12	1990–2003	40	Kamp et al. (2011)
13	Zanskar glacier 13	1992–2002	20	Kamp et al. (2011)
14	Zing-Zing Bar (Patsio)	1971–2011	220 225	Negi et al. (2013)
15	Baralacha	1971–2011	10 100	Negi et al. (2013)
16	Miyar	1961–1996	160	Sangewar and Kulkarni (2011)
17	Triloknath	1968–1996	180	Sangewar and Kulkarni (2011)
18	Panchi nala I	1963–2007	110	Sangewar and Kulkarni (2011)
19	Panchi nala II	1963–2007	120	Sangewar and Kulkarni (2011)
20	Beas Kund	1963–2003	190	Sangewar and Kulkarni (2011)
21	Sonapani	1906–1957	180	Sangewar and Kulkarni (2011)
22	Samudra Tapu	1962–2000	200	Kulkarni et al. (2006)
23	Hamtah	1961–2005	80	Sangewar and Kulkarni (2011)
24	Jobri	1963–2003	30	Sangewar and Kulkarni (2011)
25	Chhota Shigri	1962–1995	70	Sangewar and Kulkarni (2011)

(continued)

Table 1 (continued)

Glacier ID	Glacier	Period	Retreat rate (m/decade)	References
26	Sara ugma	1963–2004	410	Sangewar and Kulkarni (2011)
27	Bara Shigri	1906–1995	300	Sangewar and Kulkarni (2011)
28	Man Talai (Gl. No. 115)	1989–2004	230	Sangewar and Kulkarni (2011)
29	Bilare Bange	1962–1997	30	Kulkarni and Bahuguna (2002)
30	Shaune Garang	1962–1997	260	Kulkarni and Bahuguna (2002)
31	Janapa Garang	1962–1997	200	Kulkarni and Bahuguna (2002)
32	Tikku	1960–1999	220	Sangewar and Kulkarni (2011)
33	Jhajju Bamak	1960–1999	280	Sangewar and Kulkarni (2011)
34	Jaundar Bamak	1960–1999	370	Sangewar and Kulkarni (2011)
35	Bandarpunch	1960–1999	260	Sangewar and Kulkarni (2011)
36	Dokriani	1962–2007	170	Sangewar and Kulkarni (2011)
37	Gangotri	1935–1996 1935–1996	188 188	Sangewar and Kulkarni (2011)
37	Gangotri	1962–1999	338	Naithani et al. (2001)
37	Gangotri	1935–2004	220	Kumar et al. (2008)
37	Gangotri	1965–2006	200	Bhambri et al. (2012)
37	Gangotri	1962–2000	397	Bahuguna et al. (2007)
37	Gangotri	2004–2007	119	Sangewar and Kulkarni (2011)
38	Gl. No.3 (Arwa)	1932–1956	80	Sangewar and Kulkarni (2011)
39	Satopanth	1962–2006	220	Nainwal et al. (2008)
40	Bhagirathi Kharak	1962–2001	167	Sangewar and Kulkarni (2011)
40	Bhagirathi Kharak	1962–2006	73	Nainwal et al. (2008)
41	Trishul bank	1960–2003	220	Sangewar and Kulkarni (2011)
42	Devasthan Bank	1960–2003	260	Sangewar and Kulkarni (2011)
43	Uttari Rishi Bank	1960–2003	340	Sangewar and Kulkarni (2011)
44	Dakshini Rishi Bank	1960–2003	170	Sangewar and Kulkarni (2011)
45	Dakshini Nanda Devi Bank	1960–2003	130	Sangewar and Kulkarni (2011)
46	Milam	1948–1997	170	Sangewar and Kulkarni (2011)
47	Pindari	1906–2001	152	Sangewar and Kulkarni (2011)
47	Pindari	1845–1906	263	Cotter and Brown (1907)
47	Pindari	1906–1966	184	Tewari (1966)
47	Pindari	1966–2007	64	Bali et al. (2008)
48	Poting	1306–1957	50	Sangewar and Kulkarni (2011)
49	Burphu	1966–1997	50	Sangewar and Kulkarni (2011)
50	Shankalpa	1886–1957	70	Sangewar and Kulkarni (2011)
51	Jhulang	1962–2000	110	Oberoi et al. (2001)
52	Meola	1912–2000	190	Sangewar and Kulkarni (2011)
53	Chipa	1961–2000	270	Oberoi et al. (2001)
54	Nikarchu	1962–2002	90	Sangewar and Kulkarni (2011)
55	Adikailash	1962–2002	130	Sangewar and Kulkarni (2011)

(continued)

Table 1 (continued)

Glacier ID	Glacier	Period	Retreat rate (m/decade)	References
56	Rekha Samba	1974–1999	120	Fujita et al. (2001a)
57	AX010	1978–1989	30	Yamada et al. (1992)
58	AX030	1978–1989	0	Yamada et al. (1992)
59	DX 080	1978–1989	50	Yamada et al. (1992)
60	EB 050	1978–1989	30	Yamada et al. (1992)
61	Kongma	1978–1989	30	Yamada et al. (1992)
62	Kongma Tikpe	1978–1989	30	Yamada et al. (1992)
63	Chukhung	1978–1989	90	Yamada et al. (1992)
64	Rathong	1976–2005	180	Raina (2009)
65	Onglaklong	1976–2005	100	Raina (2009)
66	Talung	1976–2005	40	Raina (2009)
67	Tongshiong	1976–2005	140	Raina (2009)
68	Zemu	1976–2005	140	Raina (2009)
69	Changsang	1976–2005	220	Raina (2009)
70	E. Langpo	1976–2005	240	Raina (2009)
71	Jongsang	1976–2005	380	Raina (2009)
72	South Lhonak	1962–2008	420	Raj et al. (2013)
73	Lhonak	1976–2005	270	Raina (2009)
74	N. Lhonak	1976–2005	130	Raina (2009)
75	Chuma	1976–2005	80	Raina (2009)
76	Tasha	1976–2005	20	Raina (2009)
77	Tasha1	1976–2005	40	Raina (2009)
78	Yulhe	1976–2005	−10	Raina (2009)
79	Changme	1976–2005	30	Raina (2009)
80	Rulak	1976–2005	20	Raina (2009)
81	Tista	1976–2005	150	Raina (2009)
82	Kangkyong	1976–2005	80	Raina (2009)
83	Tenabawa	1976–2005	40	Raina (2009)

Table 2 Retreat in different regions of the Himalaya

Region Id	Region	Number of glaciers	Mean retreat in 40 years	Standard deviation
1	Jammu and Kashmir	15	603	698
2	Himachal Pradesh	17	694	400
3	Uttarakhand	23	722	359
4	Nepal	8	186	158
5	Sikkim	20	544	482

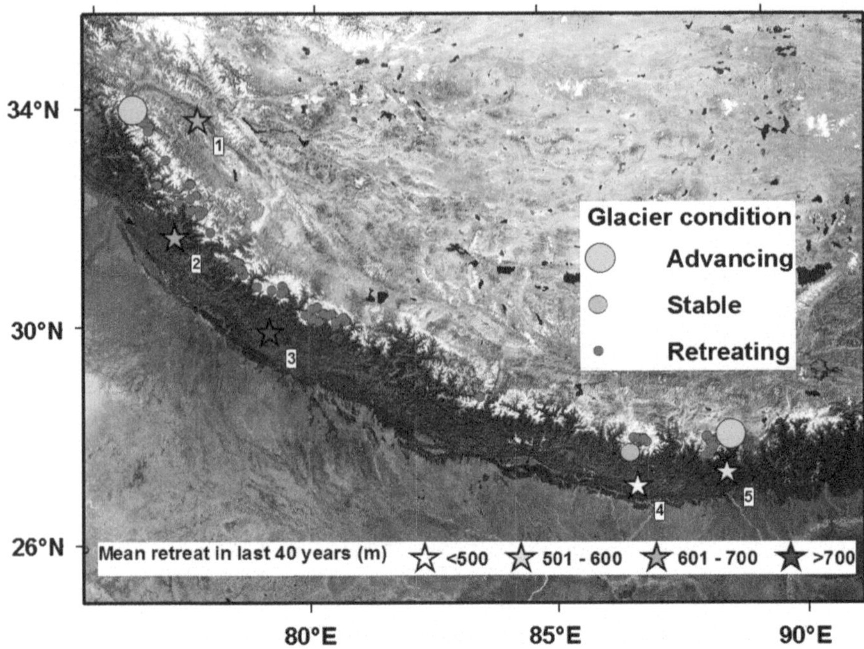

Fig. 6 The location of glaciers with retreat measurements is represented by *circle*, and the *colour of the circle* represents their condition between 1960 and 2000, indicating that most of the glaciers are retreating. The region-wise retreat of glaciers between 1960 and 2000 is represented by *star*, and their *number* represents the region. Further details are given in Table 2 (colour figure online)

addition, mean retreat of glaciers in different regions were found, varying from 186 to 722 m for years between 1960 and 2000, (Fig. 6; Table 1). This variation in retreat is due to factors such as area altitude distribution, mass balance, slope and debris cover (Deota et al. 2011). In general, loss in length is higher in the Western Himalayas than in Sikkim or north-western Himalayas. Limited data on retreat are available for Nepal, Bhutan and Arunachal Pradesh. The snout monitoring suggests that snout of more than 50 % of glaciers in the Karakoram mountain range are either advancing or are stable (Scherler et al. 2011). However, conclusions based on monitoring of only the snout could be misleading, as slope and length can influence retreat, even if loss in mass is the same. Retreat of individual glaciers may be influenced by many factors such as local geomorphic and climatic parameters. However, it may not represent regional changes in climatic condition. Therefore, it would be important to assess long-term overall changes in glacial extent.

The loss in glacial extent is predominantly carried out using remote sensing data. Topographical maps of the survey of India, aerial photographs and satellite images have been extensively used to monitor changes in glacial extent. The published data suggest that almost ~20,060 km^2 out of ~40,000 km^2 area is already studied to estimate the areal loss (Table 3; Fig. 7). It suggests an overall loss of 12.6 ± 7.5 %

Table 3 Loss in glacial area in different regions of the Himalaya

ID	Basin/Region/glacier	Period	Areal extent (km^2)	Loss in area (km^2)	Number of glaciers	References
1	Karakoram region	1990–2013	7882	4	607	Brahmabhatt et al. (2015)
2	Bhut	1962–2001/4	469	47	189	Kulkarni et al. (2011)
3	Zanskar	1962–2001/4	1023	92	671	Kulkarni et al. (2011)
4	Kang Yatze Massif	1969–2010	96	14	121	Schmidt and Nüsser (2012)
5	Warwan	1962–2001/4	847	178	253	Kulkarni et al. (2011)
6	Miyar	1962–2001/4	568	45	166	Kulkarni et al. (2011)
7	Bhaga	1962–2001/4	363	109	111	Kulkarni et al. (2011)
8	Bhaga	1963–2013	357.48 ± 5	51.49 ± 1.07	231	Birajdar et al. (2015)
9	Zing Zing Bar	1979–1989	–	8.87 ± 1.58 (%)	1	Negi et al. (2013)
10	Baralacha	1971–2011	–	16.35 ± 3 (%)	1	Negi et al. (2013)
11	Chandra	1962–2001/4	696	139	116	Kulkarni et al. (2011)
12a	Parbati	1962–2001/4	493	99	90	Kulkarni et al. (2011)
12b		1998–2009	154.3 ± 0.39	8	51	Present study
13	Baspa	1962 to 2014	173	41.3 ± 8.37	19	Present study
14	Dokriani	1962–1995	8	1	1	Dobhal et al. (2004)
15	Bhagirathi	1962–2001/4	1365	191	212	Kulkarni et al. (2011)
16	Alaknanda	1968–2006	324.7 ± 8.4	18	69	Bhambri et al. (2011)
18	Naimona'nyi Region	1976–2003	84	7	NA	Ye et al. (2006)
19	Mt Everest Region	1976–2006	3212 ± 0.019	502	NA	Yong et al. (2010)
20	South Slope Mt Everest Region	1963–2011,	–	403 ± 9m,	NA	Thakuri et al. (2015)
21	Ax010	1978–1999	1	0	1	Fujita et al. (2001b)
22	Sagarmatha National Park	1950–1990	404	19	NA	Salerno et al. (2008)

(continued)

Table 3 (continued)

ID	Basin/Region/glacier	Period	Areal extent (km^2)	Loss in area (km^2)	Number of glaciers	References
23	Kanchenjunga Region—Arun and Tomar Basin	1962–2000	323	20.8	68	Racoviteanu et al. (2015)
24a	Tista	1997–2004	403	11	57	Kulkarni et al. (2011)
24b		1989–2010	333 ± 9.1	6.9 ± 1.5	38	Basnett et al. (2014)
24c	Sikkim	1962 to 2000	634	127.7 ± 42	164	Racoviteanu et al. (2015)
25	Bhutan Himalaya	1963–1993	147	12	66	Karma et al. (2003)

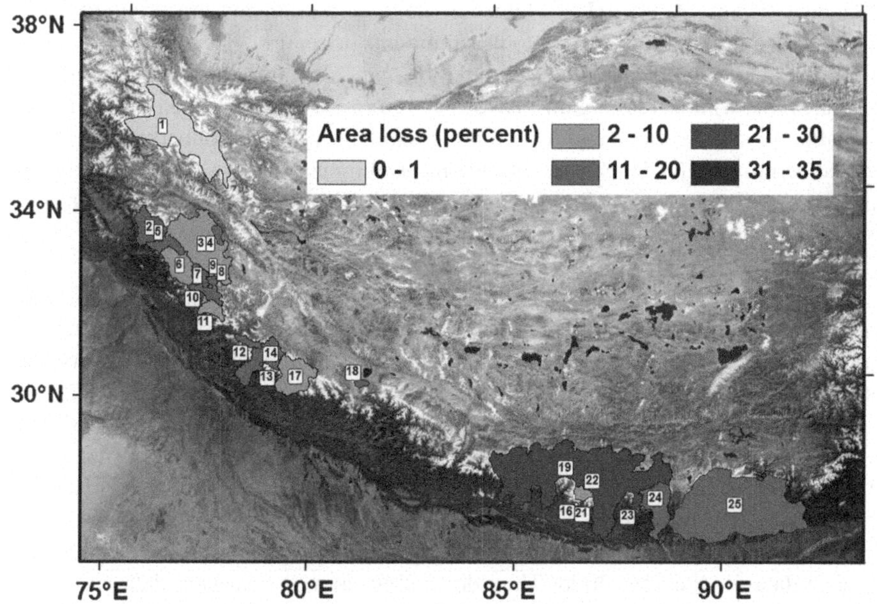

Fig. 7 Glacial area loss (%) in different regions of the Himalaya from 1960 to 2000. The *number* represents basins/regions, and the *colour* represents percent of area loss, as given in Table 3 (colour figure online)

for 40 years from 1960 onwards. Maximum loss in area is observed in (Eastern Himalaya) Tista and Mt. Everest region followed by Bhutan and Western Himalayas. Loss in area is observed to be the least in the Karakoram region (Brahmbatt et al. 2015). Area loss measurement is missing in few of the regions of the Himalaya, and this lacuna has to be filled to assess the status of glaciers in the Himalaya.

4 Mass Balance

Mass balance is the total loss or gain of mass at the end of a hydrological year (Paterson 1998) and is one of the key variables to assess the state of a glacier. Mass balance is also useful to model future changes in glacial extents and river run-off pattern (Kulkarni et al. 2004). It is more sensitive to climate change than retreat. However, measurement of mass balance is more challenging, as it needs extensive investigations on glacial ice and is measured in various ways such as traditional glaciological, hydrological, accumulation area ratio (AAR) or equilibrium line altitude (ELA), geodetic and gravimetric methods.

Traditional glaciological method involves measuring winter snow accumulation using pits and summer melt using stakes in ablation area. First field measurement of mass balance for an Indian glacier was carried out on the Gara glacier, Himachal

Himalaya (Raina et al. 1977). However, field measurements are point measurements and extrapolation of point measurements to entire glacier induces uncertainty. In addition, avalanche accumulation can also introduce uncertainties in estimates of winter mass gain. Also, in situ mass balance measurements are available for short periods and for a limited number of glaciers due to the harsh and rugged terrain of the Himalaya. In situ measurements tend to be biased towards small- to medium-sized, debris-free glaciers, as they are easy to access and thus, may not be a good representative of rest of the glaciers in the Himalayan region (Gardelle et al. 2013).

Estimates of mass loss using traditional glaciological methods are available for 14 glaciers, spread across the Himalaya. The name and location of these glaciers are shown in Fig. 8. We have estimated cumulative mean mass loss from 1975 to 2014 as 20 ± 6 m, where uncertainty in mass loss estimation is induced due to the absence of continuous mass balance data for long duration (Fig. 9). To estimate uncertainty, firstly, mean mass balance for each individual glacier from 1975 to 2014 was calculated using available field data. The year 2008 has field mass balance data for the maximum number of glaciers. Using mass balance data of 2008 and mean mass balance data for missing glaciers, we generated synthetic mass balance data for the year 2008. Then depending upon the missing data points, we estimated uncertainties in mass balance of individual years. Finally, the cumulative of uncertainties from 1975 to 2014 was deduced as the overall uncertainty in mass loss estimation.

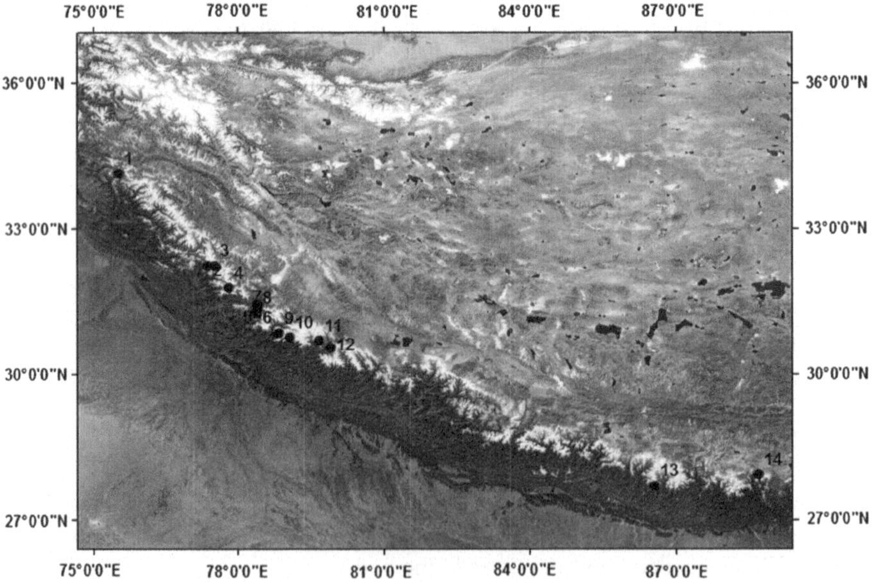

Fig. 8 Location of 14 glaciers, where field-based estimates of mass balance are available. *1* Nehnar, *2* Hamta, *3* Chhota Shigri, *4* Parbati, *5* Gara, *6* Gor Garang, *7* Shaune Garang, *8* Naradu, *9* Dokriani, *10* Chorabari, *11* Tipra bank, *12* Dunagiri, *13* AX010 and *14* Changme Khangpu

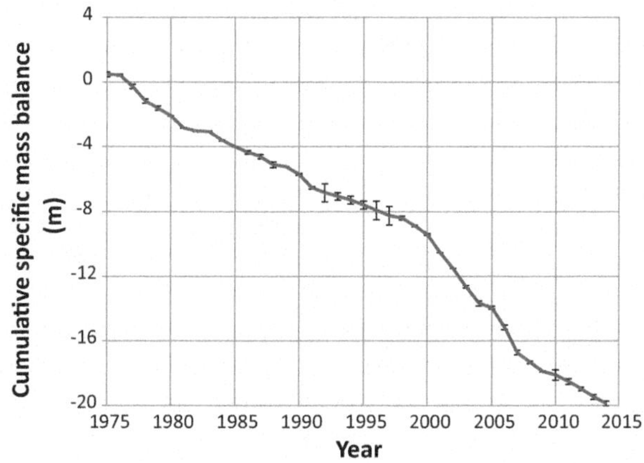

Fig. 9 Cumulative glacial mass loss from 1975 to 2014 for 14 glaciers is estimated at 20 ± 6 m, where uncertainty represents cumulative error. The *error bars* in the graph represent the uncertainty in mass loss estimation for individual year. Predominantly, field mass balance data were used to estimate the cumulative mass balance. However, for few glaciers and years, gaps in mass balance data were filled using ELA and AAR methods. The cumulative mass balance will change, if data from other regions such as Sikkim and Karakoram are available

The 14 glaciers cover only an area of ~90 km^2, compared to the vast glaciated area of the Himalayan range of ~40,000 km^2 (Kulkarni and Karyakarte 2014). Also, ELA for glaciated area of the Himalayan range is estimated at 5400 m asl (Chaturvedi et al. 2014). In the Himalaya, 57 % of the glaciated area is below ELA (Fig. 10), whereas 80 % of the area is below ELA for 14 glaciers, where mass balance data are available. This makes the mass loss estimates biased towards low-altitude glaciers. Furthermore, mass balance not only depends on the altitude, but also on the orientation, precipitation and other meteorological parameters. Hence, these 14 glaciers may not represent the glaciers of the entire Himalayan range.

Fig. 10 Area altitude distribution of 14 glaciers with field-based mass balance data in comparison with AAD of glaciers of entire Himalaya shows that they lie substantially in the lower altitude

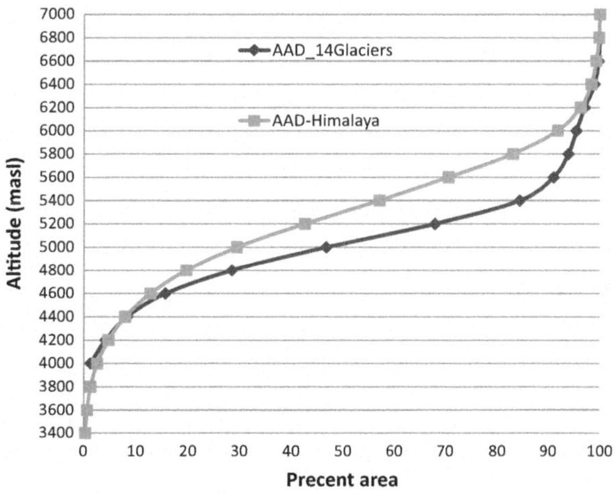

Therefore, to assess the glacial mass loss for the Himalayan region, we have combined field observations with equilibrium line altitude and accumulation area ratio method (Kulkarni 1992; Kulkarni et al. 2004). Normally at ELA, annual net mass balance is zero. AAR is the ratio of accumulation area to the total glacier area. A relationship between AAR and mass balance is already established for the Himalayan region (Kulkarni et al. 2004). The relationship was developed using field mass balance data of the Shaune and Gor Garang glaciers in the Baspa river basin (Fig. 11). ELA and AAR can also be estimated for large number of glaciers using remote sensing data, and mass balance can be derived at regional scale using the statistical relationship between AAR and mass balance. Using this relationship, mass balance was estimated over the entire Himalayan region (Chaturvedi et al. 2014). The specific mass balance for the year 2000 for the Karakoram region, Western Himalaya, Central Himalaya and Eastern Himalaya is estimated at +171 ± 63, −165 ± 80, −508 ± 69, and −809 ± 38 mm w.e.a^{-1}, respectively, and −162 ± 16 mm w.e.a^{-1} for the entire Karakoram–Himalayan region. Although, this method allows continuous monitoring of glacier mass balance and at larger spatial domain, it has limitations, as the presence of cloud cover and seasonal snow causes hindrance in accurate mapping of AAR or ELA.

Furthermore, Chaturvedi et al. (2014) used current assessment of mass balance and ensemble-mean projections of temperature and precipitation changes under different Representative Concentration Pathways (RCP) to determine mass loss in the future. There are four pathways, viz. RCP 8.5, RCP 6, RCP 4.5 and RCP 2.6 which represent time-dependent projections of atmospheric greenhouse gas (GHG) concentration under different emission scenarios. RCP 2.6 represents the

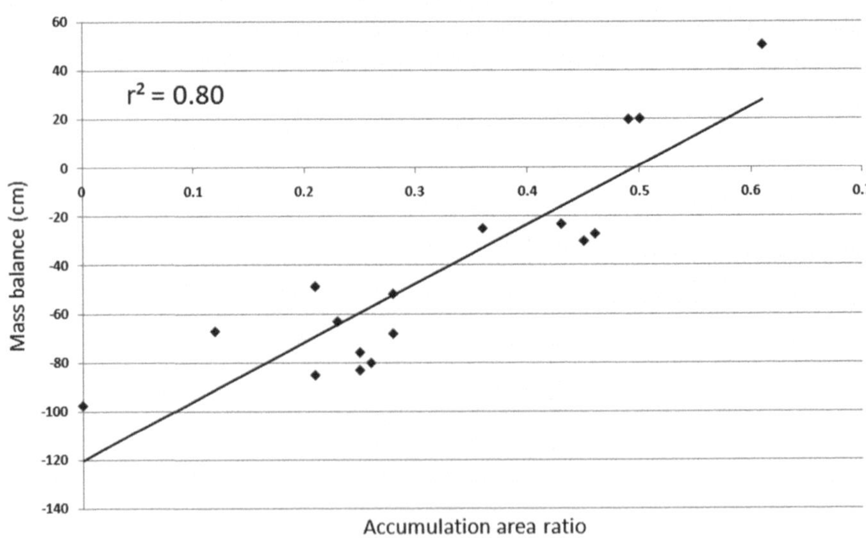

Fig. 11 Regression relationship between accumulation area ratio and mass balance for Shaune Garang and Gor Garang glaciers, in Baspa basin, Western Himalaya. *Source* Kulkarni et al. (2004)

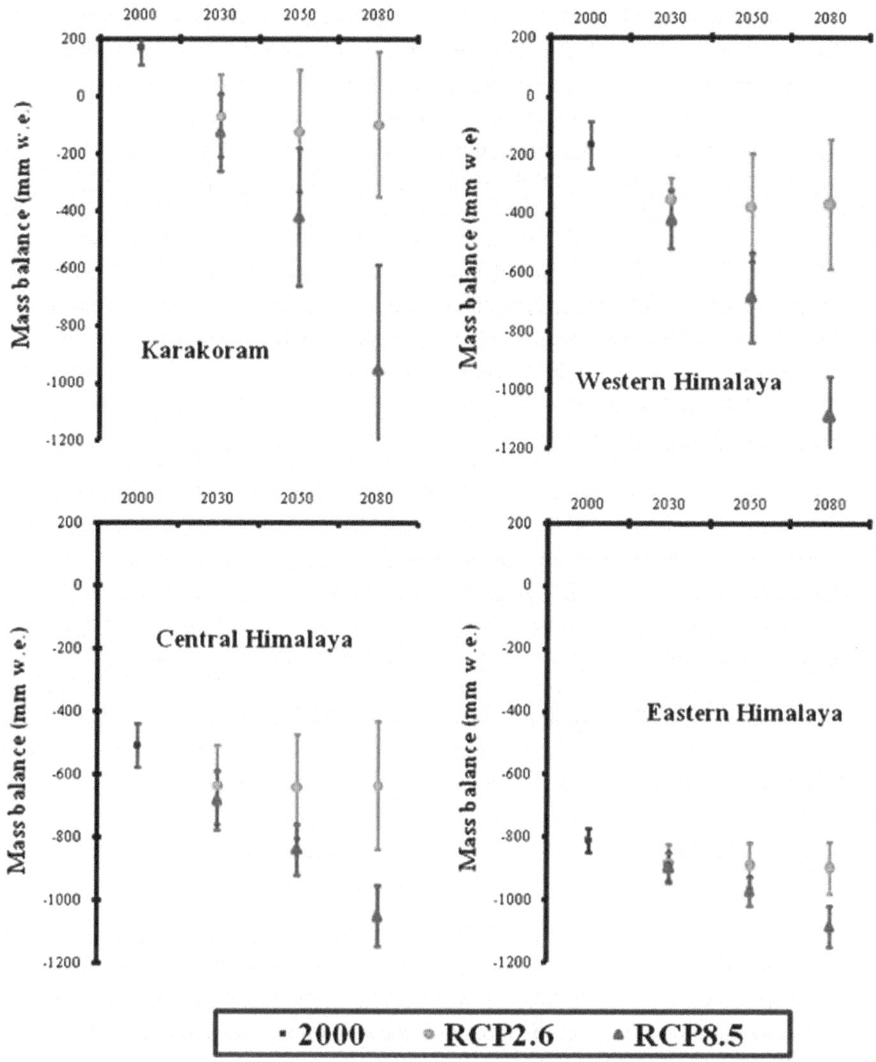

Fig. 12 Impact of climate change on the glacial mass balance of the Karakoram–Himalaya (KH) region for RCP 2.6 and RCP 8.5. *Error bars* for the Karakoram and Western Himalaya regions for the year 2000 are calculated from the standard deviation of the current ELA estimates available from Pandey et al. (2012), while for Central and Eastern Himalaya, difference in the two available studies is used to calculate the *error bars* (Wagnon et al. 2007; Owen and Benn 2005). The error bars for projections (2030, 2050 and 2080 s) are computed based on the standard deviation of the temperature and precipitation change projections under different scenarios. *Source* Chaturvedi et al. (2014)

least warming, whereas RCP 8.5 is associated with the highest warming. RCP 4.5 and RCP 6.0 represent moderate warming. It is estimated that in 2080s, the mass balance of the entire Himalaya will decrease to -342 ± 55 and -1014 ± 59 mm w.e.a^{-1}, under RCP 2.6 and RCP 8.5, respectively (Fig. 12).

At much larger spatial scale, gravimetric method is used to measure glacier mass balance using the data from Gravity Recovery and Climate Experiment (GRACE) mission launched in 2002. The GRACE mission consists of two satellites moving in tandem to each other, separated by a distance of 200 km. These satellites continuously measure the variation in distance between them, occurring due to the variation in gravity field of the earth. The earth's gravity field is a measure of mass in terms of water equivalent height (WEH) at approximately monthly intervals with a spatial resolution of a few hundred kilometres. However, as WEH reflects the changes in the integrated vertically stored water including snow cover, surface water, ground water and soil moisture, extraction of water equivalent of ice cover from the rest of the component is a challenge (Agrawal et al. 2014). Gravimetric measurements suggest mass loss of -5 ± 6 Gt^{-1} (Jacob et al. 2012) for the Hindu Kush region from 2003 to 2010 in comparison with -6.6 ± 1 Gta^{-1} (Chaturvedi et al. 2014) deduced using relationship between AAR and mass balance.

The decadal mass loss in the Himalayas has increased from 3.7 to 8.5 m/decade from 1980 to 2010. However, mass loss is not uniform throughout the Himalaya, and data of the individual glaciers may not represent the entire glaciated region. Rapid mitigation of greenhouse gases concentration from RCP 8.5 to RCP 2.6 can conserve some of the Himalayan glaciers by the end of this century. However, more detailed investigation to understand mass loss of the Himalayan glaciers and possible mitigation strategy is needed.

5 Snow Cover

Large area in the Himalayas is covered in snow. Snow in the Himalayas is source of water in North India. Seasonal snow cover is a natural reservoir that temporarily stores water and feeds rivers during the summer months. In addition, due to high albedo, snow is also an important component of earth's radiation budget and hence influences climate. Snow cover is very sensitive to climatic variations and is an indicator of climate change (Kulkarni and Rathore 2003). Therefore, monitoring of snow cover is necessary to assess water availability and climate change (Kulkarni et al. 2006). Modelling snow melt run-off will aid in the management of water resources, development of hydropower generation and flood forecasting (Gupta et al. 2005).

Measurement of snow cover area often poses challenges due to inaccessibility. In the field, point measurements of snowfall using gauges have several limitations and the information acquired is insufficient. Hence, remote sensing technique is used extensively to monitor spatial distribution of snow cover. Monitoring of snow cover using satellite images started in 1960 (Singer and Popham 1963). Since then, it has been carried out using various sensors such as IRS WiFS, AWiFS, LISS-II and LISS III, Terra ASTER, Landsat MSS and TM, National Oceanic and

Atmospheric Administration (NOAA) MODIS. Snow cover can be easily identified and assessed using remote sensing data as it has distinct reflectance characteristics. It has high reflectance in optical region but much lower in SWIR. This characteristic is used to develop Normalised difference Snow Index (NDSI) to map snow cover in the Himalaya (Kulkarni et al. 2006).

In the Himalayas, average snow cover area from 2000 to 2011 varied from ~0.3 to 0.03 million km^2. Mean depletion of 0.151, 0.034 and 0.003 million km^2 of seasonal snow area was observed in the Indus, the Ganga and the Brahmaputra, respectively, from 2000 to 2011 (Fig. 13) (Singh et al. 2014). The average volume of snow stored in the Indus, the Ganga and the Brahmaputra basin is estimated at 54.5, 9.3 and 14.5 billion cubic meters, respectively. Snow volume was estimated using climatological snowfall data from the ESSO-Indian Meteorological Department and snow cover area from Singh et al. (2014). Snow fall data from Dharmsala, Manali, Shimla and Srinagar stations in the Indus, Joshimath and Mukteshwar stations in the Ganga, and Gangtok and Pasighat stations in the Brahmaputra basin were considered.

Estimating overall trend in the snow cover for the entire Himalayan region is difficult due to the large variability in the snow cover. This variation is in spatial and temporal domain, making it difficult to assess long-term changes in snow cover. In addition, reliable estimates of seasonal snow cover for the Himalayan region are

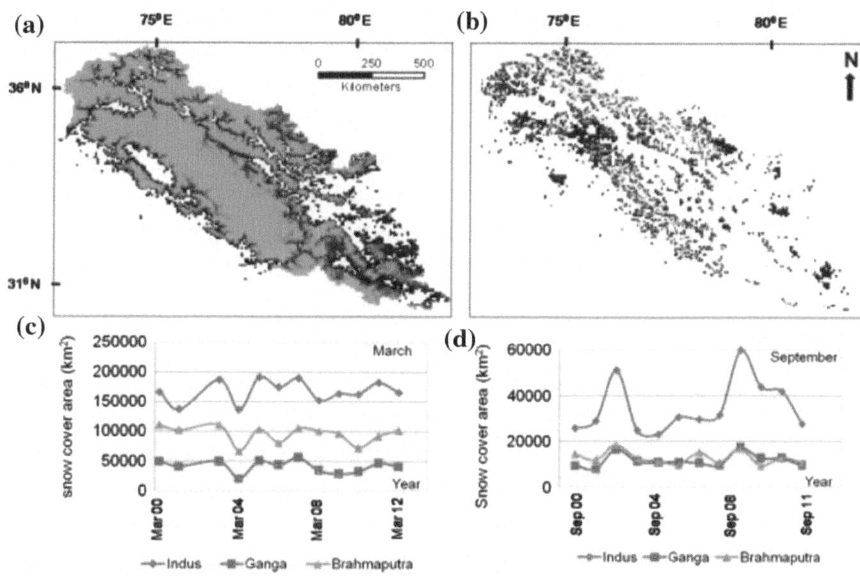

Fig. 13 Monthly March (accumulation) and September (ablation) snow cover of the period 2000–2011 for three major basins in Himalayan region. **a** Snow cover map of March (2011) for the Indus Basin; **b** Snow cover map of September (2011) for the Indus Basin; **c** Snow cover variation of all three basins for March; **d** Snow cover variation of all three basins for September. *Source* Singh et al. (2014)

available from year 2002, i.e. after MODIS data. Therefore, decadal trend in snow cover change has shown little variability. Even though no trend in mean areal extent for the entire Himalayan region was observed, declining trend was observed in the Bhutan and other parts of the Himalaya (Figure 14) (Kulkarni and Rathore 2003; Gurung et al. 2011; Shekhar et al. 2009; Kulkarni 2010; Panwar and Singh 2014). Snow cover area in Bhutan declined by -3.27 ± 1.28 % from 2002 to 2010. Further, in many low-altitude basins such as Ravi, snow ablation was observed even in mid-winter (Kulkarni and Rathore 2003; Kulkarni 2010). Therefore, to assess the changes in the seasonal snow cover, different parameters such as basins located in lower altitude should be studied.

A study was carried out at the Ravi and Bhaga basins using AWiFS data of Resourcesat satellite. The investigation was carried out from 2004 to 2007 at an interval of 5 or 10 days (Kulkarni 2010). A low-altitude basin such as the Ravi has shown a different trend of snow melting in summer and winter months than a high altitude basin such as the Bhaga. In Ravi basin, snow accumulation and ablation were continuous processes throughout winter. Even in the middle of the winter, the snow area was reduced from 90 to 55 %. In the Bhaga basin, snow melting was observed in the early part of the winter, i.e. in the month of December, and no significant melting was observed between January and April as shown in Figs. 15 and 16, indicating different snow ablation patterns in various parts of Himalayas. In addition, snow accumulation and ablation pattern using monthly snow cover distribution in the Alaknanda, Bhagirathi and Yamuna subbasins was studied from 2004 to 2012 using AWiFS data as shown in Fig. 17. Further analysis of the data has also shown small statistically insignificant increase in snow cover (Rathore et al. 2015).

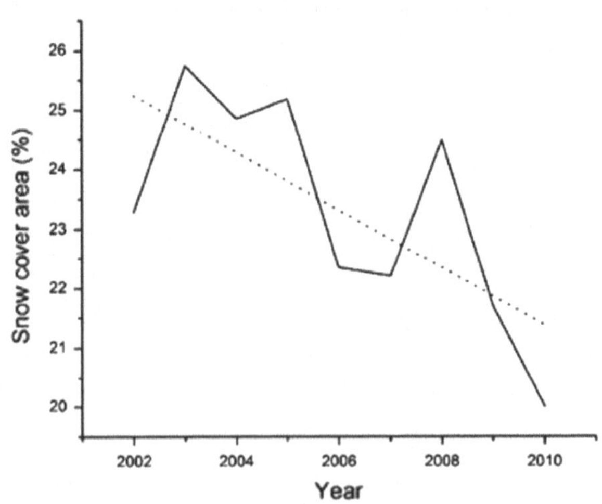

Fig. 14 Decadal trend in snow cover of Bhutan from 2002 to 2010. The *solid line* represents mean annual snow cover area, and the *dashed line* represents linear trend for the decade. *Source* Gurung et al. 2011)

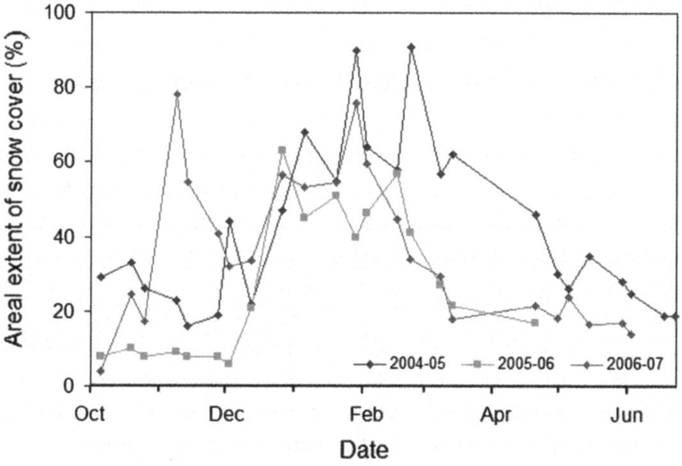

Fig. 15 Snow cover depletion curve for Ravi river basin for year 2004, 2005 and 2006. *Source* Kulkarni et al. (2010)

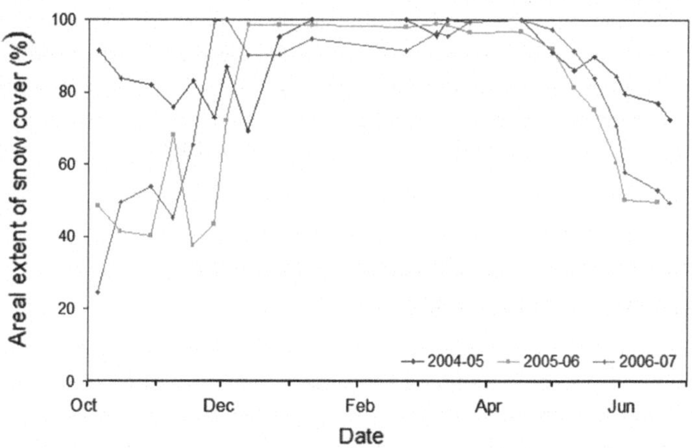

Fig. 16 Snow cover depletion curve for Bhaga river basin for year 2004, 2005 and 2006. *Source* Kulkarni et al. (2010)

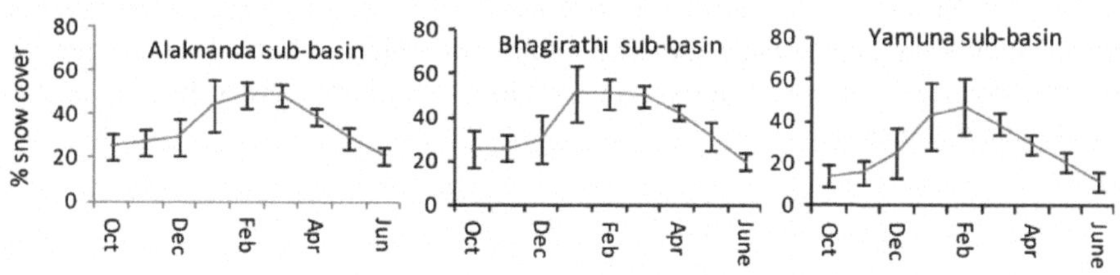

Fig. 17 Monthly average snow cover from 2004 to 2012 with $\pm 1\sigma$. *Source* Rathore et al. (2015)

6 Way Forward

6.1 *Spatial Distribution of Glacial Depth and Volume*

The spatial distribution of glacier depth and assessment of glacial-stored water is essential information for managing water resources of the Himalayas. Conventionally, several statistical techniques have been used to determine the volume or mean depth of glaciers using surface area. However, volume–area scaling technique introduces large errors in depth estimates due to the lack of field data on depth, and also small errors in delineating a glacier boundary can introduce large errors in depth estimate. Therefore, estimate of glacier-stored water in the Himalayas varies from 4000 to 12,000 Gt, depending upon the scale of glacier inventory and scaling technique. Therefore, further improvement in this assessment is necessary for better management of the Himalayan water resources.

Recent developments in the field of glaciology make it possible to improve volume estimates. The technique based on velocity, slope and flow law of ice can provide ice-thickness distribution. The technique has been successfully used in the Gangotri Glacier. Velocity and slope can be obtained using remote sensing techniques, and depth estimates can be validated using field GPR measurements. If a systematic programme is launched, the technique can be used to estimate the distribution of ice thickness of thousands of glaciers in the Himalaya. The information on spatial distribution of ice thickness can be further used to map bottom topography and assess future formation of moraine dammed lakes.

6.2 *Identification and Mapping of Potential Moraine-Dammed Lakes*

A large number of moraine-dammed lakes have been mapped in the Himalaya. This number is likely to increase as majority of the Himalayan glaciers are experiencing accelerated retreat. Glacier lakes are potential source of outburst flood, leading to destruction of life and property in the downstream. Furthermore, threats from the glacial lakes are likely to increase in the future due to the ongoing change in climate. Therefore, it is necessary to develop a systematic programme to map existing glacial lakes, and develop a systematic data base of glacial lakes, identify the potentially dangerous lakes, develop a mitigation plan and disaster management strategy including early warning system to safe guard life and property of people in the Himalayan region.

6.3 *Modelling Glacier Dynamics to Understand Future Changes in Glacial Extent Due to Changes in Climate*

Mass balance is one of the key variables to assess the state of a glacier and is useful in modelling future changes in glacial extents. Mass balance in the field is measured using traditional glaciological method. However, available in situ measurements of mass balance are sparse and biased towards small glaciers and hence not representative of the entire Himalayas. Therefore, a programme needs to be launched to measure mass balance at the basin scale. The field-based technique can be combined with remote sensing based methods such as equilibrium line altitude and photogrammetric method to estimate mass loss on basin and regional scale.

Further, our understanding of the impact of projected climate change on the Himalayan glaciers is poor. Hence, we have to launch a programme to understand future changes in mass loss due to the change in temperature and precipitation. This will help in understanding the distribution of glaciers if mass loss is combined with glacier dynamics. Hence, the programme should also include the development of better models to understand relationships between mass balance, glacial depth and any other physical parameter of glaciers. This will help us in understanding future changes in glacial area.

6.4 *Modelling Future Changes in Contribution of Snow and Glacier Melt in Stream Run-off*

Changes in glacial and snow cover can lead to changes in stream run-off pattern. This will affect not only the large rivers but also numerous tributaries. Glacial mass loss in the Himalaya is not spatially uniform, and glaciers in different regions are retreating at different rates. Therefore, influence of climate change on stream run-off is likely to differ from basin to basin. Numerous models in the Himalayas are already developed to understand contribution of snow and glaciers to stream run-off. However, lack of field meteorological data and proper estimates of changes in the Himalayan cryosphere and poor understanding of ongoing climate make it difficult to understand future changes in stream run-off. Therefore, a programme needs to be developed to better understand the food and water security in India.

Acknowledgments The work was carried out at the Divecha Centre for Climate Change, Indian Institute of Science, Bengaluru, India, and the authors are grateful to the centre for the support. Authors would also like to thank numerous colleagues working in the field of glaciology for providing published and unpublished information on glacier inventory, mass balance, glacier retreat and snow cover.

References

Ageta KY, Naito N, Iwata S and Yabuki H (2003) Glacier distribution in the Himalayas and glacier shrinkage from 1963 to 1993 in the Bhutan Himalayas. Bull. Glacier Res., 20, 29– 40

Arendt A, Echelmeyer K, Harrison W, Lingle C, Zirnheld S, Valentine V, Ritchie B, and Druckenmiller M (2006) Updated estimates of glacier volume changes in the western Chugach Mountains, Alaska, and a comparison of regional extrapolation methods. Journal of Geophysical Research, Vol. 111, F03019.

Arendt A, Bolch T, Cogley JG, Gardner A, Hagen J-O, Hock R, Kaser G, Pfeffer WT, Moholdt G, Paul F, Radić V, Andreassen L, Bajracharya S, Barrand N, Beedle M, Berthier E, Bhambri R, Bliss A, Brown I, Burgess D, Burgess E, Cawkwell F, Chinn T, Copland L, Davies B, De Angelis H, Dolgova E, Filbert K, Forester RR, Fountain A, Frey H, Giffen B, Glasser N, Gurney S, Hagg W, Hall D, Haritashya UK, Hartmann G, Helm C, Herreid S, Howat I, Kapustin G, Khromova T, Kienholz C, Köonig M, Kohler J, Kriegel D, Kutuzov S, Lavrentiev I, Bris R Le, Lund J, Manley W, Mayer C, Miles E, Li X, Menounos B, Mercer A, Mölg N, Mool P, Nosenko G, Negrete A, Nuth C, Pettersson R, Racoviteanu A, Ranzi R, Rastner P, Rau F, Raup B, Rich J, Rott H, Schneider C, Seliverstov Y, Sharp M, Sigurðsson O, Stokes C, Wheate R, Winsvold S, Wolken G, Wyatt F, Zheltyhina N (2012) Randolph Glacier Inventory – A Dataset of Global Glacier Outlines: Version 3.2. *Global Land Ice Measurements from Space, Boulder Colorado, USA*. Digital Media.

Agrawal R, Singh SK, Rajawat AS and Ajai (2014) Estimation of regional mass anomalies from Gravity Recovery and Climate Experiment (GRACE) over Himalayan region. The International Archives of the Photogrammetry, Remote Sensing and Spatial Information Sciences, Volume XL-8, 2014 ISPRS Technical Commission VIII Symposium, Hyderabad, India.

Bahr DB (1997) Global distributions of glacier properties: A stochastic scaling paradigm. Water Resources Research, Vol. 33, No. 7, pp 1669-1679.

Bahuguna IM, Kulkarni AV, Nayak S, Rathor BP, Negi HS and Mathur P (2007) Himalayan glacier retreat using IRS 1C PAN stereo data. Int. J. Remote Sensing, 28(2), 437–442

Bali R, Agarwal KK, Ali N and Rastogi SK (2008) Geomorphological framework and recessional behavior of the Pindari glacier, Kumauon Himalaya. In National Seminar on Glacial Geomorphology and Paleoglaciation in Himalaya, Lucknow, pp. 32–33.

Bajracharya SR and Shrestha B (2011) The status of glaciers in the Hindu Kush – Himalayan region. International Centre for Integrated Mountain Development, Kathmandu, Nepal.

Bhambri R, Bolch T, Chaujar R and Kulshreshtra AC (2011) Glacier changes in the Garwal Himalaya, India, from 1968 to 2006 based on remote sensing. J. Glaciol, 57(203), 543 –556

Bhambri R, Bolch T and Chaujar RK (2012) Frontal recession of Gangotri Glacier, Garhwal Himalayas, from 1965 to 2006, measured through high- resolution remote sensing data. Current Science, 102(3), 489–494.

Basnett S, Kulkarni A and Bolch T (2013). The influence of debris cover and glacial lakes on the recession of glaciers in Sikkim Himalaya, India, J. Glaciol., 59, 1035–1046.

Birajdar F, Venkataraman G, Samant H (2015) Temporal and spatial variability over the glacierized regime of Bhaga basin, western Himalaya, India, during the past half century (1963–2013). International Symposium on Glaciology in High-Mountain Asia. Kathmandu, Nepal.

Brahmbhatt RM, Bahuguna IM., Rathore BP, Singh SK, Rajawat AS, Shah RD and Kargel JS (2015) Satellite monitoring of glaciers in the Karakoram from 1977 to 2013: an overall almost stable population of dynamic glaciers. The Cryosphere Discuss., 9, 1555–1592

Cotter G de P and Brown J (1907) Notes on certain glaciers in Kumaun, Rec. Geol. Surv. India, 35 (4), 148–157.

Chen J and Ohmura A (1990) Estimation of Alpine glacier water resources and their change since the 1870s, Hydrology in Mountainous Regions. I-Hydrological Measurements; the Water Cycle, Proceedings of two Lausanne Symposia, IAHS Publ. No. 193.

Cuffey KM and Paterson WSB (2010) The Physics of Glaciers, Fourth Edition, Butterworth-Heinemann, Elsevier.

Chaturvedi RK, Kulkarni A, Karyakarte Y, Joshi J and Bala G (2014) Glacial mass balance changes in the Karakoram and Himalaya based on CMIP5 multi-model climate projections. Climatic Change 123:315–328

Dobhal DP, Gergan JT and Thayyen RJ (2004) Recession and morphogeometrical changes of Dokriani glacier (1962–1995) Garhwal Himalaya, India. Curr. Sci. 86(5), 692–696

Deota BS, Trivedi YN, Kulkarni AV, Bahuguna IM And Rathore BP (2011) RS and GIS in mapping of geomorphic records and understanding the local controls of glacial retreat from the Baspa Valley, Himachal Pradesh, India. Curr. Sci., 100(10), 1555 – 1563

Fujita K, Nakazawa F and Rana B (2001) Glaciological observations on Rikha Samba glacier in Hidden Valley, Nepal Himalayas, 1998 and 1999. Bull. Glacier Res., 18, 31 –35.

Fujita K, Kadota T, Rana B, Kayastha RB and Ageta Y (2001) Shrinkage of glacier AX010 in Shorong region, Nepal Himalayas in the 1990s. Bull. Glacier Res., 18, 51–54.

Farinotti D, Huss M, Bauder A, Funk M and Truffer M (2009) A method to estimate ice volume and ice-thickness distribution of alpine glaciers. J. Glaciol., 55(191), 422–430

Gergan JT, Dobhal DP and Kaushik R (1999) Ground penetrating radar ice thickness measurements of Dokriani bamak (glacier), Garwal Himalaya. Curr. Sci. 77, 169–173.

Gupta RP, Haritashya UK and Singh P (2005) Mapping dry/wet snow cover in the Indian Himalayas using IRS multispectral imagery Remote Sensing of Environment 97 458 – 469

Gurung DR, Kulkarni AV, Giriraj A, Aung KS and Shrestha B (2011) Monitoring of seasonal snow cover in Bhutan using remote sensing technique. Current Science, Vol. 101, No. 10, 25.

Gabbi J, Farinotti D, Bauder A and Maurer H (2012) Ice volume distribution and implications on runoff projections in a glacierised catchment. Hydrol. Earth Syst. Sci., 16(12), 4543–4556 (doi: 10.5194/hess-16-4543-2012)

Gardelle J, Berthier E, Arnaud Y, Kaab A (2013) Region wide glacier mass balances over the Pamir-Karakoram-Himalaya during 1999-2011. Cryosphere, 7 (6), pp.1885-1886.

Gantayat P, Kulkarni AV and Srinivasan J (2014) Estimation of ice thickness using surface velocities and slope: Case study at Gangotri glacier, India. Journal of Glaciology, Vol No 60 No 220

Haeberli W and Hoelzle M (1995) Application of inventory data for estimating characteristics of and regional climate-change effects on mountain glaciers: a pilot study with the European Alps. Ann. Glaciol., 21, 206–212.

Immerzeel WW, Van Beek LPH, Bierkens MFP (2010) Climate change will affect the Asian water towers. Science 328 (5984), 1382-1385

Jacob T, Wahr J, Pfeffer WT and Swenson S (2012). Recent contributions of glaciers and ice caps to sea level rise. Nature 482:514–518.

Kulkarni AV and Buch AM (1991) Glacier Atlas of Indian Himalaya, ISRO report number SAC/RSA/RSAG-MWRD/SN/05/91,

Kulkarni AV (1992) Mass balance of Himalayan glaciers using AAR and ELA Methods. Journal of Glaciology. Vol 38, No. 128

Kulkarni, A. V. and Bahuguna, I. M., (2002) Glacial retreat in the Baspa Basin, Himalayas, monitored with satellite stereo data. J. Glaciol., 48(160), 171–172

Kulkarni AV, Rathore BP (2003) Snow cover monitoring in Baspa basin using IRS WiFS data. Mausam, 54 (1), 335-340.

Kulkarni AV, Rathore BP and Alex S (2004) Monitoring of glacial mass balance in the Baspa basin using accumulation area ratio method. Current Science, 86(1), 185–190.

Kulkarni AV, Dhar S, Rathore BP, Raj, K B. G. and Kalia R (2006), Recession of Samundra Tapu glacier, Chadra river basin, Himachal Pradesh. J. Indian Soc. Remote Sensing, 34(1), 39–46

Kulkarni AV, Singh SK, Mathur P and Mishra VD, (2006) Algorithm to monitor snow cover using AWiFS data of RESOURCESAT for the Himalayan region, International Journal of Remote Sensing, Vol. 27, No. 12, 2449-2457

Kumar K, Dumka,RK, Miral MS, Satyal GS and Pant M (2008) Estimation of retreat rate of Gangotri glacier using rapid static and kinematic GPS survey. Current Science, 92(2), 258–262

Khan SB, Ahmad F, Sadaf S, and Kashif RH (2010) Crops area and production (by districts) (1981-82 to 2008-09), Volume I Food and cash crops, Government of Pakistan, Statistics Division, Federal Bureau of Statistics (EconomicWing), Islamabad.

Kulkarni, AV, Rathore BP, Singh SK and Ajai (2010) Distribution of seasonal snow cover in central and western Himalaya. Annals of Glaciology 51(54).

Kamp U, Byrne M and Bolch T (2011) Glacier fluctuations between 1975 and 2008 in the Greater Himalayan Range of Zanskar, Southern Ladakh. J. Mt. Sci., 8, 374–389

Kulkarni, AV, Rathore, BP, Singh SK and Bahuguna IM (2011) Understanding changes in Himalayan cryosphere using remote sensing technique. Int. J. Remote Sensing, 32(3), 601 – 615.

Kulkarni AV and Yogesh Karyakarte (2014) Observed changes in Himalayan glaciers. Current Science, 106 (2)

Liu CH and Sharma CK (1998) Report on first expedition to glaciers in the Pumqu (Arun) and Poiqu (Bhote-Sun Kosi) river basins, Xizang (Tibet), Science Press, Beijing, pp. 192

Naithani AK, Nainwal HC, Sati KK and Prasad C (2001) Geomorphological evidences of retreat of the Gangotri glacier and its characteristics. Current Science, 80(1), 87–94

Nainwal HC, Negi BDS, Chaudhary M, Sajwan KS And Gaurav A (2008) Temporal changes in rate of recession: evidences from Satopanth and Bhagirath Kharak glaciers, Uttarakhand, using Total Station Survey. Curr. Sci., 94(5), 653–660

Negi HS, Saravana G, Rout R and Snehmani (2013) Monitoring of great Himalayan glaciers in Patsio region, India using remote sensing and climatic observations. Current Science, Vol. 105 (10)

Oberoi LK, Siddiqui MA and Srivastava D (2001) Recession of Chipa, Meola and Jhulang (Kharsa) Glaciers in Dhauliganga Valley Between 1912 and 2000, GSI Special Publication No. 65, vol. II, pp. 57–60

Paterson W SB (1998) The Physics of Glaciers, Butterworth Heinemann, pp. 26–53

Panwar A, Singh D (2014) Monitoring of seasonal snow cover in Yamuna basin of Uttarakahnd Himalaya using remote sensing techniques International Journal of Engineering Sciences & Emerging Technologies, Volume 6, Issue 5, pp: 447-453

Raina VK, Kaul, MK and Singh S (1977), mass balance studies of Gara glacier, J glaciology 18 (80) 415-423

Raina VK (2009) Himalayan glaciers: a state-of-art review of glacial studies, glacial retreat and climate change. MoEF Discussion Paper 56

Reserve Bank of India (RBI) report (2011), State-Wise Production of Foodgrains and Major Non-Foodgrain Crops,

Raj KBG, Remya SN and Kumar KV (2013) Remote sensingbased hazard assessment of glacial lakes in Sikkim Himalaya, Curr. Sci., 104(3), 359–364.

Racoviteanu AE, Arnaud Y, Williams MW, and Manley WF (2015) Spatial patterns in glacier characteristics and area changes from 1962 to 2006 in the Kanchenjunga–Sikkim area, eastern Himalaya. The Cryosphere, 9, 505–523, 2015

Rathore BP, Singh SK, Bahuguna M, Brahmbhatt RM, Rajawat AS, Thapliyal A, Panwar A and Ajai (2015) Spatio-temporal variability of snow cover in Alaknanda, Bhagirathi and Yamuna sub-basins, Uttarakhand Himalaya. Current Science, Vol. 108, No. 7, 10 April 201 5

Singer FS and Popham RW (1963) Non-meteorological observations from satellites. Astronaut. Aerospace Eng., 1(3), 89–92.

Sangewar CV and Shukla SP (1991) Inventory of the Himalayan Glaciers, Special Publication No. 34, Geological Survey of India, Space Applications Centre, Ahmedabad,.

Swaroop S, Raina VK and Sangeswar CV (2003) Ice flow of Gangotri glacier. In Proceedings of the Workshop on Gangotri glacier, 26–28 March 2003, Lucknow, India. (Spec. Publ. 80) Geological Survey of India, Kolkata

Salerno F, Buraschi E, Bruccoleri G, Tartari G and Smiraglia C (2008) Glacier surface-area changes in Sagarmatha national park, Nepal. In the second half of the 20th century, by comparison of historical maps. J. Glaciol., 54(187), 758–752

Singh P and Jain SK (2009) Snow and glacier melt in the Satluj River at Bhakra Dam in the western Himalayan region, Hydrological Sciences, 47, 93-106.

Singh KK, Kulkarni AV, Mishra VD (2010) Estimation of Glacier Depth and Moraine Cover Study using Ground Penetrating Radar (GPR) in the Himalayan Region J. Indian Soc. Remote Sens. 38: 1– 9.

Sangewar CV and Kulkarni AV (2011) Observational studies of the recent past. In Report of the Study Group on Himalayan glaciers prepared for the Office of the Principal Scientific Adviser to the Government of India, PSA/2011/2, 25-76

Scherler D, Bookhagen B and Strecker MR (2011) Spatially variable response of Himalayan glaciers to climate change affected by debris cover. Nature Geosci., 4,156–159

Schmidt S and Nüsser M (2012) Changes of high altitude glaciers from 1969 to 2010 in the Trans-Himalayan Kang Yatze Massif, Ladakh, Northwest India. Arct., Antarct. Alp. Res., 44 (1), 107–121

Singh SK, Rathore BP, Bahuguna IM, Ramnathan AL and Ajai (2012) Estimation of glacier ice thickness using Ground Penetrating Radar in the Himalayan region Current Science, Vol. 103, No. 1, 10 July

Singh SK, Rathore BP, Bahuguna IM and Ajai (2014) Snow cover variability in the Himalayan–Tibetan region. Int. J. Climatol. 34: 446–452 (2014)

Tewari AP (1966) Recent changes in the position of the snout of the Pindari glacier (Kumaun Himalaya) Almora District, Uttar Pradesh, India. In International Symposia on the Role of Snow and Ice in Hydrology, International Association of Hydrological Science, vol. 106, pp. 1144–1149.

Thakuri S, Salerno F, Smiraglia C, Bolch T and Tartari G (2015) Scrutinizing glacier changes on the south slope of Mount Everest using optical satellite imagery. International Symposium on Glaciology in High-Mountain Asia. Kathmandu, Nepal, 2–6 March

Yamada T, Shiraiwa T, Lida H, Kadota T, Watanabe T, Rana B, Ageta Y and Fushimi H (1992) Fluctuations of the Glaciers from 1970s to 1989 in the Khumbu, Shorong and Langtang region, Nepal Himalayas. Bull. Glacier Res.,10, 11 –19

Ye Q, Yao T, Kang S, Chen F and Wang J (2006) Glacier variations in the Naimona'nyi region, western Himalaya, in the last three decades. Ann. Glaciol., 43(1), 385–389.

Yong N, Yili Z, Linshan L and Jiping Z (2010) Glacial change in the vicinity of Mt. Qomolangma (Everest), central high Himalayas since 1976. J. Geogr. Sci., 20(5), 667–686.

Chapter 13
Variability of Atmospheric Aerosols Over India

S.K. Satheesh, S. Suresh Babu, B. Padmakumari, G. Pandithurai and V.K. Soni

1 Introduction

The effect of aerosols on climate through various radiative interactions (involving the atmosphere, clouds, and the cryosphere) is well recognized by the international community (e.g., Andreae et al. 2005; Satheesh and Moorthy 2005; Bollasina et al. 2011; IPCC 2013). Atmospheric aerosols play a significant role in climate change due to their ability to scatter and absorb the incoming and outgoing radiation (direct effect). In addition to this, aerosols can also impact climate through modifying cloud properties, such as droplet size distribution and cloud lifetime, a process known as "indirect effect" (Twomey 1974; Kaufman et al. 2005; Rosenfeld et al. 2014). In addition, absorption of solar radiation by aerosols can lead to changes in cloud properties known as semi-direct effect (Lohmann et al. 2001). Of late, several reports focus on the impact of carbon-containing aerosols on monsoon, even though there have been contradictions and the processes are not understood well. Further, there have been reports that aerosols might impact crop yields, stimulate tropical cyclones, cause droughts and floods, and so on, most of these not validated yet (Chameides et al. 1999; Chung and Ramanathan 2006). Carbon-containing aerosols

S.K. Satheesh (✉)
Indian Institute of Science, Bengaluru, India
e-mail: satheesh@caos.iisc.ernet.in

S. Suresh Babu
Space Physics Laboratory, Indian Space Research Organization, Thiruvananthapuram, India

B. Padmakumari · G. Pandithurai
Indian Institute of Tropical Meteorology, Earth System Science Organization, Pune, India

V.K. Soni
India Meteorological Department, Earth System Science Organization, New Delhi, India

M.N. Rajeevan and S. Nayak (eds.), *Observed Climate Variability and Change Over the Indian Region*, Springer Geology, DOI 10.1007/978-981-10-2531-0_13

are composed of black carbon (BC) and organic carbon (OC); however, they have different climate forcing. Major sources of both BC and OC are biomass and fossil fuel burning. While OC is of light-scattering type, the most important light-absorbing aerosol species is BC. The OC through its cloud condensation nuclei (CCN) activity also contributes to indirect forcing while BC can contribute to evaporation of clouds known as "cloud burn off" (Ackerman et al. 2000). The fact that atmospheric aerosols exhibit high degree of spatial and temporal heterogeneity which makes their role in cloud microphysics very complex. Consequently, the aerosol–cloud interactions have been a subject matter of great interest to the climate science community, and efforts are being made by making use of a wide variety of experiments onboard ground-based, airborne, and space-borne platforms (Koren et al. 2004, 2008, 2014). Thus, it is imperative that relevant aerosol quantities are being measured from ground, aircraft, and space and used to carefully answer the fundamental questions related to the aerosol–climate problem.

Systematic studies on aerosols, clouds, and radiation budget were virtually nonexisting in India till the 1980s, except a few studies based on the atmospheric turbidity measurements using Volz Sun photometers by the India Meteorological Department (IMD) (Mani 1968; Mani and Huddar 1972; Mani and Chacko 1980; Rangarajan and Mani 1982). Even though aerosol research in India dates back to 1950s, systematic studies of the aerosols commenced only in the 1980s under the Indian Middle Atmosphere Programme (I-MAP) (Murthy 1988; Moorthy et al. 1988, 1999, 2009). During the I-MAP, a project to monitor the aerosol characteristics was initiated by utilizing multi-wavelength radiometers over a few selected locations (Murthy et al. 1988; Moorthy et al. 1988, 1989, 1991, 1993), following common instrumentation, data and analysis protocols, based on a network concept involving multi-institutional collaboration. This project became operational in the mid-eighties. Realizing its scientific potential and success, this project has been continued after IMAP under ISRO's Geosphere Biosphere Programme (ISRO-GBP) as Aerosol Climatology and Effects (ACE) having long-term objectives. With this modest beginning, networking at several geographically distinct locations, a few selected sites were equipped with a regional focus under ISRO-GBP since the 1990s (Moorthy et al. 1993, 1994, 1996; Moorthy and Satheesh 2011). Around the same time, vertical profiles of aerosols have been obtained using rocket-borne and balloon-borne payloads and Lidars (Parameswaran et al. 1984; Devara et al. 1994). These activities have enabled the increased recognition of the important role aerosols may have on climate forcing, and the ACE activity was converted to a national project namely the Aerosol Radiative Forcing over India (ARFI) project (Moorthy et al. 2009; Moorthy and Satheesh 2011) with the regional network of observatories known as ARFINET. This currently comprises of 41 aerosol observatories spread across India covering urban, remote, island, coastal, inland, semiarid, arid, and remote mountain regions over the mainland and the adjoining oceanic regions (Babu et al. 2013) (Fig. 1 for ARFINET locations). Around the same time, the global network of aerosols such as AErosol RObotic NETwork (AERONET) also established surface-based sites in India (Holben et al. 2001). In addition, recent years witnessed the expansion of several other regional networks also in India such as SKYNET (Kim et al. 2005).

In this chapter, long-term changes and variability of aerosol concentration and its spatial vertical distribution over the Indian region are discussed, based on the results from several experiments conducted over the Indian region. Aerosol observations carried out during various experiments/campaigns are discussed, and analysis was

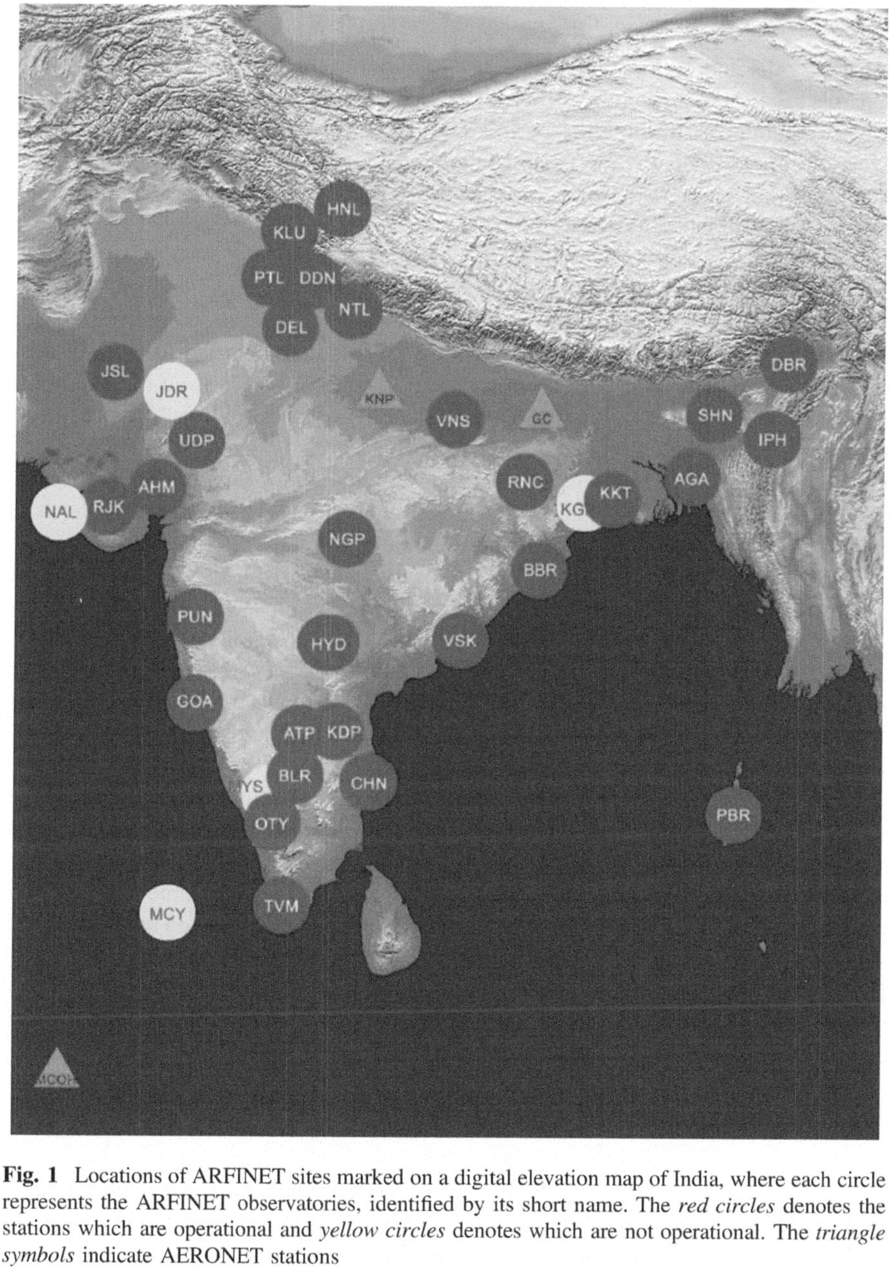

Fig. 1 Locations of ARFINET sites marked on a digital elevation map of India, where each circle represents the ARFINET observatories, identified by its short name. The *red circles* denotes the stations which are operational and *yellow circles* denotes which are not operational. The *triangle symbols* indicate AERONET stations

made on different types of aerosols with special emphasis to black carbon aerosols. Potential linkages between the changes in aerosol, radiation and clouds are also discussed.

2 Variability in Aerosol Properties

During 1940s and 1950s, climate science community was focused on hygroscopic aerosols because of their ability to act as CCN. Sea salt aerosol was the most important species of interest during this period (Woodcock 1953, 1957; Monahan 1968; Tsunogai et al. 1972; Lovett 1978; Monahan et al. 1982, 1983). Stuhlman (1932) suggested that the bursting of bubbles produced jets of water, which results in the production of small seawater droplets. Kohler (1936, 1941) proposed that strong winds can cause spray at the wave crest and are responsible for the airborne sea salt particles. Kientzler et al. (1954), using high-speed photographic study of bursting bubbles, concluded the mechanism of production of seawater droplets from bubble breaking. By early 1960s onwards, attention got gradually shifted to mineral dust aerosols due to their ability to get transported to long distances from their sources. A number of reports regarding the transport of aerosols from continents to ocean and vice versa are available in the literature (Eriksson 1959, 1960; Toba 1965a, b; Junge 1972; Delany et al. 1973; Prospero 1979; d'Almeida 1986; Bergametti et al. 1989; Arimoto et al. 1995; Gong et al. 2003; Zender et al. 2003). The presence of Saharan dust even over the remote areas of the Atlantic and Pacific Oceans is an example (Carlson and Prospero 1972; Junge 1972; Prospero and Carlson 1972; Prospero 1979; d'Almeida 1986; Bergametti et al. 1989; d'Almeida et al. 1991). Bergametti et al. (1989) have carried out chemical analysis of aerosol samples from the Atlantic Ocean, which revealed that source is Africa. By late 1970s/early 1980s, investigators have started realizing the radiative effects of mineral dust aerosols especially the fact that dust aerosols have important role in the warming of lower atmosphere due to short-wave absorption (Alpert et al. 1998; Miller and Tegen 1998). Several studies have indicated that, in general, under clear-sky conditions, dust approximately doubles the shortwave radiation absorption. Tegen and Fung (1994) have shown that dust causes a net cooling at the surface, simultaneous with an increase in atmospheric warming. By late 1980s, most likely due to the enhanced presence of anthropogenic aerosols as a consequence of industrial revolution, several studies have focused on anthropogenic sulfate aerosols and their climate impacts (Charlson et al. 1991, 1992). By early 1990s, there was an increase in the interest of black carbon aerosols and their effect on climate due to their strong absorbing characteristics (Cachier et al. 1989; Satheesh et al. 1999; Satheesh and Ramanathan 2000; Babu et al. 2002).

2.1 Trends in Aerosol Optical Depth

Of late, long-term changes (trends) in aerosols have gained increased interest due to their importance to regional and global climate. There have been several studies over the Indian region to address trends in aerosols using ground-based measurements using short-term, single-station data (Parameswaran et al. 1998; Moorthy et al. 1999; Satheesh et al. 2002; Kaskoutis et al. 2012) as well as long-term multi-station databases (Moorthy et al. 2013; Babu et al. 2013). Some investigators have also used satellite-derived products to study trends in aerosols (Ramachandran et al. 2012; Dey and Girolamo 2011). Aerosol optical depth (AOD) is a common data available on aerosols from several satellites regularly, and there are limitations arising from the large uncertainties involved in the satellite-retrieved AODs, especially over the landmass mainly due to the diverse surface reflectance and cloud contamination (Jethva et al. 2007). Studies by Zhang et al. (2005), Remer et al. (2005), Kahn et al. (2007), Levy et al. (2010), and several others have suggested that satellite data may have problems in long-term trend analysis due to the fact that noises and biases in the products can often be (mistakenly) interpreted as legitimate. Such studies are also beleaguered by the calibration drifts, which can often be (mistakenly) interpreted as trends in aerosols. Therefore, data from ground-based Sun photometers (preferably from a network), which can provide the most accurate AOD data, are the best choice for trend analysis. As on now, ARFINET is the only ground-based network with such long-term data.

The detailed regional synthesis of long-term data on AOD from the ARFINET has shown an increasing trend with a seasonal variability (Moorthy et al. 2013; Babu et al. 2013). The results are shown in Figs. 2, 3, and 4. Comparison of the turbidity coefficients in this study with those reported around late 1960s and early 1970s (Mani 1968) using Volz Sun photometer data indicates the extraordinary nature of the increase in aerosols during the last decades. The rate of increase is always high during the dry season (December–March) over the entire region, whereas the trends are rather inconsistent and weak during the pre-monsoon (April–May) and summer monsoon period (June–September) (Fig. 3). The variations in the spectral variation in AOD reveal the important contribution of anthropogenic sources on the increasing trend in AOD (Fig. 4, which reveals an increasing trend in the submicron aerosol abundance). The insignificant trend in AOD observed over the Indo-Gangetic Plain during the summer and pre-monsoon months is mainly ascribed to the competing effects of transport of dust and wash out of aerosols by the monsoon rainfall. Nonetheless, if this trend continues so, AOD at several locations would nearly double and approach unity in the next few decades. This can lead to an enhancement in aerosol-induced lower atmospheric warming by a factor of two. These observations indicate that trends in the aerosol forcing elements and their regional and global climate implications need to be better evaluated using global climate models.

Though not extensive, there are other studies on aerosol trends based on data from one or a few stations. Parameswaran et al. (1998) have reported an increasing trend in the AOD from 1989 to 1994. Satheesh et al. (2002) reported increase in AOD using two decades of ground-based Sun photometer measurements. Aerosols have an important role in determining solar radiation reaching the Earth's surface. Padmakumari et al. (2007) have made an assessment of monthly mean surface-reaching solar radiation under all-sky conditions for several stations, distributed across the Indian region, for the period 1981–2004 (see Chap. 9). They have shown that all the stations showed a decline in S ranging from −0.17 to −1.44 W m^2 per year. The consequent annually averaged solar dimming over India for the period 1981–2004 is estimated to be −0.

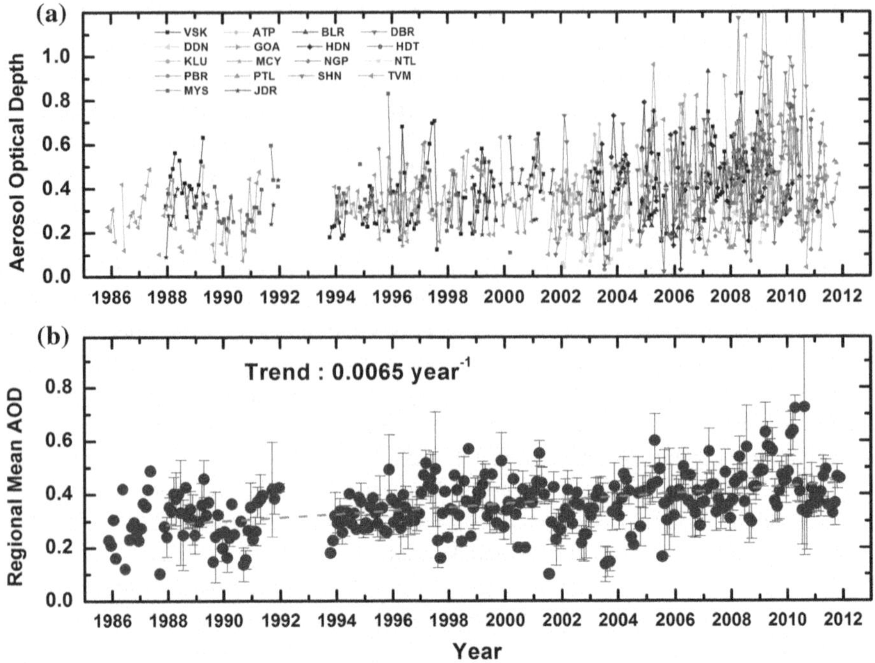

Fig. 2 **a** Long-term trends in aerosol optical depth at 500 nm, derived from measurements at ARFINET stations where *different colors* and *symbols* differentiate the stations. **b** Long-term trend in regional mean aerosol optical depth at 500 nm (color figure online)

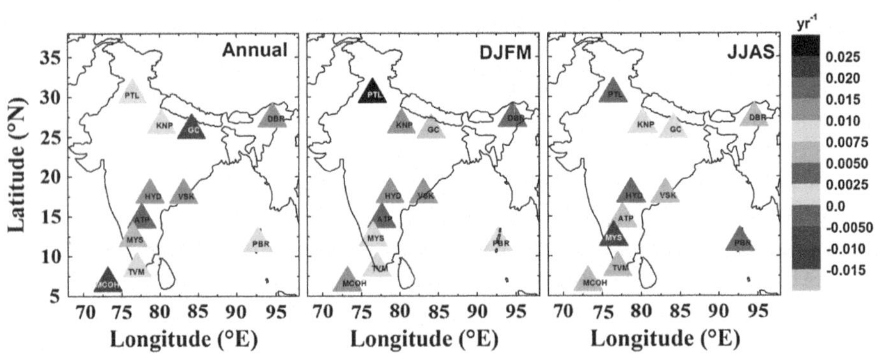

Fig. 3 Seasonal changes in the spatial variation in AOD trend over Indian region

86 W m^{-2}, while the seasonal mean values for winter, pre-monsoon, and monsoon seasons are pegged at −0.94, −1.04, and −0.74 W m^2 per year, respectively. Dey and Girolamo (2011) have used ten years (2000–2010) of observations from MISR to quantify seasonal linear trends of AOD and showed that many regions have statistically significant increasing trend in the range 0.1–0.4 in the last decade. Kaskoutis et al. (2012) discussed aerosol loading variability and trends at Kanpur using AERONET

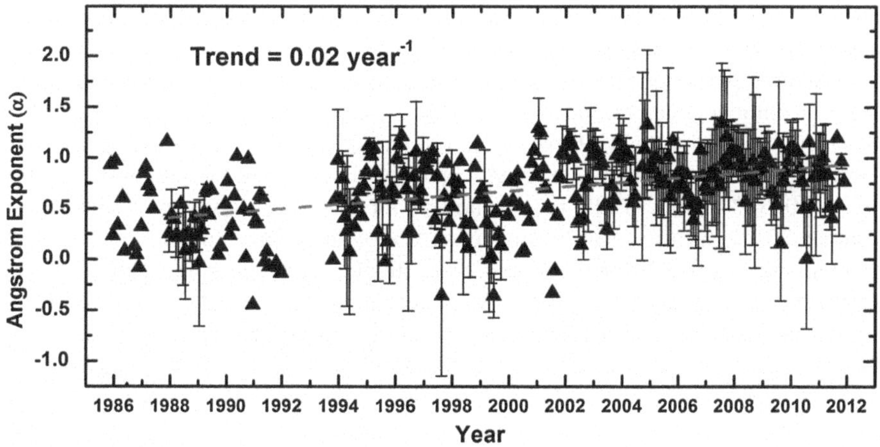

Fig. 4 Long-term trend in the regional mean values of Angstrom wavelength exponent, an indicator of anthropogenic impact of aerosol column loading

data. Their studies show an increase in AOD and attributed to an increase in the seasonal/monthly averaged AOD during the winter and post-monsoon seasons. Aerosol sources during these seasons are dominated by anthropogenic emissions. In contrast, a weak decreasing trend is observed during pre-monsoon and monsoon seasons. Soni et al. (2012) have studied the variability in annually averaged irradiance (global and diffuse) and bright sunshine duration over 12 stations of solar radiation network of IMD for the period 1971–2005 (see Chap. 9). They have reported that the annually averaged all-sky global irradiance decreased between 0.1 % (0.3 W m^{-2}) and 3.6 % (9. 0 W m^{-2}) per decade at these stations. Examining the seasonal and annual mean trends in AODs for the last decade using MODIS data over different locations in India, Ramachandran et al. 2012 have reported an increasing trend. Dani et al. (2012) have reported an increasing trend of 45 % per decade over Pune, based on Sun photometer-derived multispectral AOD measurements. Pathak et al. (2012) examined the variability in aerosol loading over the central region of the Indo-Gangetic Plains (IGP), during a decade (2001–10). An overall increase in AOD (ground-based radiometric measurements) was observed on yearly basis. Lu et al. (2013) presented the first inter-annual comparison of SO_2 emissions and those retrieved from satellite data [ozone monitoring instrument (OMI)] for Indian coal-fired power plants during 2005–2012. The results show that SO_2 emissions increased by 71 % during the study period.

Horizontal surface visibility is one of the simplest measures of local atmospheric pollution. Jaswal et al. (2013) have studied trends of morning poor visibility days (PVD, visibility <4 km) and afternoon good visibility days (GVD, visibility >10 km) based on 279 surface meteorological stations, well distributed over India, for the period 1961–2008. Their studies show that during the last 5 decades, all India averaged range of annual morning PVD has increased from 6.7 to 27.3 % days, while the range of afternoon GVD has decreased from 76.1 to 30.6 % days. Sreekanth (2013) has used the MODIS data to study trends in AOD over Bangalore.

They have shown an overall increasing trend, due to sustained increase in the seasonal averaged AOD during summer. Ramachandran and Kedia (2013) have used satellite-derived AOD to study inter-annual and regional variations in aerosols over six homogeneous rainfall zones in India for the month of July from 2000 to 2010 and reported that AOD over India in a drought year (2002) was higher when compared to normal monsoon years. Investigations using MODIS data for the period 2003–2012 over Delhi region show an increase in AODs by more than 25 % (Kumar 2014). Yearly mean Terra/Aqua AOD values have shown an increasing trend at a rate of 0.005/0.009 per year, respectively, with higher rates during winter (0.012/0.007 per year, respectively). An increasing trend in AOD is also reported by Soni et al. (2014) over Raniganj (7.31 %) in eastern and Korba (5.0 %) in south-east, and Godavari Valley (32 %) in the south coalfield region in India using MODIS data from 2000 to 2012 during winter and post-monsoon periods. Thus, overall, AOD over India shows an increasing trend, which is steady and stronger during dry season, and weak and inconsistent during the monsoon season.

2.2 *Black Carbon Aerosols: Spatial and Temporal Variability*

Major sources of black carbon (BC) are fossil fuels, biofuel, and biomass burning. Due to its ability to absorb, BC aerosols warm the atmosphere. In addition, BC, when deposited on snow and ice, can reduce the albedo. BC can contaminate other aerosols substantially, thereby altering the radiative properties of the entire aerosol system. Though there have been several reports on the climate impact of BC, several of them are contradicting to each other, for example impact of aerosols on monsoon. Recently, brown carbon, resulting from biomass burning, has attracted the attention of global climate change community because of its strong absorption characteristics in blue and UV region, with nearly no absorption in the mid-visible, unlike BC.

The ISRO-GBP annual review meeting in 1998 recognized the importance of BC aerosols on climate system, and it was decided to pursue studies of BC in subsequent years (Moorthy et al. 1999). Details of this research activity is also available in "IGBP In India 2000 A status report on projects," edited by Narasimha et al. (2000) and published on behalf of the Indian National Science academy (INSA). Later, the Indian Ocean Experiment (INDOEX), an Indo-US project, takes measurements of BC over the Indian Ocean. During the INDOEX field campaign, extensive measurements of BC were taken over the Indian Ocean, probably the first BC measurements over the Indian Ocean region. Based on these and other associated measurements, Satheesh et al. (1999) have developed an aerosol model for tropical Indian Ocean, which demonstrated that BC contributes 11 % to composite AOD. Later, using several calibrated satellite radiation measurements and five independent surface radiometers, Satheesh and Ramanathan (2000) quantified that even though BC contributes 11 % to optical depth, its contribution to radiative forcing can be as much as 60 %. Over the Indian mainland, continuous and long-term measurements of BC were initiated under GBP. Babu and Moorthy (2001) published the first report of the anthropogenic impact on aerosol black carbon mass concentration based on a case study from the coastal station, Thiruvananthapuram. Thereafter, several investigators reported BC measurements at various locations in India (Babu et al. 2002, 2004; Vinoj et al. 2010; Padithurai et al. 2004; Moorthy et al. 2004; Gangulay et al. 2005; Parashar et al. 2005; Dey et al. 2006; Satheesh et al. 2006, 2008, 2013; Pant et al. 2006; Dumka et al. 2006; Ramachandran et al. 2006; Safai et al. 2007; Sreekanth et al. 2007; Niranjan et al. 2007; Rengarajan et al. 2007; Beegum et al. 2009; Rastogi and Sarin 2009; Gogoi et al. 2014), and extensive station-specific as well regional synthesis of BC is available.

In a first of the kind attempt to generate a regional synthesis of aerosol properties, a mobile land campaign (LC-I) was conducted during February to March 2004 under the support of the ISRO-GBP, where simultaneous aerosol measurements were taken over spatially separated locations, using identical instruments, mounted on a mobile van, and operated following a common protocol. The LC-I has generated a wealth of information on black carbon as well as important aerosol parameters and covered an area of more than a million square kilometers (Moorthy et al. 2004, 2005; Ganguly et al. 2005). During this campaign, aircraft-based measurements were taken on the vertical profiles of BC for the first time in India, over Hyderabad. Using these data, Moorthy et al. (2004) have shown a rapid decrease in BC concentration within the atmospheric boundary layer (ABL) up to about 500 m and a nearly steady level above. As a continuation of this experiment, land campaign II (LC-II) was organized under ISRO-GBP during December 2004 by concurrent measurements from a number of observatories across the IGP. The analysis of the comprehensive database thus generated provided deeper insights on the microphysical and optical properties of aerosols over the IGP, the roles of mesoscale and synoptic meteorology in modulating them as well the significant role of the dynamics of the ABL in controlling the aerosol concentration near the surface as well as the role of long-range transport (Tare 2006; Ganguly et al. 2006; Niranjan et al. 2006, 2007; Nair et al. 2007; Rengarajan et al. 2007). Results from LC-II showed extremely high BC concentration, often exceeding similar to 20 $\mu g\ m^{-3}$, prevailing during December over the Indo-Gangetic basin. Simultaneous measurements of the local ABL height and wind fields revealed a very close association between the BC concentration and the ventilation coefficient (defined as the product of the boundary layer height and the transport wind) (Nair et al. 2007). Despite their significances, all the above field campaigns had several limitations. They did not provide adequately spatially resolved data over the oceans around India. Measurements were limited to only a few of the aerosol parameters and mostly to one season or even less, simultaneous measurements over land and oceans were not taken, and vertical profiles of aerosol characteristics over the land and ocean (especially on the continental outflows) were not made. The first and perhaps the only field experiment, integrating different observation platforms and variety of scientific instruments, has been the Integrated Campaign for Aerosols, gases and Radiation Budget (ICARB). This was a multi-platform field campaign with participation from numerous institutions, where integrated observation and measurements of aerosols, trace gases, and radiation were taken simultaneously. The details of these campaigns and the major findings have been reported in the literature

(Moorthy et al. 2008, 2009; Babu et al. 2008; Vinoj et al. 2008; Beegum et al. 2008; Nair et al. 2008; Satheesh et al. 2008, 2009, 2010). All these studies showed the persistence of high AOD and black carbon concentrations near the surface.

The only long-term data on BC aerosols are from ARFINET observatories, which is now a network spanning the entire country and surrounding oceans. Some of its stations have databases of more than a decade for BC, enabling quantifying of its long-term trends. Data from Thiruvananthapuram, a remote coastal location in the southern peninsula, on the concentration of BC, normally considered as a tracer for human impact, show a decreasing trend of ~ 250 ng m^{-3} per year. This is particularly perceptible after 2004. Decreasing trend in BC is not a unique feature of Thiruvananthapuram. Another site, Bangalore, showed decreasing trend of ~ 330 ng m^{-3} per year. This finding has several implications. The reduction in surface BC could be considered as an indicator of the impact of stringent emission control strategies, while the increase in AOD indicates an overall increase in columnar abundance. It remains to be seen whether this increase is occurring at higher levels, above the ABL. An extensive analysis of seasonal trends in BC concentration over India is available in Babu et al. (2013). Wang et al. (2014) estimated global BC emissions from 1960 to 2007 and reported that the BC emission intensity in power plants, the residential sector, and transportation shows decreasing trend for all the regions, especially China and India. Improvements in combustion technology and changes in fuel composition are attributed to the decline of BC emission intensities.

2.3 *Evolution of Size Distributions*

Aerosol size distribution is one of the key characteristics required for estimating the spectral optical and CCN properties of aerosols. It is also important in determining the scattering phase function, which in turn determines the up-scatter fraction and eventually the radiative forcing. In early 1960s, Angstrom wavelength exponent was used extensively to infer on aerosol size distributions (Ångström 1961), especially when they followed a simple inverse power law relation. Using spectral variation in AOD, following the inverse power law representation, Angstrom wavelength exponent can be derived. The value of Angstrom wavelength exponent depends on the ratio of the concentration of large to small aerosols. Early measurements of Angstrom wavelength exponent was mostly taken over the continental locations where spectral variation in AOD was best represented following Angstrom power law (Mani and Huddar 1972; Mani and Chacko 1980; Rangarajan and Mani 1982). It is known that aerosol spectral optical depths contain information pertaining to their size distribution (King et al. 1978). Making use of this property, it is possible to infer the columnar size distribution of aerosols by numerically inverting the spectral optical depth measurements, following linear inversion technique by King (1982), which involves the numerical inversion of the Mie integral equation. Adapting this technique, it has been observed that over the coastal regions influenced by both land and ocean environments, the aerosol size distributions were shown to depict changes associated with synoptic meteorology, showing the signatures of advected sea salt aerosols during marine air mass periods (Moorthy et al. 1991, 1993). By 1990s, several measurements were taken over oceanic regions where bimodal distributions are observed (Satheesh et al. 1998). Recent measurements show that a simple power law (even at two different wavelength ranges) or even bimodal distribution is not sufficient to get a best fit to spectral optical depths indicating multimodal size distributions indicating the influence of several sources (Kompalli et al. 2014).

Though near-surface aerosol size distribution measurements are being taken as part of various field campaigns, such information is still not adequate. There have been only a few attempts to develop climatology of aerosol size distribution (e.g., Gogoi et al. 2014; Kompalli et al. 2014; Kanawade et al. 2014). There have been improvements in the algorithms to derive column-averaged volume size distribution from Sun photometer measurements (Dubovik et al. 2006). Limited validation studies show agreement for the derived size distribution against in situ (surface as well as aircraft-based) measurements (Haywood et al. 2011; Smirnov et al. 2011; Gogoi et al. 2014; Kanawade et al. 2014), but these inversion products have not been comprehensively validated. There has also been some progress in measuring the aerosol size distribution down to a few nanometer size (e.g., Feldpausch et al. 2006) which enabled studies of new particle formation.

Khemani et al. (1982) have taken aerosol size distribution measurements based in Pune, and their report is among one of the earliest reports of size distribution over India. Measurements of size-distributed aerosol mass concentrations were taken at Bombay using a quartz crystal microbalance cascade impactor (Sharma and Patil 1992), and a power-function fit was applied to the size distributions. Extensive characterization of the seasonality of near-surface aerosol size distributions and their association with columnar AOD over the coastal regions is found in Pillai and Moorthy (2004). Dhanorkar and Kamra (1993) have reported that size distribution of aerosol particles is bimodal in shape based on measurements taken in Pune. Niranjan et al. (1997) have used spectral optical depths following the constrained linear inversion algorithms to derive aerosol size distributions and yielded a bimodal aerosol size distribution in Visakhapatnam. Size-differentiated concentrations of atmospheric aerosols measured in a suburban area of Agra city during 1992–1993 have revealed a bimodal distribution (Kulshrestha et al. 1998). Pandithurai et al. (1997) have approximated aerosol size spectrum measured in Pune with a composite power law distribution function.

New particle formation from precursor gases has been proposed as a source of nucleation aerosols under different environments, from polluted urban to pristine high-altitude Himalayas (Chate and Devara 2005; Moorthy et al. 2011; Kompalli et al. 2014). Several mechanisms for new particle formation have been proposed. Condensation of a binary mixture of sulfuric acid and water; secondary organic aerosol formation involving condensation of low- or nonvolatile organic compounds; and homogeneous nucleation of iodine oxides are examples. Laboratory studies have shown these mechanisms are significantly affected by temperature, humidity, and the surface area of preexisting particles. Many investigators measured high concentrations of ultra-fine particles, which indicate efficient new particle formation in the troposphere under low condensation sinks (Chate and Devara 2005; Kompalli et al. 2014). This is in contrast to the prevailing assumption that aerosol nucleation takes place only during the daytime and typically from sulfuric acid. Carefully examining the changes in the size distribution associated with new particle bursts, estimates of the growth rate of aerosol sizes due to coagulation and condensation processes were obtained under contrasting environments (Kompalli et al. 2014). Since not many such observations are available in the literature, more future studies are required to understand the nucleation mechanisms especially during nighttime. One of the major problems hampering our current understanding of the process of new particle formation is that these new particles are smaller than the lower size detection limit of the most instruments available as of today. They are only observed after some particle growth has occurred.

2.4 Vertical Distribution

When the amounts of absorbing aerosols such as BC are significant, AOD and chemical composition are not the only determinants of aerosol radiative effects, but the altitude of the aerosol layer relative to clouds (if present) is also important. Aerosols that are transported to higher altitudes are much more likely to travel long distances. Thus, as compared to AODs, aerosol vertical profile retrievals are very important as they provide more insights on the aerosol impacts on climate such as warming of the atmosphere and the impact on the thermal structure and stability of the atmosphere. Over India, aerosol vertical distribution was studied using rocket- and balloon-borne instruments (Jayaraman et al. 1987; Ramachandran and Jayaraman 2003), ground-based lidar (Parameswaran et al. 1984; Devara et al. 1995; Jayaraman et al. 1995; Satheesh et al. 2006, 2009; Niranjan et al. 2007; Raj et al. 2008), twilight photometry (Ashok et al. 1984; Padmakumari et al. 2003, 2005, 2006), airborne lidar (Gadhavi and Jayaraman 2006; Satheesh et al. 2008), and space-borne lidar (Padmakumari et al. 2012; Prabha et al. 2012; Gautam et al. 2011). There also exist a few studies using aircraft (Moorthy et al. 2004; Tripathi et al. 2005; Babu et al. 2008; Safai et al. 2012; Padmakumari et al. 2013a). The advent of CALIPSO (McGill et al. 2007; Huang et al. 2008) also improved our knowledge about vertical as well as spatial distribution of aerosols, which enabled to study the 3D structure of desert dust transport and biomass burning aerosols (Liu et al. 2008; Uno et al. 2008). ICARB aircraft campaign has shown the presence of elevated aerosol layers during the pre-monsoon season (Satheesh et al. 2008). A substantial fraction (as much as 50–70 %) of aerosols were found above (reflecting) clouds (Moorthy et al. 2009; Satheesh et al. 2008, 2009). Aerosol extinction coefficient at higher atmospheric layers (>2 km) was found much larger compared to that near the surface (by as much as a factor of 2–3). The aerosol-induced warming was mostly confined below 2 km over the southern Indian region, whereas it is found up to 4 km over the central Indian regions. The results showed a strong meridional gradient in warming ($\sim$4 K) at atmospheric levels above 2 km (Satheesh et al. 2008, 2009, 2010). The data from CALIPSO also paved way to extensive investigations of the assessment of the impact of aerosols above clouds (Chand et al. 2009).

Ground-based and satellites are proven to be suitable platforms for the measurements of atmospheric aerosol. Recently, aircraft is also proven to be an effective platform for aerosol measurements over a reasonably large spatial domain. Over India, measurements of aerosol vertical distribution during the monsoon season are very sparse to understand aerosol–cloud interactions. Aerosols are found to inhibit the cloud growth (Albrecht 1989) and also found, under suitable conditions, to invigorate deep convection (Koren et al. 2005). Most of the earlier airborne measurements were taken during the pre-monsoon season. CAIPEEX, a national experiment, conducted during the pre-monsoon and monsoon seasons with an instrumented research aircraft led a pathway to understand the vertical distribution of aerosols and their interaction with monsoon clouds (Kulkarni et al. 2012). CAIPEEX was conducted over different parts of India in 2009, while over central India in 2010 and 2011. Aerosol vertical variability and spatial distribution over different locations in continental India was studied using the airborne observations during CAIPEEX from May to September 2009 (Padmakumari et al. 2013a). The instruments used onboard for aerosol measurements were Passive Cavity Aerosol Spectrometer Probe (PCASP) and Aethalometer (see Kulkarni et al. 2012 for more details). CAIPEEX data during 2009 revealed the presence of elevated aerosol layers during the pre-monsoon as well as during the monsoon period (Padmakumari et al. 2013a, b). These elevated layers are mostly influenced by the depth of ABL, the origin of air mass trajectories, and also the presence of clouds (Padmakumari et al. 2013a). Figure 5 reveals that during the monsoon, aerosol number concentration showed strong vertical gradient and a transition is clearly observed between the boundary layer and the free troposphere as compared to that during the pre-monsoon (Fig. 5a) and just before the monsoon onset (Fig. 5b). It is also noted that during pre-monsoon, aerosols showed significant variation spatially at the elevated layers as compared to that in the boundary layer, while during monsoon high variability is observed only in the boundary layer. The surface-level aerosol number concentration and the height of boundary layer were found to influence the AODs significantly.

Vertical profiles of BC mass loading over Hyderabad and Bengaluru during monsoon 2009 showed decreasing trend from surface to an altitude of 7 km, but anomalously high BC loadings were encountered at different altitudes (Safai et al. 2012). CAIPEEX measurements over northeast India also revealed the formation of a second layer of BC in the upper atmosphere, which generates heating of $\sim$2 K/day (Rahul et al. 2014).

During the pre-monsoon, near the foot hills of west of the Himalayas, fine mode as well as coarse mode aerosols showed elevated layers. Characteristics of aerosols in the elevated layer and their interaction with clouds were also studied. During CAIPEEX, clouds profiled above the elevated aerosol layers showed increase in droplet number concentration (200–1400 cm^{-3}) and small effective radius (3.5–5 μm) at the cloud base indicating the observational evidence of aerosol–cloud interactions. Figure 6 shows the presence of ice phase above 6 km at temperatures

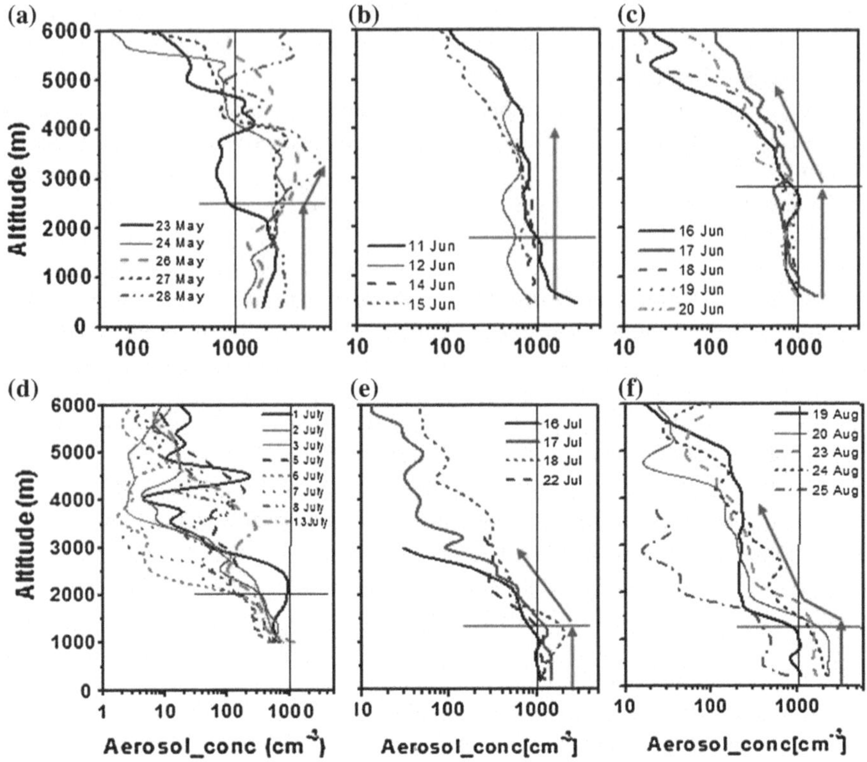

Fig. 5 Aerosol vertical distribution at different regions during different phases of monsoon. **a** Pathankot (pre-monsoon). **b** Hyderabad (just before monsoon onset at Hyderabad). **c** Hyderabad (after monsoon onset at Hyderabad). **d** Bengaluru (monsoon). **e** Bareilly (peak monsoon in IGP). **f** Bareilly (peak monsoon in IGP). The *horizontal lines* represent mean height of boundary layer obtained from radiosonde profiles. The *vertical* and *slant thick lines* represent the aerosol gradient and transition from boundary layer to free troposphere during different phases of monsoon

lower than −15 °C, representing the presence of dust aerosol acting as a potential ice-forming nucleus (Padmakumari et al. 2013b). The presence of dust aerosol in the elevated layers affects the ice microphysics. The vertical profiles of aerosols and cloud parameters obtained from aircraft were also compared with the simultaneous satellite measurements of CALIPSO and CloudSat (Padmakumari et al. 2012).

2.5 Aerosol Direct Radiative Forcing: Past and Present

There have been several efforts in India to estimate aerosol radiative forcing under clear-sky conditions (Satheesh and Ramanathan 2000; Babu et al. 2002; Pandithurai et al. 2004; Ganguly et al. 2005; Pant et al. 2006; Ramanchandran et al. 2006 and so on). Over the Arabian Sea and the Indian Ocean, using simultaneous measurements of aerosols and surface-reaching solar fluxes, Jayaraman et al. (1998) have estimated with the change in surface solar fluxes with AOD. Satheesh and Ramanathan (2000) quantified the aerosol forcing simultaneously at the Earth's surface and at the top of the atmosphere over the tropical northern Indian Ocean. They reported that aerosols are responsible for the decrease in shortwave radiation at the ocean surface by 12–30 W m^{-2} and increase in that top of the atmosphere by 4–10 W m^{-2}. They argued that this threefold difference and the large magnitude of the observed surface forcing are largely due to the absorption of solar radiation by black carbon.

Babu et al. (2002) made simultaneous measurements of aerosols, black carbon, size-segregated aerosol mass concentrations over a continental urban location. Their estimated aerosol forcing was −23 W m^{-2} (at the surface) and +5 W m^{-2} [at the top of the atmosphere (TOA)] even during relatively cleaner winter months. The consequent atmospheric heating was 0.8 K day^{-1} for cleaner period and 1.5 K day^{-1} for moderately turbid conditions. Pandithurai et al. (2004) estimated that during the dry season, aerosol forcing at the surface, the TOA, and the atmosphere over Pune (an urban site) were −33, 0, and 33 W m^{-2}, respectively.

Since detailed chemical composition was not available during 1980s, often a hybrid approach is being used to estimate aerosol radiative forcing from spectral optical depths and other available measurements (Satheesh et al. 1999, 2002; Babu et al. 2002). We follow a modified version of this approach by Satheesh and Srinivasan (2006). This method can be used to derive simple aerosol models that are "optically equivalent" and can simulate the observed aerosol optical properties and radiative fluxes, from spectral optical depth measurements. Thus, aerosol single-scattering albedo and hence aerosol radiative forcing can be estimated. The purpose of the proposed method is to estimate clear-sky aerosol radiative forcing and not to obtain the exact chemical composition of the aerosols. By incorporating this information in radiative transfer models, aerosol radiative forcing can be estimated. This method is not a substitute for the measurement of aerosol chemical composition, but a method to estimate aerosol radiative forcing over regions where aerosol chemical composition data are not available. The spectral AOD and Angstrom wavelength exponent data from ARFINET observations are used as input, and derived aerosol composition for 1985 and 2012 is shown in Fig. 7. Aerosol single-scattering albedo thus estimated for 1985 is ~0.97 and for 2012 is ~0.84.

The aerosol spectral optical depth, Angstrom wavelength exponent, and single-scattering albedo as obtained above are incorporated in a Discrete Ordinate Radiative Transfer code, which was designed and developed by the University of California, Santa Barbara (Ricchiazzi et al. 1998). A radiative forcing at the surface, TOA, and atmosphere for 1985 and 2012 is shown in Fig. 8. The TOA forcing changed sign from −4.6 to +1.4 W m^{-2} from 1985 to 2012 due to the increased presence of absorbing aerosols. The estimated surface and atmospheric forcing changed from −8.8 and +4.2 W m^{-2}, respectively, in 1985 to −35.7 and +37.1 W m^{-2} in 2012, respectively.

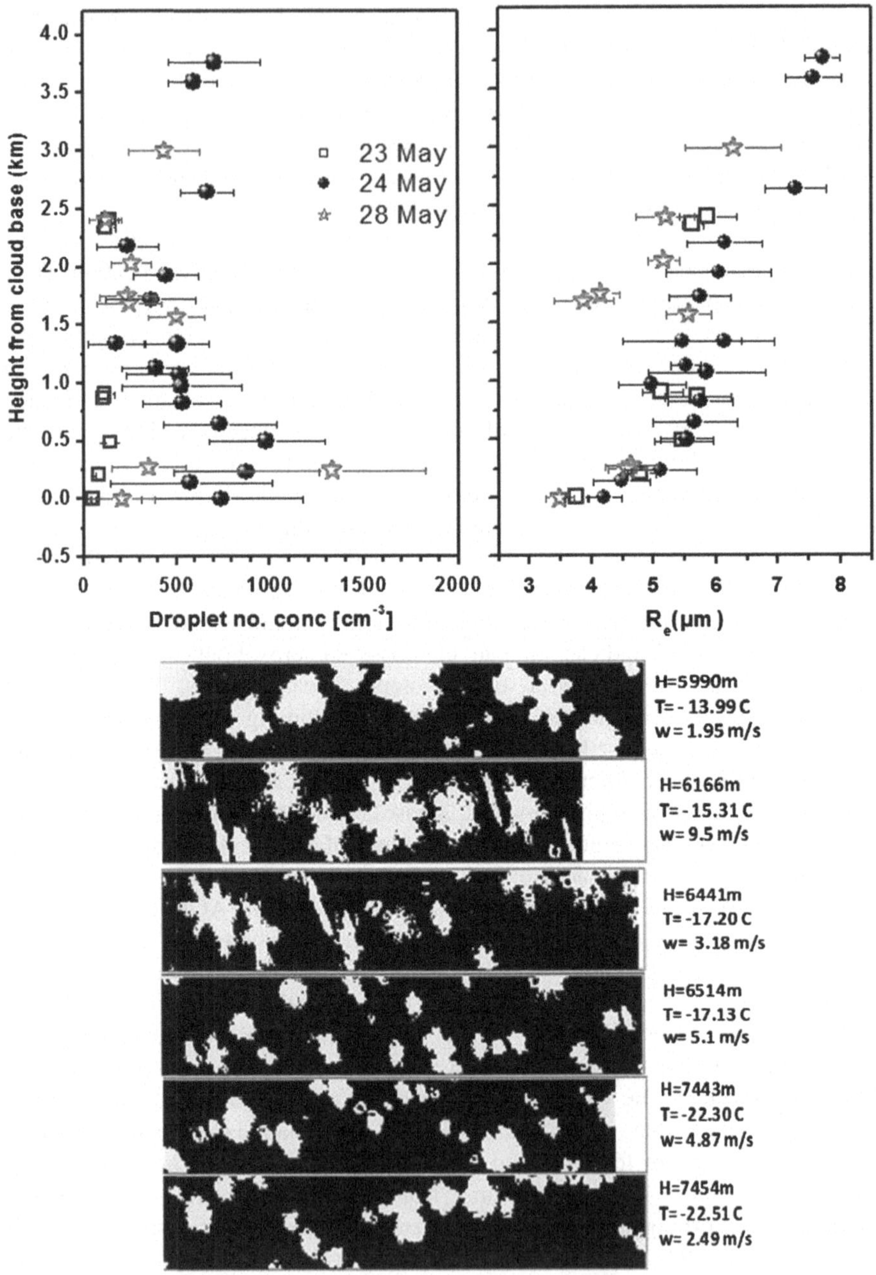

Fig. 6 *Upper panel* shows the droplet number concentration and effective radii of the clouds profiled above the elevated pollution layers. May 24 and 28 are more polluted as compared to May 23. *Lower panel* shows the ice phase of the cloud observed above 6 km on May 28 measured by Cloud Imaging Probe. These measurements are first of its kind near the foot hills of the Himalayas during the pre-monsoon season

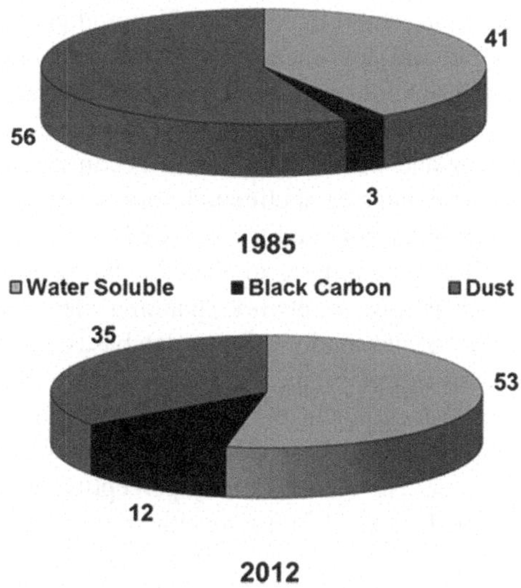

Fig. 7 Aerosol composition for 1985 and 2012 derived using spectral AOD and Angstrom wavelength exponent data from ARFINET observations as input

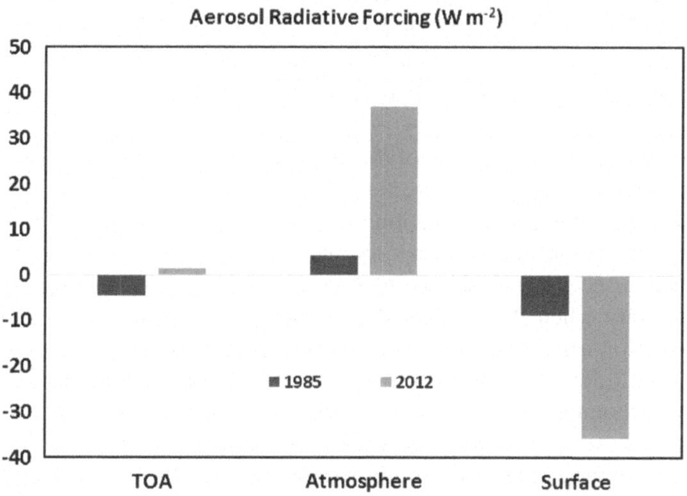

Fig. 8 Radiative forcing at the surface, TOA, and atmosphere for 1985 and 2012

3 Conclusions

The first regional synthesis of long-term primary data from the ARFINET has revealed a statistically significant increasing trend with a significant seasonal variability. Comparison of the data for 50 years reveals the phenomenal increase in aerosol loading. The rate of increase is consistently high during the dry months (December to March) over the entire region. However, the trends are inconsistent and weak during the pre-monsoon and summer monsoon period. The trends in the spectral variation of AOD reveal the significance of anthropogenic activities on the increasing trend in AOD. If this trend persists, AOD at several locations will nearly double and approach unity in the next few decades. This can lead to an amplification of aerosol-induced lower atmospheric warming by a factor of two. However, a regionally averaged scenario can be ascertained only in the coming years, when longer and denser data would become available. These observations indicate that regional and global climate implications of such trends in the forcing elements need to be better assessed using GCMs.

The only long-term data on BC aerosols are from ARFINET observatories, which is now a network spanning the entire country and the surrounding oceans. Some of its stations have databases of more than a decade for BC, enabling quantification of its long-term trends. Data from Thiruvananthapuram, a remote coastal location in the southern peninsula, on the concentration of BC, normally considered as a tracer for human impact, show a decreasing trend of ~ 250 ng m^{-3} per year. This is particularly perceptible after 2004. Decreasing trend in BC is not a unique feature of Thiruvananthapuram. Another site, Bengaluru, showed decreasing trend of ~ 330 ng m^{-3} per year. This finding has several implications. The reduction in surface BC could be considered as an indicator of the impact of stringent emission control strategies, while the increase in AOD indicates an overall increase in columnar abundance. It remains to be seen whether this increase is occurring at higher levels, above the ABL. Improvements in combustion technology and changes in fuel composition are attributed to the decline of BC emission intensities.

Aerosol size distribution is one of the key characteristics required for estimating the spectral optical and CCN properties of aerosols. It is also important in determining the scattering phase function, which in turn determines the up-scatter fraction and eventually the radiative forcing. Early 1960s, Angstrom wavelength exponent was used extensively to infer on aerosol size distributions (Ångström 1961), especially when they followed a simple inverse power law relation. New particle formation from precursor gases has been proposed as a source of nucleation aerosols under different environments, from polluted urban to pristine high-altitude Himalayas. Laboratory studies have shown these mechanisms can occur in the atmosphere and are significantly affected by temperature, humidity, and the surface area of preexisting particles. Carefully examining the changes in the size distribution associated with new particle bursts, estimates of the growth rate of aerosol sizes due to coagulation and condensation processes were obtained under contrasting environments. Since not many such observations are available in the literature, more future studies are required to understand the nucleation mechanisms especially nighttime. One of the major problems hampering our current understanding the process of new particle formation is that these new particles are smaller than the lower size detection limit of most instruments available as of today and are only observed after some particle growth has occurred.

Aircraft is also proven to be an effective platform for in situ measurements of atmospheric aerosol, apart from ground-based and satellite measurements, over a reasonably large spatial domain. In situ measurements are very much essential as aerosols are found to inhibit the cloud growth and also found to invigorate deep convection under suitable conditions. Most of the earlier airborne measurements were taken during the pre-monsoon. A national experiment called CAIPEEX conducted during pre-monsoon and monsoon seasons with an instrumented aircraft led a pathway to understand the aerosol vertical distribution and their interaction with monsoon clouds in India. CAIPEEX data revealed the presence of elevated aerosol layers during pre-monsoon as well as during monsoon and mostly influenced by the depth of ABL, origin of air mass trajectories and the presence of clouds. Such studies also revealed that during the monsoon period, aerosol number concentrations exhibit strong vertical gradient and a transition between the ABL and the free troposphere. The spatial distribution shows significant variation at the elevated layers as compared to that in the ABL during pre-monsoon, while high variability in the ABL during monsoon. The surface-level number concentration and the height of ABL were found to influence AOD significantly.

References

Ackerman, A. S et al., 2000, Reduction of tropical cloudiness by soot, Science, 288, 1042–1047.

Albrecht, B. A. 1989, Aerosols, cloud microphysics, and fractional cloudiness, Science, 245, 1227–1230, doi:10.1126/science.245.4923.1227.

Alpert, P et al., 1998, Quantification of dust -forced heating of the lower troposphere, Nature, 395, 367–370.

Andreae, MO, 2005, Jones, CD; Cox, PM, Strong present-day aerosol cooling implies a hot future, NATURE, 435, 1187-1190.

Ångström, A. (1961). Techniques of Determining the Turbidity of the Atmosphere, Tellus, 8:214-223.

Arimoto, R., et al., 1995, Trace elements in the atmosphere over the north Atlantic. Journal of Geophysical Research 100, 1199–1213.

Ashok, N. M., H. C. Bhatt, T. Chandrasekar, J. N. Desai, and D. B. Vaidya, 1984, Twilight optical studies of the El Chichon volcanic dust over Ahmedabad, India. J. Atmos. Terr. Phys., 46, 411-418.

Babu, S. S., and K. K. Moorthy, Anthropogenic impact on aerosol black carbon mass concentration at a tropical coastal station: A case study, Current Science, 81, 1208–1214, 2001.

Babu, S. S., K. K. Moorthy, and S. K. Satheesh (2004), Aerosol black carbon over Arabian Sea during intermonsoon and summer monsoon seasons, Geophys. Res. Lett., 31, L06104, doi:10.1029/2003GL018716.

Babu, S.S., Moorthy, K.K., Manchanda, R.K., Puna Ram Sinha, Satheesh, S.K., Dinkar Prasad Vajja, Srinivasan, S., Arun Kumar, V.H., 2011, Free tropospheric black carbon aerosol measurements using high altitude balloon: Do BC layers build "their own homes" up in the atmosphere?. Geophysical Research Letters 38, L08803, doi:10.1029/2011GL046654.

Babu, S.S., S.K. Satheesh and K. Krishna Moorthy, 2002, Enhanced aerosol radiative forcing due to aerosol black carbon at an urban site in India, Geophys. Res. Lett., 29 (18), 1880, doi:10.1029/2002GL015826.

Babu, S.S., S. K. Satheesh, Moorthy,K.K., C. B. S. Dutt, Vijayakumar S Nair, Denny P Alappattu and P K Kunhikrishnan, Aircraft Measurements of Aerosol Black carbon from a coastal location in the north-east part of peninsular India during ICARB, Journal of Earth System Science, 117, S1, 263-271, 2008.

Babu, S.S., M. R. Manoj, K. Krishna Moorthy, Mukunda M. Gogoi,Vijayakumar S. Nair, Sobhan Kumar Kompalli, S. K. Satheesh, K. Niranjan, K. Ramagopal, P. K. Bhuyan, and Darshan

Singh, 2013, Trends in aerosol optical depth over Indian region: Potential causes, and impact indicators, J. Geophys. Res , 118, 11,794–11,806, doi:10.1002/2013JD020507.

Barrie, LA; Yi, Y; Leaitch, WR; et al., 2001, A comparison of large-scale atmospheric sulphate aerosol models (COSAM): overview and highlights, TELLUS SERIES B-CHEMICAL AND PHYSICAL METEOROLOGY, 53, 615-645.

Beegum, S. N., Moorthy, K. K., Babu, S. S., Satheesh, S. K., Vinoj, V., Badarinath, K. V. S., Safai, P. D., Devara, P. C. S., Singh, S., Vinod, Dumka, U. C., and Pant, P.: Spatial distribution of aerosol black carbon over India during pre-monsoon season, Atmos. Environ., 43, 1071–1078, doi:10.1016/j.atmosenv.2008.11.042, 2009.

Beegum, S.N., Moorthy,K.K., V. S. Nair, S. S. Babu, S K Satheesh, V Vinoj, R. R. Reddy, K Rama Gopal, K V S Badarinath, K Niranjan, S. K. Pandey, M Behera, A Jeyaram, P K Bhuyan, M M Gogoi, Sacchidanand Singh, P Pant, U C Dumka, Yogesh Kant, J C Kuniyal, Darshan Singh. Characteristics of Spectral Aerosol Optical Depths over India during ICARB, Journal of Earth System Science, 117, S1, 303-313, 2008.

Bergametti, G., et al., 1989, African dust observed over Canary islands: source regions identification and transport pattern for some summer situations. Journal of Geophysical Research 94, 14,855–14,864.

Bollasina, M.A., Y. Ming, V. Ramaswamy, 2011, Anthropogenic Aerosols and the weakening of the South Asian Summer Monsoon, *Science*, 334, 502-505.

Cachier, H, M. P. Brémond and P. Buat-Ménard (1989), Carbonaceous aerosols from different tropical biomass burning sources, Nature, 340, 371–373, doi:10.1038/340371a0.

Carlson, T.N., Prospero, J.M., 1972, The large scale movement of Saharan air outbreaks over the equatorial North Atlantic. Journal of the Atmospheric Science 11, 283–297.

Chameides et. al., 1999, Case study of the effects of atmospheric aerosols and regional haze on agriculture: An opportunity to enhance crop yields in China through emission controls?, Proc. of National Academy of Sciences, 96, 13626-13633.

Chand, D., R. Wood, T.L. Anderson, S.K. Satheesh and R.J. Charlson, 2009, Satellite-derived direct radiative effect of aerosols dependent on cloud cover, Nature Geoscience, DOI: 10.1038/NGEO437, 181-184.

Charlson, R. J, J. Lagner, H. Rodhe, C. B. Leovy, and S. G. Warren (1991), Perturbation of the northern hemisphere radiative balance by backscattering from anthropogenic sulfate aerosols, Tellus, Ser. AB, 43, 152–163

Charlson, R. J, S. E. Schwartz, J. M. Hales, R. D. Cess, J. A. Coakley, J. E. Hansen, and D. J. Hoffman (1992), Climate forcing by anthropogenic aerosols, Science, 255, 423–430.

Chate, D. M., and P. C. S. Devara (2005), Parametric study of scavenging of atmospheric aerosols of various chemical species during thunderstorm and nonthunderstorm rain events, J. Geophys. Res., 110, D23208, doi:10.1029/2005JD006406.

Chung, CE; Ramanathan, V, 2006, Weakening of North Indian SST gradients and the monsoon rainfall in India and the Sahel, JOURNAL OF CLIMATE, 19, 2036-2045, 2006.

Dani, KK, Raj, PE, Devara, PCS, Pandithurai, G, Sonbawne, SM, Maheskumar, RS, Saha, SK Jaya Rao, Y. 2012. Long-term trends and variability in measured multi-spectral aerosol optical depth over a tropical urban station in India. International Journal of Climatology 32: 153–160, DOI:10.1002/joc.2250.

Delany, A.C., Pollock, W.H., Shedlovsky, J.P., 1973, The tropospheric aerosol: the relative contribution of marine and continental components 78, 6249–6265.

Devara, P.C.S, Raj, P.E and S. Sharma (1994), Remote sensing of atmospheric aerosol in the nocturnal boundary layer using lidar, Environ. Pollut., 85, 97–102.

Devara, P.C.S., Raj, P.E., Pandiduari, G 1995, Aerosol-profile measurements in the lower troposphere with four-wavelength bistatic argon-ion lidar. Applied Optics, 34, pp. 51–65.

Dey, S. and L. Girolamo, 2011, A decade of change in aerosol properties over the Indian subcontinent, *Geophys. Res. Lett.*, 38, L14811, doi:10.1029/2011GL048153.

Dhanorkar, S. and Kamra, A.K. Diurnal and seasonal variations of the small-, intermediate-, and large-ion concentrations and their contributions to polar conductivity, Journal of Geophysical Research, 98: doi: 10.1029/93JD00464, 1993.

Dipu, S., T.V. Prabha, G. Pandithurai, J. Dudhia, G. Pfister, K. Rajesh and B.N. Goswami, 2013, Impact of elevated aerosol layer on the cloud microphysical properties prior to monsoon onset, Atmos. Environ, 70, 454-467.

d'Almeida, G.A., 1986, A model for Saharan dust transport. Journal of Climate and Applied Meteorology 25, 903–916.

d'Almeida, G.A., Koepke, P., Shettle, E.P., 1991, Atmospheric Aerosols-Global Climatology and Radiative Characteristics, In: Deepak, A., (Ed.) Hampton, VA.

Dubovik, O., A. Sinyuk, T. Lapyonok, B.N. Holben, M. Mishchenko, P. Yang, T.F. Eck, H. Volten, O. Muñoz, B. Veihelmann, W.J. van der Zande, J.-F. Leon, M. Sorokin, and I. Slutsker, 2006: Application of spheroid models to account for aerosol particle nonsphericity in remote sensing of desert dust. J. Geophys. Res., 111, D11208, doi:10.1029/2005JD006619.

Dumka, U.C. et al., 2006, Surface changes in solar irradiance due to aerosols over central Himalayas, J. Geophys. Res., 33 (20): Art. No. L20809, doi:10.1029/2006GL027814,

Eriksson, E., 1959, The yearly circulation of chloride and sulphur in nature. Meteorological, geological and pedological implications, Part-I. Tellus 11, 375–403.

Eriksson, E., 1960, The yearly circulation of chloride and sulphur in nature Meteorological, geological and pedological implications, Part-II. Tellus 12, 63–109.

Feingold, G., W. R. Cotton, S. M. Kreidenweis, and J. T. Davis, 1999, The impact of giant cloud condensation nuclei on drizzle formation in stratocumulus: Implications for cloud radiative properties. J. Atmos. Sci., 56: 4100-4117.

Feldpausch, P., Fiebig, M., Fritzsche, L., and Petzold, A.: A Measurement of ultrafine size distributions by a combination of diffusion screen separators and condensation particle counters, J. Aerosol. Sci., 37, 577–597, 2006.

Ganguly D, Jayaraman A, Rajesh TA, et al., 2006,Wintertime aerosol properties during foggy and nonfoggy days over urban center Delhi and their implications for shortwave radiative forcing, J. Geophys. Res.,111, Article Number: D15217.

Ganguly D., et al., 2005, Single scattering albedo of aerosols over the central India: Implications for the regional aerosol radiative forcing, Geophys. Res. Lett., 32, L18803.

Gadhavi, H., and Jayaraman, A., 2006, Airborne lidar study of the vertical distribution of aerosols over Hyderabad, an urban site in central India, and its implication for radiative forcing calculations. Ann. Geophys., 24, 2461–2470.

Gautam, R., et al., 2011. Accumulation of aerosol over the Indo-Gangetic plains and southern slopes of the Himalayas: Distribution, properties and radiative effects during the 2009 premonsoon season. Atmospheric Chemistry and Physics 11, 12,841–12,863, doi:10.5194/acp-11-12841-2011.

Gogoi, M.M., K.K. Moorthy, Kompalli, Sobhan Kumar, J.P. Chaubey, S.S. Babu, M.R. Manoj, V. S. Nair, T.P. Prabhu, "Physical and optical properties of aerosols in a free tropospheric environment: Results from long-term observations over western trans-Himalayas", Atmospheric Environment, 84, 262-274, 2014. (http://dx.doi.org/10.1016/j.atmosenv.2013.11.029)

Gong, S.L., Zhang, X.Y., Zhao, T.L., McKendry, I.G., Jaffe D.A., Lu, N.M., 2003, Characterization of soil dust aerosol in China and its transport and distribution during 2001 ACE-Asia. 2: Model simulation and validation. Journal of Geophysical Research 108 (D9).

Haywood, J. M., Johnson, B.J., Osborne, S.R., Mulcahy, J., Brooks, M.E., Harrison, M., Milton, S.F., Brindley, H., 2011. Observations and modeling of the solar and terrestrial radiative effects of Saharan dust: a radiative closure case-study over oceans during the GERBILS campaign, Q. J. R. Meteorol. Soc., 137: 1211–1226, DOI: 10.1002/qj.770.

Holben, B., D. Tanré, A. Smirnov, et al. (2001), An emerging groundbased aerosol climatology: aerosol optical depth from AERONET, J. Geophys. Res., 106, 12067–12098.

Huang et al., 2008, Long-range transport and vertical structure of Asian dust from CALIPSO and surface measurements during PACDEX. J. Geophys. Res., 113, D23, doi:10.1029/2008JD010620.

Intergovernmental Panel on Climate Change (2007), Climate Change 2007, The Physical Science Basis. Contribution of Working Group I to the Fourth Assessment Report of the

Intergovernmental Panel on Climate Change, edited by S. Solomon et al., Cambridge Univ. Press, Cambridge, U. K.

Intergovernmental Panel on Climate Change (2013), Climate Change 2013: The Physical Science Basis. Contribution of Working Group I to the Fifth Assessment Report of the Intergovernmental Panel on Climate Change .In: Stocker, T.F., D. Qin, G.-K. Plattner, M. Tignor, S.K. Allen, J. Boschung, A. Nauels, Y. Xia, V. Bex and P.M. Midgley (eds.). Cambridge University Press, Cambridge, United Kingdom and New York, NY, USA, 1535 pp, doi:10.1017/CBO9781107415324.

Jaswal, A.K., Kumar, N., Prasad, A.K. et al., Decline in horizontal surface visibility over India (1961–2008) and its association with meteorological variables, Nat Hazards, 68:929. doi:10. 1007/s11069-013-0666-2, 2013.

Jayaraman, A., Lubin, D., Ramachandran, S., Ramanathan, V., Woodbridge, E., Collins, W.D. and Zalpuri, K.S. (1998). Direct observations of aerosol radiative forcing over the tropical Indian Ocean during the January-February 1996 pre-INDOEX cruise. Journal of Geophysical Research 103: doi: 10.1029/98JD00559

Jayaraman, A., Ramachandran, S., Acharya, Y.B., Subbaraya, B.H., 1995, Pinatubo volcanic aerosol layer decay observed at Ahmedabad (23 N) India using Nd:YAG backscatter lidar. Journal of Geophysical Research Letters, **100**, pp. 23209–23214.

Jayaraman, A., Subbaraya, B. H., and Acharya Y. B., 1987, The vertical distribution of aerosol concentration and their size distribution function over the tropics and their role in radiation transfer. Physica Scripta, 36, 358–361.

Jethva, H., S. K. Satheesh, and J. Srinivasan (2007), Assessment of second-generation MODIS aerosol retrieval (Collection 005) at Kanpur, India, Geophys. Res. Lett., 34, L19802, doi:10. 1029/2007GL029647

Junge, C.E., 1972, Our knowledge of the Physico-Chemistry of aerosols in the undisturbed marine environment. Journal of Geophysical Research 77, 5183–5200.

Kahn, R. A., M. J. Garay, D. L. Nelson, K. K. Yau, M. A. Bull, B. J. Gaitley, J. V. Martonchik, and R. C. Levy (2007), Satellite-derived aerosol optical depth over dark water from MISR and MODIS: Comparisons with AERONET and implications for climatological studies, J. Geophys. Res., 112, D18205, doi:10.1029/2006JD008175.

Kanawade VP, Tripathi SN, Siingh D, Gautam AS, Srivastava AK, Kamra AK, Soni VK, Sethi V. 2014. Observations of new particle formation at two distinct Indian subcontinental urban locations, Atmos. Environ. 96 :370 – 379, doi: 10.1016/j.atmosenv.2014.08.001

Kaskoutis D. et al. 2011, Variability and trends of aerosol properties over Kanpur, northern India using AERONET data (2001–10), *Environ. Res. Lett.*, 7 (2012) 024003.

Kaufman, YJ; Remer, LA; Tanre, D; et al., A critical examination of the residual cloud contamination and diurnal sampling effects on MODIS estimates of aerosol over ocean. IEEE Transactions on Geoscience and Remote Sensing, 43, 2886–2897, 2005.

Khemani, L. T., Momin, G. A., Naik, M. S., Vijayakumar, R. and Ramana Murty, Bh. V., Chemical composition and size distribution of atmospheric aerosols over the Deccan Plateau, India. Tellus, v. 34, pp. 151–158, 1982.

King et al., 1978, M.D. King, D.M. Byrne, B.M. Herman, J.A. Reagan, Aerosol size distribution obtained by inversion of spectral optical depth measurements, J. Atmos. Sci., 35 (1978), pp. 2153–2167

King, M.D. (1982). Sensitivity of Constrained Linear Inversion to the Selection of Lagrange Multiplier., J. Atmos. Sci., 39: 1356–1369

Kohler, H., 1936, The nucleus in and the growth of hygroscope droplets. Transactions of the Faraday Society 32, 1152–1161.

Kohler, H., 1941, An experimental investigation on sea water nuclei. Nova Acta Regional Society, Upsaliensis 4, 1–55.

Kientzler, C.F., et al., 1954, Photographic investigation of the projection of droplets by bubbles bursting at a water surface. Tellus 6, 1–7 IEEE TRANSACTIONS ON GEOSCIENCE AND REMOTE SENSING, 43, 2886-2897, 2005.

Kim, D. H., B. J. Sohn, T. Nakajima, and T. Takamura, 2005, Aerosol radiative forcing over East Asia determined from ground-based solar radiation measurements, J. Geophys. Res., 110, D10S22, doi:10.1029/2004JD004678.

Kompalli, Sobhan Kumar, S. Suresh Babu, K. Krishna Moorthy, Mukunda M. Gogoi, Vijayakumar S. Nair and Jai Prakash Chaubey, "The formation and growth of ultrafine particles in two contrasting environments: A case study", Annales Geophysicae, 32, 817–830, doi:10.5194/angeo-32-817-2014, 2014.

Konwar, M., R. S. Maheskumar, J. R. Kulkarni, E. Freud, B. N. Goswami, and D. Rosenfeld, 2012, Aerosol control on depth of warm rain in convective clouds, J. Geophys. Res., 117, D13204, doi:10.1029/2012JD017585.

Koren, I; Kaufman, YJ; Remer, LA; et al., 2004, Measurement of the effect of Amazon smoke on inhibition of cloud formation, SCIENCE, 303, 1342-1345,

Koren, I., Kaufman, Y. J., Rosenfeld, D.L., Remer, A., and Rudich, Y., 2005, Aerosol invigoration and restructuring of Atlantic convective clouds, Geophys. Res. Lett., 32, L14828, doi:10.1029/2005GL023187.

Koren, Ilan; Martins, J. Vanderlei; Remer, Lorraine A.; et al., 2008,Smoke invigoration versus inhibition of clouds over the Amazon, SCIENCE, 321, 946-949.

Koren, Ilan; Dagan, Guy; Altaratz, Orit, 2014, From aerosol-limited to invigoration of warm convective clouds, SCIENCE, 344, 1143-1146.

Kulkarni, J.R., Maheskumar, R.S., Morwal, S.B., Padmakumari, B., Konwar, M., Deshpande, C. G., Joshi, R.R., Bhalwankar, R.V., Pandithurai, G., Safai, P.D., Narkhedkar, S.G., Dani, K.K., Nath, A., Sathy Nair, Sapre, V.V., Puranik, P.V., Kandalgaonkar, S.S., Mujumdar, V.R., Khaladkar, R.M., Vijayakumar, R., Prabha T.V. , Goswami, B.N., 2012, The Cloud Aerosol Interactions and Precipitation Enhancement Experiment (CAIPEEX): Overview and Preliminary Results. Current Science 102, 413-425.

Kulshrestha, U.C., Saxena, A., Kumar, N., Kumari, K.M. and Srivastava, S.S., Chemical Composition Association of Size Differentiated Aerosols at a Suburban Site in a Semi Arid Tract of India, J. Atmo. Chem.29: 109-118, 1998.

Lal, D.M., S.D. Patil, H.N. Singh, S.D. Ghude, S. Tiwari and M.K. Srivastva, 2013, Influence of aerosol on clouds over Indo-Gangetic Plain, India, Clim. Dyn, 41, 601-612.

Levy, R. C., Remer, L. A., Kleidman, R. G., Mattoo, S., Ichoku, C., Kahn, R., and Eck, T. F.: Global evaluation of the Collection 5 MODIS dark-target aerosol products over land, Atmos. Chem. Phys., 10, 10399-10420, doi:10.5194/acp-10-10399-2010, 2010.

Liu, D., Wang, Z., Liu, Z., Winker, D., and Trepte, C., 2008, A height resolved global view of dust aerosols from the first year CALIPSO lidar measurements, J. Geophys. Res., 113, D16214, doi:10.1029/2007JD009776.

Lohmann, U. and Feichter, J., 2001, Can the direct and semi-direct aerosol effect compete with the indirect effect on a global scale?, Geophys. Res. Lett., 28, doi:10.1029/2000GL012051

Lovett, R.F., 1978, Quantitative measurement of airborne sea- salt in the North Atlantic. Tellus 30, 358–364.

Lu, Z.; Streets, D.G.; de Foy, B.; Krotkov, N.A. Ozone Monitoring Instrument observations of interannual increases in SO2 emissions from Indian coal-fired power plants during 2005–2012; Environ. Sci. Technol., 47, 13993–14000, 2013.

Miller, R.L., and I. Tegen, 1998: Climate response to soil dust aerosols. J. Climate, 11, 3247–3267, doi:10.1175/1520-0442.

Mani, A., 1968, Aspects of maintenance and calibration, Bull. Amer. Met. Society, 49, 1115.

Mani, A., Huddar, B.B., 1972, Studies of surface aerosols and their effects on atmospheric electric parameters, Pune and Applied Geophysics, 100, 154–166.

Mani, A., and Chacko.O., 1980, Attenuation of Solar Radiation in the Atmosphere, Solar Energy, 24, 347–349, 1980.

McGill, M. J., et al., 2007, Airborne validation of spatial properties measured by the CALIPSO lidar, J. Geophys. Res., 112, D20201, doi:10.1029/2007JD008768.

Monahan, E.C., 1968, Sea spray as a function of low elevation wind speed. Journal of Geophysical Research 73, 1127–1137.

Monahan, E.C., Davidson, K.L., Spiel, D.E., 1982, White cap aerosol productivity deduced from simulation tank measurements. Journal of Geophysical Research 87, 8898–8904.

Monahan, E.C., Fairall, C.W., Davidson, K.L., Jones Boyle, P., 1983, Observed inter-relations between 10 m winds, Ocean white caps and marine aerosols. Quarterly Journal of the Royal Meteorological Society 109, 379–392.

Moorthy, K. K., Babu, S. S., Sunilkumar, S. V., Gupta P. K., and Gera, B. S., 2004, Altitude profiles of aerosol BC, derived from aircraft measurements over an inland urban location in India, Geophys. Res. Lett., 31, L22103.

Moorthy, K. K, P.R. Nair, B.V.K. Murthy, S.K. Satheesh, S.K., 1996, Time Evolution of the Optical Effects and Aerosol Characteristics of Mt. Pinatubo Origin from Ground Based Observation, *J. Atmos. Terr. Phys., 58(10)*, 1101–1116.

Moorthy, K.K., S. K. Satheesh and B.V. Krishna Murthy, 1997, Investigations of Marine Aerosols over the Tropical Indian Ocean, *J. Geophys. Res., 102*, 18,827-18,842, doi: 10.1029/97JD01121.

Moorthy K.K. et al. (1999), Aerosol Climatology over India. 1 - ISRO GBP MWR network and database, ISRO/GBP, SR-03- 99.

Moorthy, K.K., S.K. Satheesh, S. Suresh Babu, C.B.S. Dutt, Integrated Campaign for Aerosols, gases and Radiation Budget (ICARB): An Overview, Journal of Earth System Science, 117, S1, 243-262, 2008.

Moorthy, K. K. S.S. Babu, S.K. Satheesh, et al., 2009, Climate implications of atmospheric aerosols and trace gases: Indian Scenario, *"Climate Sense"*, pp157-160, Published by WMO (ISBN: 978-92-63-11043-5), Tudor Rose, UK.

Moorthy, K.K., S.K. Satheesh, 2011, Black carbon aerosols over India, *UNEP's Black Carbon e-Bulletin, 3*, 1-3.

Moorthy, KK; Nair, PB; Murthy, BVK, 1988, A study on Aerosol optical depth at a coastal station, Trivandrum, Indian Journal of Radio & Space Physics, 17, 16-22

Moorthy, KK; Nair, PR; Murthy, BVK, 1989, Multiwavelength solar radiometer network and features of Aerosol spectral optical depth at Trivandrum, Indian Journal of Radio & Space Physics, 18, 5-6, 194-201.

Moorthy, KK; Nair, PR; Murthy, BVK, 1991, Size distribution of coastal aerosols - Effects of local-sources and sinks, Journal of Applied Meteorology, 30, 844-852.

Moorthy, KK; Nair, PR; Prasad, BSN; et al., 1993, Results from the MWR network of IMAP. Indian Journal of Radio & Space Physics, 22, 243-258, 1993.

Moorthy, K.,K., S. S.Babu, M. R. Manoj, and S. K. Satheesh (2013), Buildup of aerosols over the Indian Region, Geophys. Res. Lett., 40, 1011–1014, doi:10.1002/grl.50165.

Moorthy, K.K., S.V. Sunilkumar, P S Pillai, K. Parameswaran, P R Nair, Y N Ahmed, K. Ramgopal, K. Narasimhulu, R R Reddy, V. Vinoj, S.K. Satheesh, K. Niranjan, B. M Rao, P.S. Brahmanandam, A Saha, K.V.S. Badarinath, T.R. Kiranchand and K. M Latha, Wintertime Spatial Characteristics of Boundary Layer Aerosols over Peninsular India., J. Geophys. Res., 110, D08207, doi: 10.1029/2004JD005520, 1-11, 2005.

Moorthy,K.K., Vijayakumar S. Nair, S. Suresh Babu and S. K. Satheesh, Spatial and vertical heterogeneities in aerosol properties over oceanic regions around India: Implications for radiative forcing, Quarterly Journal of Royal Meteorological Society, 135, 2131–2145, DOI: 10.1002/qj.525, 2009.

Moorthy, K.K., V. Sreekanth, J.P. Chaubey, M.M. Gogoi, S.S. Babu, S.K. Kompalli , S.P. Bagare, B.C. Bhatt, V. Gaur, T.P. Prabhu, and S.N. Singh, "Fine and ultra fine particles at near free-tropospheric environment over the high altitude station Hanle, in Trans- Himalayas: New particle formation and size distribution", Journal of Geophysical Research, 116, D20216, 1-12, 2011. (http://dx.doi.org/doi:10.1029/2011JD016343)

Murthy, Krishna, B.V., 1988, Radiation budget of the middle atmosphere. Indian Journal of Radio Space Physics 17, 203–219.

Nair VS, Moorthy KK, Alappattu DP, et al., 2007, Winter-time aerosol characteristics over the Indo-Gangetic Plain (IGP): Impacts of local boundary layer processes and long-range transport, Journal of Geophysical Research, 112, D13205.

Nair, V.S., K.K. Moorthy, S. S. Babu, K Narasimhulu, Siva Sankara Reddy, R Ramakrishna Reddy, K Rama Gopal, V Sreekant, B L Madhavan, and K Niranjan, Size segregated aerosol mass concentration measurements over the Arabian Sea during ICARB, Journal of Earth System Science, 117, S1, 2315-323, 2008.

Narasimha, R. et al., 2000, IGBP in India 2000: A status report on project, INSA report, pp. 1–496

Niranjan, K., Ramesh Babu, Y., Satyanarayana, G. V., and Thulasiraman, S., Aerosol spectral optical dephs and typical size distributions at a coastal urban location in India, Tellus, 49B, 439–446, 1997.

Niranjan,K., V. Sreekanth, B. L. Madhavan, and Moorthy, K.K., Aerosol physical properties and Radiative forcing at the outflow region from the Indo-Gangetic plains during typical clear and hazy periods of wintertime, Geophys. Res. Lett, 34, L19805, doi:10.1029/2007GL031224, 2007.

Niranjan,K., V. Sreekanth, B. L. Madhavan, and Moorthy,K.K., Wintertime aerosol characteristics at a north Indian site Kharagpur in the Indo-Gangetic plains located at the outflow region into Bay of Bengal, J. Geophys. Res., 111, D24209, doi:10.1029/2006JD007635, 2006

Pandithurai, G., T. Takamura, J. Yamaguchi, K. Miyagi, T. Takano, Y. Ishizaka, and A. Shimizu, 2009, Aerosol effect on cloud droplet size as monitored from surface remote sensing over East China Sea region, Geophys. Res. Lett., 36, L13805, doi:10.1029/2009GL038451.

Pandithurai, G., S. Dipu, T. V. Prabha, R. S. Maheskumar, J. R. Kulkarni, and B. N. Goswami, 2012, Aerosol effect on droplet spectral dispersion in warm continental cumuli, J. Geophys. Res., 117, D16202, doi:10.1029/2011JD016532.

Parameswaran, K., K.O. Rose, and B.V. Krishna Murthy, Aerosol characteristics from bi-static Lidar observation, J. Geophys. Res., 89, No. D2, 2541-2552, 1984.

Padmakumari, B., J. M. Trigo-Rodriguez, A. L. Londhe, H. K. Trimbake, and D. B. Jadhav, 2005, Optical observations of meteoric dust in the middle atmosphere during Leonid activity in recent years 2001-2003 over India. Geophys. Res. Lett., **32**, L16807 doi: 10.1029/2005GL023434.

Padmakumari, B., A. L. Londhe, D. B. Jadhav, and H. K. Trimbake, 2006, Seasonal variability in the stratospheric aerosol layer in the current volcanically- quiescent period over two tropical stations in India using the twilight sounding method. Geophys. Res. Lett., **33**, L12809 doi: 10. 1029/2006GL026087.

Padmakumari, B., A. L. Londhe, D. B. Jadhav, 2007, Observational evidence of solar dimming: Offsetting surface warming over India. Geophys Res Lett 34: L21810. doi:10.1029/ 2007GL031133.

Padmakumari, B., R.S. Maheskumar, G. Harikishan, J.R. Kulkarni, and B.N. Goswami, 2012, Comparative study of aircraft- and satellite-derived aerosol and cloud microphysical parameters during CAIPEEX-2009 over the Indian region, International Journal of Remote Sensing, Vol. 34, No. 1, 358–373.

Padmakumari, B., Maheskumar, R.S., Harikishan, G., Morwal, S.B., Prabha, T.V., Kulkarni, J.R., 2013a. In situ measurements of aerosol vertical and spatial distributions over continental India during the major drought year 2009. Atmos. Environ. 80, 107–121.

Padmakumari, B., Maheskumar, R.S., Morwal, S.B., Harikishan, G., Konwar, M., Kulkarni, J.R., Goswami, B.N., 2013b. Aircraft observations of elevated pollution layers near the foothills of the Himalayas during CAIPEEX-2009. Quarterly Journal of the Royal Meteorological Society 139, 625-638, doi:10.1002/qj.1989.

Padmakumari, B., Trimbake, H. K., Londhe, A.L., and Jadhav, D. B., 2003, A case study of twilight probing of the atmosphere during Leonid meteor shower 2001. Current Science, 84, 1238–1241, 2003.

Pandithurai, G et al., 2004, Aerosol radiative forcing over a tropical urban site in India, Geophysical research Letters, 31, L12107.

Pandithurai, G., P. C. S. Devara, P. E. Raj, and S. Sharma, Retrieval of aerosol size index from high spectral resolution radiometer observations, Aerosol Sci. & Tech., 26, 154–162, 1997.

Pant, P., et al., 2006, Study of aerosol black carbon radiative forcing at a high altitude location, J. Geophys. Res., 111 (D17): Art. No. D17206, doi:10.1029/2005JD006768.

Parameswaran, K., Rajan, R., Vijayakumar, G., Rajeev, K., Moorthy, K.K., Nair, P.R., Satheesh, S.K., 1998. Seasonal and long term variations of aerosol content in the atmospheric mixing region at a tropical station on the Arabian Sea-coast. J. Atmos. Sol.-Terr. Phys. 60 (1), 17–25.

Parashar D C, Gadi R, Mandal T K, et al., Carbonaceous aerosol emissions from India, Atmos. Environ., 39, 7861–7871, 2005.

Parashar et al., Dey,S. and S.N.Tripathi, 2007, Estimation of aerosol optical properties and radiative effects in the Ganga basin, northern India, during the wintertime, Journal of Geophysical Research, 112, D03203.

Pathak, B., P.K. Bhuyan, M. Gogoi and K. Bhuyan, Seasonal heterogeneity in aerosol types over Dibrugarh-North-Eastern India, Atmospheric environment 47, 307-315, 2012.

Pillai, P.S., and K. K. Moorthy, Size distribution of near surface aerosols and its relation to the columnar Aerosol Optical Depths - Response to airmass types, Ann. Geophys., 22, 3347-3351, 2004

Prabha T.V., Karipot, A., Axisa, D., Padmakumari, B., Maheskumar R.S., Konwar M., Kulkarni, J.R., Goswami, B.N, 2012, Scale interactions near the foothills of Himalaya during CAIPEEX. Journal of Geophysical Research 117, D10203, doi:10.1029/2011JD0167.

Prospero, J.M., Carlson, T.N., 1972, Vertical and area distribution of Saharan dust over the western equatorial North Atlantic Oceans. Journal of Geophysical Research 77, 5255–5265.

Prospero, J.M., et al., 1970, Dust in the Caribbean Atmosphere Traced to an African Dust Storm. Earth and Planetary Science Letters 9, 287–293.

Prospero, J.M., 1979, Mineral sea salt aerosol concentrations in various Oceans regions. Journal of Geophysical Research 84, 725–731.

Rahul, P.C.R., Bhawar, R.L., Ayantika, D.C., Safai, P.D., Thara, V., Padmakumari, B., and Raju, M.P., 2014, Double blanket effect caused by two layers of black carbon aerosols escalates warming in the Brahmaputra River Valley. Nature Scientific Reports, 4:3670, 2014, DOI:10. 1038/srep03670, 1-7.

Raj, P.E., Saha, S.K., Sinbawne, S.B., Deshpande, S.M., Devara, P.C.S., Jaya Rao, Y., Dani, K.K., Pandithurai, G., 2008, Lidar observation of aerosol stratification in the lower troposphere over Pune during pre-monsoon season of 2006. Journal of Earth System Science 117(S1), 293–302.

Ramachandran, S. and Jayaraman, A., 2003, Balloon-borne study of the upper tropospheric and stratospheric aerosols over a tropical station in India. Tellus, 55(3)B, 820–836.

Ramachandran, S., and Kedia, S., Aerosol, clouds and rainfall: inter-annual and regional variations over India, Climate Dynamics, 40, 1591-1610, 2013.

Ramachandran S., S.Kedia, R.Srivastava (2012), Aerosol optical depth trends over different regions of India, *Atmos. Environ.,* 49, 338–347.

Ramachandran S, Rengarajan R, Jayaraman A, et al., 2006, Aerosol radiative forcing during clear, hazy, and foggy conditions over a continental polluted location in north India, J. Geophys. Res., 111, Article Number: D20214.

Ramanathan, V., P.J. Crutzen, J. T. Kiehl, and D. Rosenfeld, 2001, Aerosol, climate, and hydrological cycle, Science, 294, 2119–2124, doi:10.1126/science.1064034.

Ramana, M.V., Ramanathan, V., Podgorny, I.A., Pradhan, B.B., Shrestha, B., 2004, The direct observations of large aerosol radiative forcing in the Himalayan region. Geophysical Research Letters 31, L05111, doi:10.1029/2003GL018824.

Rangarajan, S., Mani, A., 1982, Total Precipitable Water in the Atmosphere over India, Proc. Indian.Acad.Sciences, 91, 189-207.

Rastogi, N. and Sarin, M.M., 2009. Quantitative chemical composition and characteristics of aerosols over western India: One year record of temporal variability, Atmospheric Environment 43, 3481-3488, DOI: 10.1016/j.atmosenv.2009.04.030.

Ravi Kiran, V., M. Rajeevan, S. Vijaya Bhaskara Rao, and N. Prabhakara Rao, 2009, Analysis of variations of cloud and aerosol properties associated with active and break spells of Indian summer monsoon using MODIS data, Geophys. Res. Lett., 36, L09706, doi:10.1029/ 2008GL037135.

Remer, L. A., et al. (2005), The MODIS algorithm, products and validation, J. Atmos. Sci., 62, 947 – 973.

Rengarajan, R., M.M. Sarin, A.K., 2007, Sudheer Carbonaceous and inorganic species in atmospheric aerosols during wintertime over urban and high-altitude sites in North India, Journal of Geophysical Research, 112, D21307

Ricchiazzi, P., S. Yang, C. Gautier, and D. Sowle, 1998, SBDART, A research and teaching tool for plane-parellel radiative transfer in the Earth's atmosphere, *Bull. Am. Meteorol. Soc., 79*, 2101– 2114.

Rosenfeld, Daniel; Andreae, Meinrat O.; Asmi, Ari; et al., 2014, Global observations of aerosol-cloud-precipitation-climate interactions, REVIEWS OF GEOPHYSICS, 52, 750-808.

Safai, P.D., Kewat, S., Praveen, P.S., Rao, P.S.P., Momin, G.A., Ali, K., Devara, P.C.S., Seasonal variation of black carbon aerosols over a tropical urban city of Pune, India. Atmospheric Environment, 41, 2699 – 2709, 2007.

Safai, P.D., Raju, M.P., Maheskumar, R.S., Kulkarni, J.R., Rao, P.S.P., Devara, P.C.S. , 2012, Vertical profiles of black carbon aerosols over the urban locations in South India, Science of the Total Environment , 431, 323-331.

Sarkar, S., R. Chokngamwong, G. Cervone, R.P.Singh and M . Kafatos, 2006, Variability of aerosol optical depth and aerosol forcing over India. Adv. Space.Res., 37, 2153-2159.

Satheesh, S.K. and J. Srinivasan, A Method to Estimate Aerosol Radiative Forcing from Spectral Optical Depths, J. Atmos. Sci., 63 (3): 1082-1092, 2006.

Satheesh, S.K., and K. Krishna Moorthy, 2005, Radiative Effects of Natural Aerosols: A Review, Atmos. Environ., 39 (11): 2089-2110.

Satheesh, S.K., and K. Krishna Moorthy, 1996, Atmospheric Total Ozone Content from Spectral Extinction Measurements, Ind. J. Radio and Space Phys., 25, 204-210.

Satheesh, S.K., Vinoj, V., Suresh Babu, S., Krishna Moorthy, K., S. Nair, V., Vertical distribution of aerosols over the east coast of India inferred from airborne LIDAR measurements, 2009, Annales Geophysicae, 27 (11), 4157-4169.

Satheesh, S.K., V. Vinoj and K. Krishnamoorthy, 2010, Improved assessment of aerosol radiative forcing over oceans adjacent to Indian subcontinent, Advances in Meteorology, 2010, 1-13, doi:10.1155/2010/139186.

Satheesh, S.K., and V. Ramanathan, 2000, Large Difference in Tropical Aerosol Forcing at the Top of the Atmosphere and Earth's Surface, Nature, 405, 60-63.

Satheesh, S.K., V. Ramanathan, X.L. Jones, J.M. Lobert, I.A. Podgorny, J.M. Prospero, B.N. Holben and N.G. Loeb, 1999, A Model for Natural and Anthropogenic Aerosols over the Tropical Indian Ocean Derived from INDOEX data, J. Geophys. Res., 104, D22, doi: 10.1029/1999JD900478, 27, 42127, 440.

Satheesh, S.K.,V. Ramanathan, B. N. Holben, K. K. Moorthy, N. G. Loeb, H. Maring, J. M. Prospero and D. Savoie, 2002, Physical, Chemical and Radiative Properties of Indian Ocean Aerosols, J. Geophys. Res., 107 (D23), 4725, doi: 10.1029/2002JD002463.

Satheesh, S.K., Vinoj, V., Moorthy, K.K., 2006, Vertical distribution of aerosols over an urban continental site in India inferred using a micro pulse lidar. Geophysical Research Letters 33 (20), L20816, doi:10.1029/2006GL027729.

Satheesh, S.K., K. Krishna Moorthy and B.V. Krishna Murthy, 1998, Spatial Gradients in Aerosol Characteristics over the Arabian Sea and Indian Ocean, J. Geophys. Res., 103, D20, 26,183-26,192, doi: 10.1029/98JD00803.

Satheesh, S. K., K. Krishna Moorthy, S. Suresh Babu, V. Vinoj and C.B.S. Dutt, 2008, Climate implications of large warming by elevated aerosol over India, 1008, Geophysical Research Letters, Vol. 35, 1-6.

Sharma V K and Patil R S. (1992), Chemical composition and source identification of Bombay aerosol, Environ Technol. 13: 1043-1052.

A. Smirnov, B.N. Holben, D.M. Giles, I. Slutsker, N.T. O'Neill, T.F. Eck, A. Macke, P. Croot, Y. Courcoux, S.M. Sakerin, T.J. Smyth, T. Zielinski, G. Zibordi, J.I. Goes, M.J. Harvey, P.K. Quinn, N.B. Nelson, V.F. Radionov, C.M. Duarte, R. Losno, J. Sciare, K.J. Voss, S. Kinne, N. R. Nalli, E. Joseph, K.K. Moorthy, D.S. Covert, S.K. Gulev, G. Milinevsky, P. Larouche, S. Belanger, E. Home, M. Chin, L.A. Remer, R.A. Kahn, J.S. Reid, M. Schulz, C.L. Heald, J. Zhang, K. Lapina, R.G. Kleidman, J. Griesfeller, B.J. Gaitley, Q. Tan, T.L. Diehl, Maritime aerosol network as a component of AERONET – First results and comparison with global aerosol models and satellite retrievals, Atmospheric Measurement Techniques, 4 , pp. 583–597, 2011.

Sreekanth, V., Satellite derived aerosol optical depth climatology over Bangalore, India, Advances in Space Research, 51, 2297 – 2308, doi.org/10.1016/j.asr.2013.01.022, 2013.

Sreekanth V, Niranjan K, Madhavan BL, 2007, Radiative forcing of black carbon over eastern India, Geophys. Res. Lett., 34, Article Number: L17818,

Stuhlman Jr., O., 1932, The mechanics of effervescence. Journal of Applied Physics 2 (6), 457–466.

Tare,V., 2006, Measurements of atmospheric parameters during Indian Space Research Organization Geosphere Biosphere Program Land Campaign II at a typical location in the Ganga Basin: 2. Chemical properties, Journal of Geophysical Research, 111, D23210.

Tegen, I., and I. Fung, 1994, Modeling of mineral dust in the atmosphere: Sources, transport, and optical thickness. J. Geophys. Res. ,99, doi:10.1029/94JD01929.

Toba, Y., 1965a, On the giant sea salt particles in the atmosphere I. General features of the distribution. Tellus 17, 131–145.

Toba, Y., 1965b, On the giant sea salt particles in the atmosphere II. Theory of the vertical distribution in the 10 m layer over the ocean. Tellus 11, 365–382.

Tripathi, S. N., Sagnik Dey, Vinod Tare, S. K. Satheesh, Shyam Lal and S. Venkataramani, 2005, Enhanced layer of black carbon in a north Indian industrial city, Geophy. Res. Letts, 32, L12802, doi:10.1029/2005GL022564.

Tsunogai, O., et al., 1972, Chemical composition of oceanic aerosol. Journal of Geophysical Research 77, 5283–5292.

Twomey, S. 1974, Pollution and the planetary albedo, Atmos. Environ., 8, 1251–1256, doi:10.1016/0004-6981(74)90004-3.

Uno, I., et al., 2008, 3D structure of Asian dust transport revealed by CALIPSO lidar and a 4DVAR dust model, Geophys. Res. Lett., 35, L06803, doi:10.1029/2007GL03232

Vinoj V, S. K. Satheesh and KK Moorthy, 2010, Optical, radiative and source characteristics of aerosols at Minicoy, a remote island in the southern Arabian Sea, J. Geophys. Res., 115, D01201.

Vinoj,V., S.K. Satheesh and K.K.Moorthy, Aerosol Characteristics at a Remote Island: Minicoy in Southern Arabian Sea, Journal of Earth System Science, 117, S1, 389-397, 2008.

Wang, X., C. L. Heald, D. A. Ridley, J. P. Schwarz, J. R. Spackman, A. E. Perring, H. Coe, D. Liu, and A. D. Clarke (2014), Exploiting simultaneous observational constraints on mass and absorption to estimate the global direct radiative forcing of black carbon and brown carbon, Atmos. Chem. Phys., 14, 10,989–11,010, doi:10.5194/acp-14-10989-2014.

Woodcock, A.H., 1953, Salt nuclei in marine air as a function of altitude and wind force. Journal of Meteorology 10, 362–371.

Woodcock, A.H., 1957, Atmospheric salt nuclei data for project shower. Tellus IX, 521–524.

Zender, C.S., Bian, H., Newman, D., 2003,. Mineral Dust Entrainment and Deposition (DEAD) model: description and 1990s dust climatology. Journal of Geophysical Research 108 (D14), 4416.

Zhang, J., Reid, J. S., and Holben, B. N.: An analysis of potential cloud artifacts in MODIS over ocean aerosol optical thickness products, Geophys. Res. Lett., 32, L15803, doi:10.1029/2005GL023254, 2005.

Chapter 14
Variability of Ozone and Related Trace Gases Over India

Shyam Lal, S.K. Peshin, M. Naja and S. Venkataramani

1 Introduction

The earth's atmosphere contains a variety of trace gases in addition to the major gases like nitrogen, oxygen, and argon. Most of the trace gases are of natural and anthropogenic origin. They contribute significantly to the air quality and climate change apart from their chemical roles. Ozone (O_3) is of mostly natural origin and protects life on the earth by absorbing the biologically active UV-B (280–310 nm) radiation in the middle atmosphere (20–30 km range). However, man-made CFCs, transported to the stratospheric altitudes from surface, destroy the precious ozone layer thereby depriving earth of its natural protection. This effect is highly pronounced in the Antarctic and to a lesser extent in the Arctic regions. However, in the troposphere (ground to about 18 km) ozone acts as a greenhouse gas and is referred to as a short-lived climate pollutant. Ozone, in the lower atmosphere, is produced from the ozone precursors, namely CO and hydrocarbons in the presence of oxides of nitrogen. Thus, ozone is greatly affected by the pollution levels. It is found to increase in several polluted regions and their outflow affecting the regions away from the major pollutant regions. Global climate change is real, and its effects are being observed world over as well as in India. The climate change effects are clearly reflected over India in the changes in surface air temperature, melting of the Himalayan glaciers, unusual seasonal climate change patterns, frequent occurrence

S. Lal (✉) · S. Venkataramani
Physical Research Laboratory (PRL), Ahmedabad, India
e-mail: shyam@prl.res.in

S.K. Peshin
India Meteorological Department (IMD), Earth System Science Organization, New Delhi, India

M. Naja
Aryabhatta Research Institute of Observational Sciences (ARIES), Nainital, India

M.N. Rajeevan and S. Nayak (eds.), *Observed Climate Variability and Change Over the Indian Region*, Springer Geology, DOI 10.1007/978-981-10-2531-0_14

of extreme weather events, etc. (Kothawale and Kumar 2005; Kulkarni et al. 2007; Mallik and Lal 2011, Chaps. 3 and 4 of this book).

Tropospheric ozone is also recognized as a threat to human health and vegetation. Ozone-related deaths are estimated to make up about 5–20 % of all those related to air pollution, and India has been shown to be among top ten polluted countries (Lelieveld et al. 2013). Its deleterious impact on vegetation through plant damage impedes the uptake of carbon into the biosphere (Sitch et al. 2007). Hence, ozone plays key roles in atmospheric chemistry, radiation balance, and air quality, including human health. This chapter presents some of the studies related to ozone and associated trace gases.

2 Atmospheric Total Ozone

Ozone was discovered in the atmosphere by Schonbein in 1839, and its observations started in the late 1800s and early 1900s, using semiquantitative Schonbein ozonoscope. Later, extensive ozone observations were initiated in 1950s over different global sites during the International Geophysical Year. During this period, an Ehmert chemical sensor (in situ concentration measurement based on the principle of chemical reaction of ozone) and Dobson spectrometer (total columnar ozone based on principle of optical absorption by ozone) acquired through International Ozone Commission were operated at Physical Research Laboratory (PRL), Ahmedabad and at Mt. Abu in 1954 (Dave 1957). Additionally, India Meteorological Department (IMD) initiated observations of total columnar ozone at Srinagar (1957 up to 1989 only), Delhi (1957), Varanasi (1963), Pune (1973), and Kodaikanal (1957) using Dobson spectrometers. These spectrometers have been inter-compared/calibrated from time to time (Peshin and Singh 2013). Total ozone over the Indian region is mostly in the range of 235–300 DU (Dobson Unit*) (Fig. 1). The highest ozone is observed over Delhi (260–290 DU) and the lowest over Kodaikanal (235–265 DU). Further, the highest ozone is observed during April–June period and the lowest during winter. After June, the total column ozone is affected by mixing of lower ozone air from the lower heights (<3 km). Lower ozone levels during winter are caused by lesser photochemical production. A small shift in the pattern of monthly changes in ozone from Delhi to Kodaikanal is also noted due to the latitudinal change.

An analysis of the total ozone data up to 1996 at six sites including Ahmedabad shows increasing trend at all the sites except at Varanasi, where a decreasing trend was found (Chakrabarty et al. 1998). The rates of increase in ozone during this period were found to be 1.98, 2.33, 1.85, and 0.68 % per decade at Kodaikanal, Ahmedabad, New Delhi, and Srinagar, respectively, while rate of decrease in ozone at Varanasi is 1.02 % per decade. Another study of this data for a shorter period (1981–1998) by Londhe et al. (2003) showed decreasing trends of 8 (2.8 %), 11 (4.2 %), 8 (3.1 %), and 9 (3.4 %) DU per decade for Delhi, Varanasi, Pune, and Kodaikanal, respectively. Recently, a rigorous analysis of the total ozone over

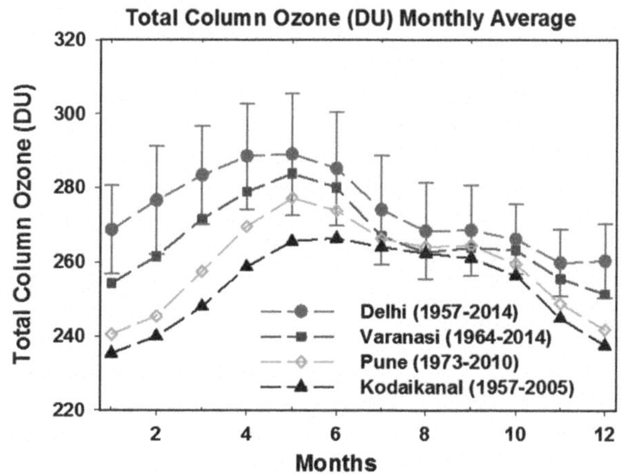

Fig. 1 Seasonal variations in total columnar ozone at Delhi (1957–2014), Varanasi (1964–2014), Pune (1973–2010), and Kodaikanal (1957–2005), which are parts of IMD's network. One sigma spread is shown only for Delhi as it is similar for other stations

various cities in India by Tandon and Attri (2011) using satellite data for the period of 1979–2008 provided statistically significant decreasing trends at Delhi, Varanasi, and Pune, which are 1.1, 0.8, 0.2 % per decade, respectively.

3 Vertical Distribution of Ozone

Scientists at IMD have been making measurements of vertical distribution of ozone using balloon-borne electrochemical concentration cell (ECC) sensor developed in India since 1971. The maximum ozone concentration occurs over Delhi, Pune, and Trivandrum at about 23–24, 25–26, and 26–27 km, respectively. Measurements of vertical distribution of ozone were also made using rocket-borne sensors launched from Thumba in campaign modes starting from 1977 (Subbaraya et al. 1994).

Ozonesonde data (1972–2001) from IMD over Delhi, Pune, and Trivandrum show statistically insignificant trends over Trivandrum. But the trend over Pune is close to statistical significance at 9.7 ± 6.1 % per year in the planetary boundary layer and somewhat insignificant above 600 hPa (Fig. 2) (Saraf and Beig 2004). A statistically significant positive trend throughout the troposphere is observed over Delhi. However, over the last decade, variations in tropospheric ozone have become statistically insignificant even over Delhi in spite of high levels of ozone precursors. The regressed tropospheric ozone residual (TOR) pattern during monsoon season shows large trend over the entire Indo-Gangetic region and is largest, 6–7.2 % per decade, over the northeastern Indo-Gangetic plain (IGP) of India. Annually, trend of about 0.4 ± 0.25 (1σ) % per year was seen over the northeastern Gangetic region (Lal et al. 2012).

Trend analysis of tropical (30°S–30°N) tropospheric columnar ozone from Nimbus7 and Earth Probe satellite—Total Ozone Mapping Spectrometer (TOMS) data for the period of 1979–2005 using multifunctional regression model indicate

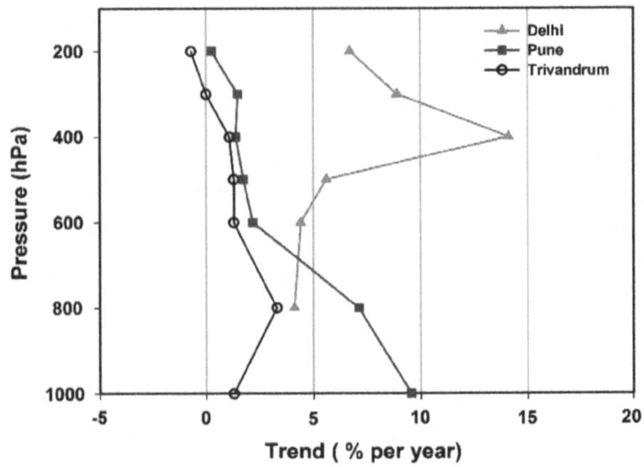

Fig. 2 Vertical distribution in the tropospheric ozone trends over Delhi, Pune, and Trivandrum (Adapted from Saraf and Beig 2004)

the highest increasing trend (7–9 % per decade) over some parts of south Asian continental region which is attributed to increasingly high emissions of ozone precursors over the region. Among different marine regions, ozone trend is found to be highest (4–6 % per decade) over the Bay of Bengal (BoB) (Beig and Singh 2007).

Role of trace gases especially those of chlorofluorocarbons (CFCs) in depleting stratospheric ozone became important in early 1980s, and PRL initiated measurements of vertical distributions of CFCs and other trace gases from Hyderabad using cryogenic air sampler flown on high-altitude balloons in collaboration with Max Plank Institute of Chemistry, Mainz, Germany in 1987 and 1990. Later, a similar cryogenic air sampler was indigenously developed by PRL and Indian Space Research Organisation (ISRO) and flown successfully from Hyderabad in 1994 and 1998 (Patra et al. 2000).

4 Effects of Dynamics on Ozone in the Troposphere

Distributions of trace gases are strongly affected by advection, convection and long-range transport as well as the movement, both diurnal and seasonal, of the planetary boundary layer (PBL). Ozone in the atmosphere has a highly variable lifetime from hours near the surface to months in the free troposphere and again hours in the upper stratosphere. Hence, long-range transport can affect ozone distributions only in the free troposphere and lower stratosphere, due to its longer lifetime. Recently, few more groups, including PRL, ARIES, Space Physics Laboratory (SPL), and National Atmospheric Research Laboratory (NARL), initiated observations of ozone vertical distribution using balloon-borne ECC sensors.

Scientists at the Physical Research Laboratory have measured ozone profiles from Ahmedabad during 2003–2007. The results show ozone levels to be lower from ground to 3–4 km height. The winds over Ahmedabad in the free troposphere are from the western region during March to May and October to November. These winds bring effects of biomass burning in the northern African region and northern parts of the South America. Ahmedabad also experiences winds from southern Europe via northwest India during winter months (Lal et al. 2014). The summer monsoon winds coming from the Indian Ocean and the Arabian Sea during June to September bring cleaner marine air and sweep away the pollutants. During other seasons, pollutants are transported over the Arabian Sea (AS) from the western region of India (Lal et al. 2013). Vertical distribution of ozone over the BoB is affected by transport from the IGP in the lower region (0–4 km), but in the free troposphere it is affected by convection. However, the AS region is more affected by downward transport from the upper troposphere and lower stratosphere (Lal et al. 2013).

Balloon-borne measurements of ozone vertical distribution and meteorological parameters over Nainital in the central Himalayan region show influence of sub-tropical jets (wind speed ~80 m/s) in the middle/upper troposphere, particularly during winter (Ojha et al. 2014). Influences of springtime biomass burning are observed in ozone enhancement (~20 ppbv) in 2–4 km altitude region. These data showed that the tropospheric ozone levels over the central Himalayas are considerably (~30 ppbv) higher than those over the western India, particularly in spring (Fig. 3). This ozone enhancement is attributed mainly to the regional pollution of the IGP supplemented with the northern Indian biomass burning.

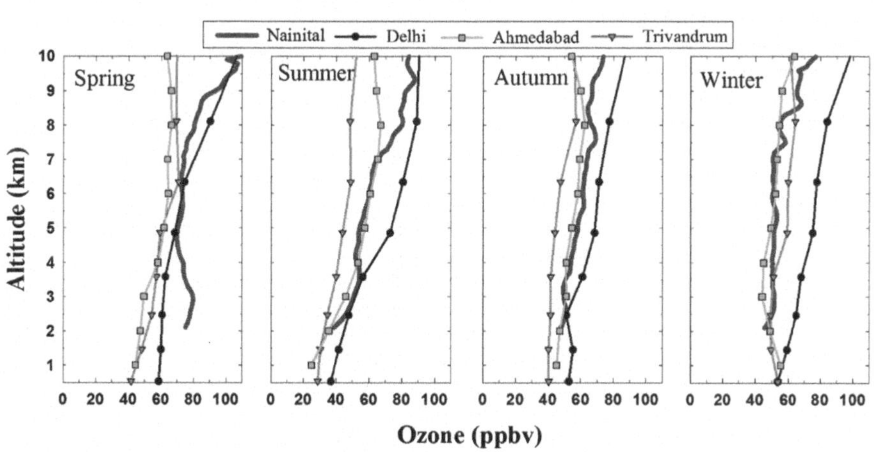

Fig. 3 Average vertical distributions of ozone over Nainital, Delhi, Ahmedabad, and Trivandrum during the four seasons (Kumar et al. 2012; Ojha et al. 2014; Lal et al. 2014)

5 Surface Ozone and Its Precursors

IMD has been making measurements of surface ozone at Delhi, Pune, Kodaikanal and Trivandrum since 1971 using a Brewer bubbler. Prior to this, measurements of surface ozone were made at Mt. Abu and Ahmedabad during 1954–55. Later in early 1990s, first systematic observations of ozone together with its precursor gases were initiated at Ahmedabad, Gadanki, Trivandrum and Mt Abu (Lal et al. 2000; Nair et al. 2002; Naja et al. 2003) by PRL under ISRO-Geosphere Biosphere program (ISRO-GBP). Now, several groups are making surface-based study of ozone and its precursors under this project. Indian Institute of Tropical Meteorology (IITM) Pune under the Ministry of Earth Sciences (MoES) has also initiated such studies at several places in India. Surface observations from various sites (Fig. 4) under different observational network have started during the last 10 years with an aim to study variability and transport effects.

There are 16 observational sites, representing different regions of India, under ISRO-ATCTM (Atmospheric Trace gases-Chemistry, Transport and Modeling) project. Almost half of the sites cover the IGP, which is considered to be highly polluted region. There are also three mountain sites, representing northern, western, and southern free tropospheric region in India. Figure 5 shows average seasonal variations from various locations in India. The monthly average ozone shows strong

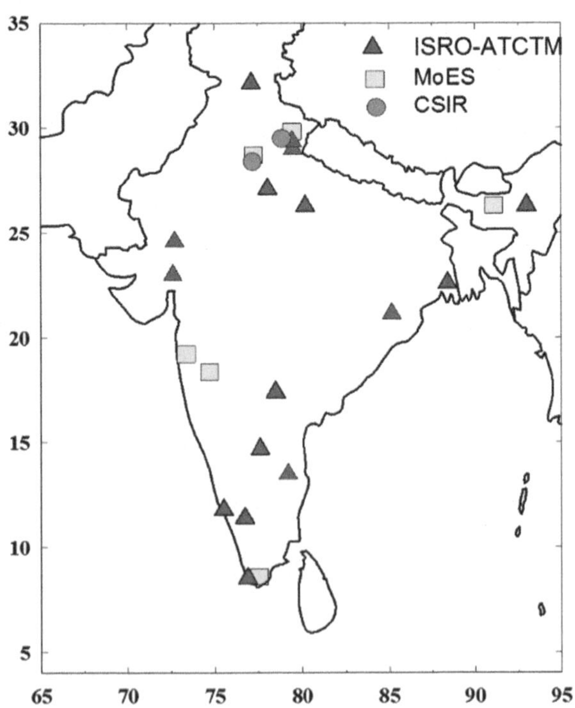

Fig. 4 Observational sites (a sample of) for surface-based and balloon-borne measurements under different networks [ISRO, MoES, and Council of Scientific and Industrial Research (CSIR)] in India

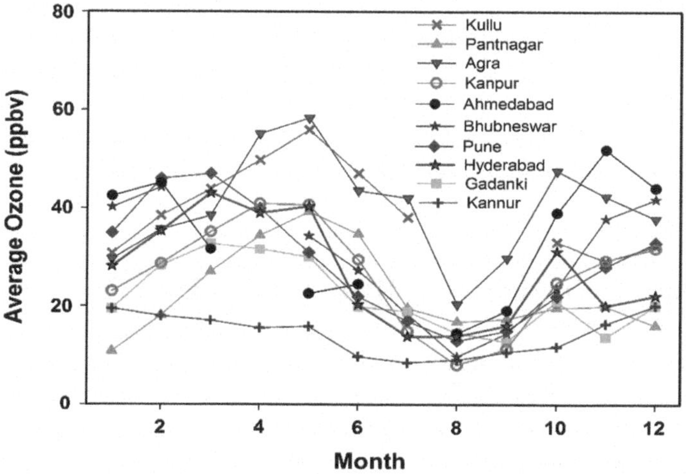

Fig. 5 Average monthly variations of surface ozone at different sites in India. Kullu—Sharma et al. (2013); Pantnagar—Ojha et al. (2012); Agra—Verma et al. (2015); Bhubaneshwar—Mahapatra et al. (2012); Gadanki—Renuka et al. (2014); Pune—Beig et al. (2008); Kanpur—Gaur et al. (2014); Kannur—Nishant et al. (2012); Hyderabad—Venkanna et al. (2014)

effect of large-scale dynamics, apart from role of photochemistry. The lowest ozone levels are observed during the summer monsoon at all the sites, and the highest ozone levels occur during spring (MAM) at most of the sites due to strong photochemistry. However, some sites also show higher ozone during autumn/winter. The average monthly ozone values vary from place to place but are in the range of 10–60 ppbv. Few sites also report hourly ozone levels higher than 80 ppbv and cross the air-quality guidelines. Generally, seasonal amplitude is greater at sites located in northern India, when compared with sites in southern India.

In general, poor correlations among ozone, CO, and NO_x (NO and NO_2) are seen over these sites suggesting the possibility of incomplete photochemical processes. There is a weak (correlation coefficient of 0.3) positive relationship between ozone and NO_x at Gadanki, a site in southern India, with ozone production efficiency of 3.3 (Naja and Lal 2002). This value is slightly higher (4.9) for spring season, when more measurements are available; nevertheless, ozone production efficiency is shown to be much higher [6–7 O_3 (ppbv) per unit NO_x (ppbv)] for mid-latitude (Kelly et al. 1984; Glavas 1999). Ozone production per molecule of CO is also observed to be lower (less than 0.13) (Naja et al. 2003), when compared (0.27–0.33) to those observed at locations in North America (Chin et al. 1994; Parrish et al. 1998). It has been suggested that lower ozone production efficiency is an indicator of incomplete photochemical process (e.g., Parrish et al. 1998). It has been shown that the ratio of combined anthropogenic NO_x sources (SN) and total CO and hydrocarbon sources (SC), i.e., SN/SC, is more than four times lower in Asia than in North America (Lelieveld et al. 2001). This also suggests that ozone photochemistry in Asia is strongly NO_x limited and regeneration of OH by NO is inefficient. The slope ($\Delta CO/\Delta NO_x$) is higher in India, when compared to USA and Europe (Lal et al. 2008; Mallik et al. 2015). This indicates that biomass/biofuel burning emissions (higher CO levels) dominate over the fossil-fuel-related emissions (low NO_x levels) in India. These observations also showed that ozone production efficiency and photochemical ozone production is much lower over the Indian region when compared to other sites in USA or Europe.

6 Surface Observations at High-Altitude Sites

Observations from high-altitude sites are very important for regional representativeness and also to study different processes. First ever surface ozone observations from a high-altitude site in India were made at Mount Abu for a short period in 1954. Later, systematic observations of ozone, CO, NO_x, and CH_4, initiated in 1993, from this site showed the absence of daytime higher ozone levels at this relatively cleaner site. Observations from this site provided information on regional background and continental ozone with levels of 33.4 and 48.1 ppbv, respectively (Naja et al. 2003). Presently, measurements of surface ozone and related trace gases are also being made at other high-altitude sites like Nainital, Darjeeling, and Ooty, representing different regions of India.

Nainital, a mountain site in the central Himalayas, is found to be a good representative site for the northern region, and regional pollution is shown to have maximum contribution (16.5 ppbv) to the ozone levels during May–June, while contribution of long-range transport is greatest during January–March (8–11 ppbv). Springtime ozone values in the central Himalayas are significantly higher than Mt. Abu, the high-altitude site in western India. Contrary to the central Himalayan region, where background ozone is estimated to be 30–35 ppbv, the background ozone is about 28 ppbv in western India. It was shown that biomass burning, which is significant over the northern Indian region during spring, could lead to enhancement in ozone levels by 19 ppbv ($\sim$34 %) (Kumar et al. 2010). Collocated observations from Pantnagar and Nainital show that emissions and photochemically processed air masses in the IGP region can influence the air quality over the cleaner Himalayan region via the boundary layer mixing process, particularly in spring when the mixing depth is higher. Analysis of relations among ozone, CO, and NO_y (all oxides of nitrogen including HNO_3) confirms incomplete photochemistry with minimal role of fresh emissions while greater role of dynamical processes (Sarangi et al. 2014).

7 Hydrocarbons

Measurements of light (C2-C5) non-methane hydrocarbons (NMHCs) from Ahmedabad are being made by PRL scientists to study their variability and relations with ozone and its other precursors using gas chromatography (Sahu and Lal 2006).

Later, this technique has been used to measure NMHCs from many different locations (Mt. Abu, Hissar, Kanpur, Nainital, Kolkata and from marine regions) by collecting air samples. A proton transfer reaction mass spectrometer (PTR-MS) is being employed for measurements of various ozone precursors such as NMHCs and VOCs including benzenoids and acetonitrile (a biomass burning tracer) at IISER Mohali. These observations, supported with back-air trajectories, showed that massive amounts of carcinogenic benzenoids are released from post-harvest paddy residue burning over northwest India, which aggravates urban smog at downwind urban centers. Subsequently a 300 % increase in ambient levels of VOCs due to wheat residue burning in northwest India was also documented (Sinha et al. 2014). Measurements of several key VOCs such as methanol, acetone, acetaldehyde, and isoprene are also made for the first time in India. The daytime ozone increase anti-correlates with acetaldehyde (Fig. 6), indicating that oxidation of acetaldehyde which occurs very rapidly produces acetyl peroxy radicals that are convert nitrogen monoxide to nitrogen dioxide, which on photolysis produces ozone rapidly. Similar measurements are now being made at PRL, Ahmedabad, and IITM, Pune, also.

A comparison of contribution of different NMHCs at Nainital, Haldwani/Pantnagar, Kanpur, and Hissar shows nearly similar composition between Kanpur and Hissar, but different than those at Nainital or Haldwani/Pantnagar (Fig. 7) (Sarangi et al. 2016). The overall differences in the composition of NMHCs are suggested to be a manifestation of the relatively aged air sampled over Nainital as compared to the IGP sites having fresh emissions. Observations at Agra showed benzene and toluene to be most abundant, among BTX (Benzene, Toluene and Xylene), with an average value of 49 ± 1.8 and 36 ± 2.6 μg m^{-3}. Concentrations of p-xylene and o-, m-xylene are found to be 6.2 ± 1.4 and 16.1 ± 2.6 μg m^{-3}, respectively. Although the concentrations of xylenes are observed to be minimum among the BTX, p-xylene is found to be the most dominant contributor to ozone formation. Benzene being the most abundant and hazardous species among BTX showed minimum potential to ozone formation. Toluene was found to have second largest contribution to ozone formation (Singla et al. 2012). Additionally, O_3/HNO_3

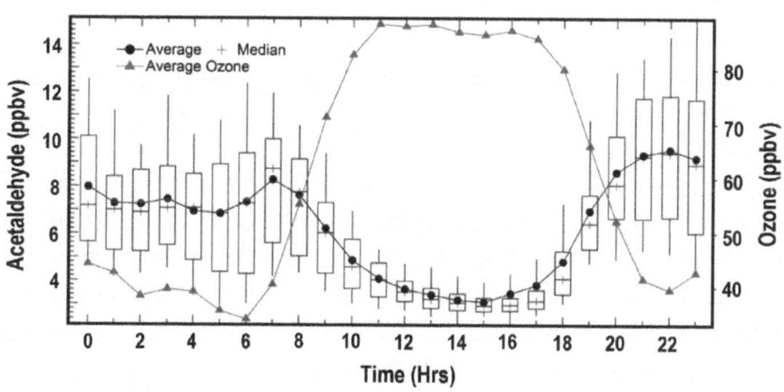

Fig. 6 Average diurnal variations in ozone and acetaldehyde at IISER, Mohali, in May, 2012 (Sinha—Personal communication)

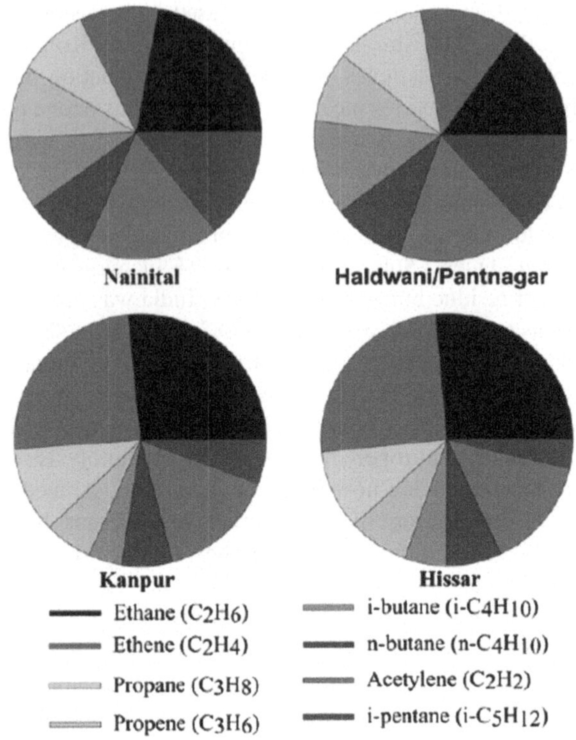

Fig. 7 Comparison of composition of different NMHCs at Kanpur, Hissar, Nainital, and Haldwani/Pantnagar

ratios range from 10 to 56, with an average value of 35.3 ± 2.6 (summer and winter) at Agra, which is mostly NO_x-sensitive region and have also been confirmed at Pantnagar.

8 Sulfur Gases

Natural sources of sulfur gases include emissions from soil, oceans, and volcanic emissions. The biogeochemical cycling of various sulfur species between various ecosystems has important implications to the climate, pH of rainwater, human health, etc. The most conspicuous form of sulfur in the atmosphere is SO_2 and its several precursors, which are called reduced sulfur compounds (RSCs). The primary source of SO_2 is combustion of coal and oil. Major atmospheric RSCs are dimethyl sulfide (DMS), hydrogen sulfide (H_2S), carbonyl sulfide (COS), carbon disulfide (CS_2), etc. Many groups in India have been making measurements of SO_2 using chemical techniques. However, these are not very accurate, and continuous measurements are limited. There has been renewed interest in the measurements of sulfur gases particularly SO_2. It is being measured using online analyzers from many sites in India in the last decade.

SO_2 observations at Ahmedabad revealed clear influence of different local sources in and around the city like a power plant in the northeast of the study location and a few industries in the east as corroborated by SO_2/NO_x value of 0.41 (ppbv/ppbv) (Mallik and Lal 2014). Further, the scientists at PRL, Ahmedabad, have been able to document the levels of atmospheric COS and CS_2 for the first time over an Indian location using a new GC setup, which shows anthropogenic contribution to COS (Mallik et al. 2016). COS is a major sulfur-containing gas in the atmosphere and is also a major component of stratospheric aerosol layer. A coherent picture of atmospheric COS distributions is still elusive, and measurements made over Ahmedabad would help to improve this understanding and better reconcile with known sources and sinks. High values of COS in a diluted boundary layer during March indicate substantial contribution of oceans. Enhanced oceanic emissions of COS and its precursors (CS_2 and $(CH_3)_2S$), which can quickly photooxidize into COS, occur during summer because COS and CS_2 are photochemically produced in the aqueous phase from dissolved organic matter. For CS_2 this aqueous source, and ultimately the production of COS, is thought to be enhanced in warm waters (enhanced hydrolysis) of the tropics.

Simultaneous measurements of SO_2 and hydrocarbons made over Kolkata revealed that despite being a megacity with strong local emissions, regional transport from the IGP plays a major role in enhancing its levels over Kolkata during winter (6.7 ppbv) and post-monsoon (4.1 ppbv). Due to a plethora of local and regional sources, the 1 sigma values can be very high, e.g., for winter, 1 sigma is about 73 % of the mean concentration. The SO_2 levels over Nainital, a mountain site, are found to be influenced by transport of air masses from the lower IGP during summer with highest average levels ($\sim$340 pptv) as compared to winter ($\sim$70 pptv) (Naja et al. 2014).

9 Trace Gases over the Marine Regions

Extensive measurements of trace gases have been made over the Arabian Sea, the Bay of Bengal, and the Indian Ocean using ORV Sagar Kanya and other ships (Fig. 8). These measurements of ozone, CO, NO_x, NMHCs, and SO_2 revealed large variability over these marine regions (Lal et al. 1998; Chand et al. 2003; Naja et al. 2004; David and Nair 2011; Mallik et al. 2013). Estimated south to north latitudinal gradients in O_3 (3.95 ppbv/°) and CO (16.56 ppbv/°) were significantly higher than those observed during earlier campaigns. Principle component analysis indicated contributions of ship emissions to NO_x levels over the BoB. Influences of fire from the Myanmar and Thailand regions are shown to be the potential contributor to enhanced CO levels (>250 ppbv) over the BoB during 14–15 November 2010 (Mallik et al. 2013). Diurnal variations in surface O_3 revealed effects of advection, entrainment, and photochemistry.

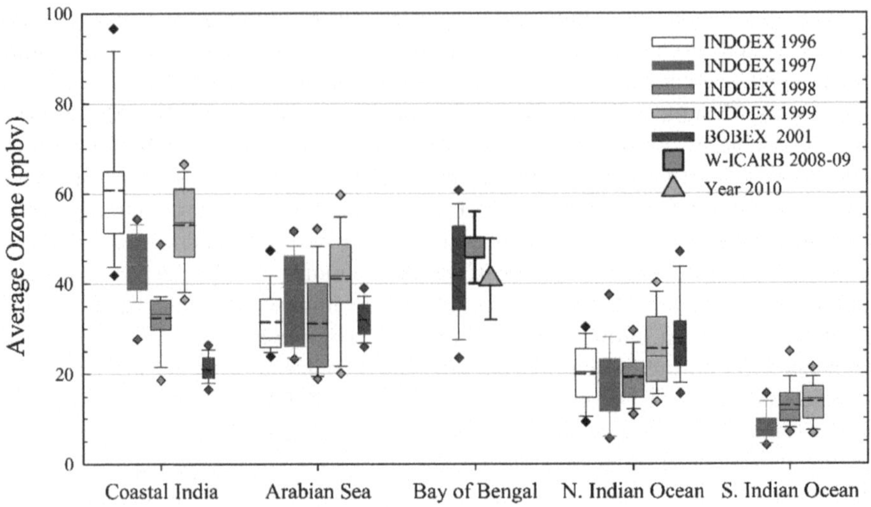

Fig. 8 Variations in surface ozone over the Arabian Sea, Bay of Bengal, Indian Ocean, and coastal region

10 Long-Term Trends in Surface Ozone

A comparison of surface ozone observations at Ahmedabad during 1954–55 and 1991–93 show increase of 1.45 % $year^{-1}$ (Naja and Lal 1996). These observations also show change in seasonal variations with maximum ozone in winter–spring during 1950s, while maximum ozone has been in autumn–winter during 1990s. Surface ozone observations made at the National Physical Laboratory, Delhi, show positive trends in daily maximum and daytime (1000–1700 h) ozone levels with a rate of about 1.7 (±0.7) and 1.3 (±0.6) ppbv $year^{-1}$, during 1997–2004, respectively. Observations of surface ozone at Dayalbagh educational site, Agra, show a tendency of ozone increase during 2009–2014, and some possibilities of decrease in average ozone (all data) are seen at Nainital. But data set at Nainital and Agra are for a short period to conclude on their trends.

11 Long-Term Trends in NO_x

Nitrogen dioxide (NO_2) plays a key role in the chemistry of the atmosphere and is emitted mainly by combustion processes. These emissions have been increasing over India over the past few years due to rapid economic growth. The GOME (Global Ozone Monitoring Experiment) and SCIAMACHY data show an increasing trend for the NO_2 tropospheric column for some of the Indian regions considered, but sometimes not as rapid as is expected from the urban growth rate in those cities. Delhi and Ahmedabad show very high growth rates of 11.3 and 10 %

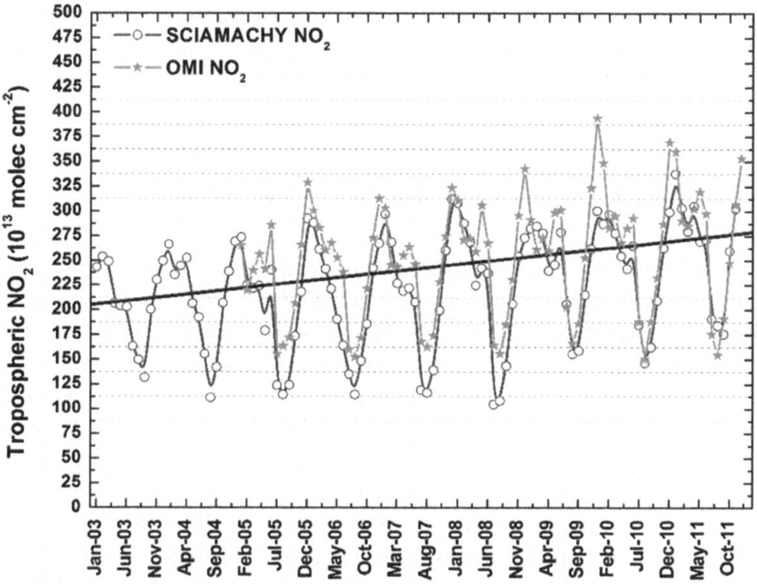

Fig. 9 Trends in tropospheric NO_2 column from OMI and SCIAMACHY over India (Adapted from Ghude et al. 2013)

per year, respectively, during 1996–2006 period (Sheel et al. 2010). However, Bengaluru and Pune show much lower growth rates of 3.5 and 1.5 %, respectively. Gurjar et al. (2008) have shown that transport emissions are responsible for most of the NO_x. The number of vehicles is increasing very fast all over India. The vehicular population in Delhi rose from 1.9 million vehicles in 1990 to nearly 3.6 million in the year 2001 (an increase of nearly 8.7 % per annum). Satellite (OMI and SCIAMACHY) data of total column NO_2 for India show an increasing trend of 1.65 % in anthropogenic NO_2 emission for the same period (Fig. 9) (Ghude et al. 2013). These observed trends are comparable to the growth rate in NO_x emission estimated in bottom-up inventories and can be attributed to the growth in power, transportation, and industrial sectors in India. Coal and oil consumption in India has also increased significantly in the last decade.

12 NO_x Emission Inventories

Work on NO_x emission inventories over India has been carried out for different fuel combustion and industrial activities by incorporating the most recently available microlevel activity data as well as country-specific emission factors (EFs) at high resolution (Sahu et al. 2012). Recently, top-down NO_x emission inventory has also been made for India (Ghude et al. 2013). A strong growth of NO_X is found during 2000s as compared to 1990s. All major cities remain as top emitters of NO_X.

13 Modeling Using Global and Regional Models

Chemistry of the atmosphere has been simulated using various types of photochemical models starting from a simple box model to coupled 3D chemistry-transport models. There are several groups in India using such models. A box model is very simple but very useful to understand chemical process, e.g., a study on changes in surface ozone during a solar eclipse (Naja and Lal 1997). Two-dimensional photochemical models have been used to simulate vertical distribution of trace gases and ion chemistry in the atmosphere by groups at PRL and IITM. Coupled chemistry-transport models have become more popular due to advancement in computing power (Beig and Brasseur 2006; Sheel et al. 2010). While these models can incorporate full chemistry and dynamical processes, the resolution is generally limited. Now, Weather Research and Forecasting (WRF) model coupled with Chemistry (WRF-Chem) is being used to simulate the emission, transport, mixing, and chemical transformation of trace gases and aerosols simultaneously with the meteorology (Kumar et al. 2012). These models are online and are able to run on higher resolution.

ARIES has made an annual simulation of tropospheric ozone and related species using the WRF-Chem model over South Asia, for the first time. The model simulated ozone, CO, and NO_x are evaluated using various observations (TES, OMI, and MOPITT). The comparison of model results indicates the capability of the model in reproducing seasonal variations of ozone and CO, but shows some differences in NO_x. Largest differences between model simulations and observations are found during spring when intense biomass burning activity occurs in this region and indicate large uncertainties in anthropogenic and biomass burning emission estimates. Additionally, it is shown that inclusion of aerosols and related heterogeneous chemistry can lead to loss of significant amount of ozone (Kumar et al. 2014). The model results indicate that ozone production in this region is mostly NO_x limited.

14 Heterogeneous Chemistry

The Indian region has high levels of aerosols due to industrial, construction, and agricultural activities, biomass burning and fossil fuel combustion as well as due to the Thar Desert. The average total suspended particulates (TSP) are found to be in the range of 100–300 $\mu g\ m^{-3}$ in different regions of India (Ram and Sarin 2012; Satsangi et al. 2013). There are high levels of black carbon also (Ramachandran and

Rajesh 2007; Ram and Sarin 2012). Ground-based aerosols observations over India showed increase in aerosols loading by 2.3 % per year since 1985 (Moorthy et al. 2013). These high levels of aerosol particles may contribute to loss of ozone as well as its precursors such as NO_x (oxides of nitrogen), HO_x (OH and HO_2), etc. (Jacob 2000). Observed levels of HO_2 and OH (No measurements of these radicals in India so far) are found to be lower than model results by 10–40 % due to uptake by aerosols and better agreement if loss of HO_2 or OH on aerosols are included in the models (Jacob 2000). HONO (HNO_2) is produced (to several ppbv) heterogeneously at night in high NO_x environments and provides HO_x by its photolysis during early morning hours. Field observations show that hydrolysis of N_2O_5 in aerosols is a major sink of NO_x, which is a precursor to ozone formation (Jacob 2000).

Simulations using a global 3D model of tropospheric chemistry show that aerosols decrease O_3 through O(1D) photolysis (5–20 %) at the surface throughout the Northern Hemisphere (largely due to mineral dust) and by a factor of 2 in biomass burning regions mostly due to black carbon (Martin et al. 2003). Aerosol uptake of HO_2 accounts for 10–40 % of total loss of HO_x radicals in the boundary layer over polluted continental regions (largely due to sulfate and organic carbon) and for more than 70 % over tropical biomass burning regions due to organic carbon (Martin et al. 2003).

Loss of ozone and other trace gases on dust particles has been found from aircraft observations and modeling (Tang et al. 2004). They found that in the near-surface layer, the modelled heterogeneous reactions indicated loss of O_3, SO_2, NO_2, and HNO_3 by up to 20, 55, 20, and 95 %, respectively. Model simulation for the South Asia shows that dust storm can reduce many trace gases by 5–100 %, and more than 80 % of this reduction is due to heterogeneous chemistry (Kumar et al. 2014). Hence, aerosols can affect levels of these species. However, there are no such measurements in India.

15 Effects on Crops and Human Health

High levels of surface ozone can damage plants and reduce crop yields such as wheat, rice mustard, black gram, soybean, cotton, and potato. Ozone negatively affected the yield and quality of crops. However, sensitivity of crops differed among species and cultivars. These effects have been studied for wheat and other crops (Rai et al. 2010; Sinha et al. 2015). The yields of major crops of wheat and rice are found to decrease by 10–15 % depending upon the cultivars for both the crops at this place. This type of study has been extended covering entire India using model estimated values of surface ozone. Ghude et al. (2014) have estimated a loss of about 9 % to the total production of cereals in India. However, the experimental work needs to be carried out at other places to confirm such results. Additionally, models depend on several assumptions and findings from such models have to be supported by observations.

The air quality in many of the major cities is reaching hazardous levels very often particularly during the winter season. Extremely high levels of various pollutants such as PM2.5 are often occurring in Delhi and many other cities of India during winter months. These cause severe health effects. The global mean per capita mortality caused by air pollution is about 0.1 % year^{-1}. The highest premature mortality rates are found in the Southeast Asia and Western Pacific regions (about 25 and 46 % of the global rate, respectively) where more than a dozen of the most highly polluted megacities are located (Lelieveld et al. 2013).

16 Concluding Remarks and Future Direction

Measurements of total ozone, its vertical distribution, and surface-level measurements are being pursued in different regions of India. The Dobson instrument is now supported by Brewer instrument at some of the sites by IMD. These are providing long-term total ozone measurements in India. Regular ozone profilings by IMD are being made from New Delhi, Pune, and Trivandrum covering the Indian region. There are additional measurements by a few groups in campaign mode to study with a focus on short-term changes in the troposphere. Surface-level measurements of ozone and its various precursors are being made not only by IMD but also by many other groups all over India. These are providing valuable information on the changing air chemistry in different regions. Most of such measurements have been initiated only in the last 10–15 years. The air chemistry in the Indian region is also being studied using available satellite data and modeling.

The Asian region and in particular China and India are having accelerated economic growth. This is causing increased anthropogenic emissions of various trace gases related to air chemistry, air pollution, and climate change. While the interest in studying air chemistry has increased significantly in the Indian region, there is a need for more high-quality calibrated and detailed measurements. Such in situ measurements will help in identifying changes due to anthropogenic activities. There is a need to understand detailed chemistry by making measurements of intermediate products and radicals also. Ozone and its related trace gases (NO_x, CO, CH_4, NMHCs, VOCs, sulfur compounds) contribute significantly to the changing air chemistry, air quality as well as in chemistry–monsoon and chemistry–climate interactions (Ravishankara et al. 2015; Monks et al. 2015). The air quality in many of the major cities is reaching hazardous levels very often particularly during the winter season. The premature mortality caused by air pollution is increasing (Lelieveld et al. 2013). Secondary aerosols are formed by many of the trace species (NO_x, SO_2, VOCs, etc.). Aerosols can also affect levels of various species through heterophase reactions. Many of these species are also affected due to changes in meteorology, monsoon circulation and climate (Brasseur et al. 2006; Worden et al. 2009; Randel et al. 2011; Vogel et al. 2015). The Indian summer monsoon may be affected by these species and other greenhouse gases directly or indirectly by the formation of secondary aerosols, changes in radiation balance, and the temperature structure (Ueda et al. 2006; Rosenfeld et al. 2008; Ramanathan and Feng 2009; Bollasina et al. 2011; Monks et al. 2015). Hence, measurements of trace gases along with aerosols and their physical and chemical properties at urban and remote sites representing different parts of India will be very useful to study these changes.

We also lack simultaneous measurements of ozone along with its important precursors and aerosols in the free troposphere. Aircraft-based measurements of trace gases, aerosols, and radiation in the free troposphere along with ground-based measurements are needed to delineate anthropogenic and natural effects.

In addition to the requirements of quality measurements, we also need dedicated modeling efforts. Emission scenarios have to be developed for the Indian/Asian region as the emissions in this region are fast increasing/changing. Detailed chemistry-transport models can only predict changes in the future and effort to be made to identify the actions to be taken to control the changes. The changing air chemistry can potentially have deleterious effects on health, crop yields, and monsoon and can perturb the climate.

Acknowledgements We are thankful to ATCTM project of ISRO-GBP and various PIs and others for providing necessary inputs. We are thankful to Dr. C. Mallik of MPIC, Mainz, Germany, for his valuable suggestions.

References

Beig, G. and G. P. Brasseur, Influence of anthropogenic emissions on tropospheric ozone and its precursors over the Indian tropical region during a Monsoon, Geophysical Research Letters, 33, L07808, doi:10.1029/2005GL024949, 2006.

Beig, G., and V. Singh (2007), Trends in tropical tropospheric column ozone from satellite data and MOZART model, Geophys. Res. Lett., 34, L17801, doi:10.1029/2007GL030460.

Beig, G., S. D. Ghude, S. D. Polade, and B. Tyagi (2008), Threshold exceedances and cumulative ozone exposure indices at tropical suburban site, Geophys. Res. Lett., 35, L02802, doi:10.1029/2007GL031434.

Bollasina M. A. Yi Ming, V. Ramaswamy, Anthropogenic Aerosols and the Weakening of the South Asian Summer Monsoon, Science 334, 502 (2011);

Brasseur G. P., M. Schultz, C. Granier, M. Saunois, T. Diehl, M. Botzet, and E. Roeckner (2006), Impact of Climate Change on the Future Chemical Composition of the Global Troposphere, J. Climate, 19, p. 3932

Chakrabarty, D. K., Peshin, S. K.,Pandya, K. V., Shah, N. C., 1998. Long-term trend of ozone column over the Indian region, J. Geophys. Res. 103, 19245-19249.

Chand D., Lal S., Naja M. (2003), Variations of ozone in the marine boundary layer over the Arabian Sea and the Indian Ocean during the 1998 and 1999 INDOEX campaigns. J. Geophys. Res., Vol. 108, p.doi:10.1029/2001JD001589.

Chin, M., Jacob, D.J., Munger, J.W., Parrish, D.D., Doddridge, B.G., 1994. Relationship of ozone and carbon monoxide over North America. Journal of Geophysical Research 99, 14565–14573.

Dave J. V., (1957), Studies on twilight and atmospheric ozone, Ph. D. thesis, Gujarat University, Sept, 1957.

David, L. M., and P. R. Nair (2011), Diurnal and seasonal variability of surface ozone and NO_x at a tropical coastal site: Association with mesoscale and synoptic meteorological conditions, J. Geophys. Res., 116, D10303, doi:10.1029/2010JD015076.

Gaur A., S. N. Tripathi, V. P. Kanawade, V. Tare and S. P. Shukla, (2014) Four-year measurements of trace gases (SO2, NO_x, CO, and O3) at an urban location, Kanpur, in Northern India, J Atmos Chem, DOI 10.1007/s10874-014-9295-8

Ghude, S. D., C. Jena, D. M. Chate, G. Beig, G. G. Pfister, R. Kumar, and V. Ramanathan (2014), Reductions in India's crop yield due to ozone, Geophys. Res. Lett., 41, 5685–5691, doi:10.1002/2014GL060930.

Ghude, S.D., Kulkarni, S.H., Jena, C., Pfister, G.G., Beig, G., Fadnavis, S., and R. J. van der A., 2013. Application of satellite observations for identifying regions of dominant sources of nitrogen oxides over the Indian Subcontinent. J. Geophys. Res. Atmos., 118, 1075-1089.

Glavas, S., Surface ozone and NO_x concentrations at a high altitude Mediterranean site, Greece, Atmos. Environ., 33, 3813–3820, 1999.

Gurjar B. R. et al., (2008) Evalution of emissions and air quality in megacities, Atmos. Env, 42, 1593-1606.

Jacob D. J. (2000), Heterogeneous chemistry and tropospheric ozone, Atmospheric Environment 34,p. 2131.

Kulkarni, A.V., Bahuguna, M, Rathore, B.P. & Singh, S.K. (2007) : Glacial retreat in Himalaya using Indian Remote Sensing satellite data, Current Science, Vol. 92, No. 1

Kelly, N. A., G. T. Wolff, and M. A. Ferman, Sources and sinks of ozone in rural areas, Atmos. Environ., 18, 1251– 1266, 1984.

Kothawale, D. R., and K. Rupa Kumar (2005), On the recent changes in surface temperature trends over India, Geophys. Res. Lett., 32, L18714, doi:10.1029/2005GL023528

Kumar R., M. Naja, S. Venkatramani, and O. Wild, Variations in surface ozone at Nainital, a high altitude site in the central Himalayas, J. Geophys. Res. 115, D16302, doi:10.1029/2009JD013715, 2010

Kumar R., M. Naja, G. G. Pfister, M. C. Barth, C. Wiedinmyer, G. P. Brasseur, Simulations over South Asia with the Weather Research and Forecasting Model with Chemistry (WRF-Chem): Chemistry evaluation and initial results, Geoscientific Model Development, 5, 619-648, doi:10.5194/gmd-5-619-2012, 2012.

Kumar, R., M. C. Barth, S. Madronich, M. Naja, G. R. Carmichael, G. G. Pfister, C. Knote, G. P. Brasseur, N. Ojha, T. Sarangi, Effects of dust aerosols on tropospheric chemistry during a typical pre-monsoon season dust storm in northern India, ACP, 14, 6813-6834, 2014.

Lal, D. M., Ghude, Sachin D. ,S. D. Patil, Santosh H. Kulkarni, Chinmay Jena, S. Tiwari, Manoj K. Srivastava, "Tropospheric ozone and aerosol long-term trends over the Indo-Gangetic Plain (IGP) region, India", Atmospheric Research, 13: 748–755, doi:10.1016/j.atmosres.2012.02.014 (2012).

Lal S., L. K. Sahu, S. Gupta, S. Srivastava, K. S. Modh, S. Venkataramani, T. A. Rajesh (2008) Emission characteristic of ozone related trace gases at a semi-urban site in the Indo-Gangetic plain using intercorrelations, J. Atmos. Chem. Vol. 60, p.189–204, doi:10.1007/s10874-008-9115-0

Lal, S., M. Naja, and A. Jayaraman, Ozone in the marine boundary layer over the tropical Indian Ocean, J. Geophys. Res., 103, 18907-18917, 1998.

Lal, S., M. Naja, and B. H. Subbaraya, Seasonal variations in surface ozone and its precursors over an urban site (Ahmedabad) in the tropics, Atmospheric Environment, 34, 2713-2724, 2000.

Lal, S., S. Venkataramani, S. Srivastava, S. Gupta, M. Naja, T. Sarangi, X. Liu, Transport effects on the vertical distribution of tropospheric ozone over the tropical marine regions surrounding India, J. Geophys. Res., 118, 1513-1524, 2013.

Lal, S., S. Venkataramani, N. Chandra, O. R. Cooper, J. Brioude, M. Naja, Transport effects on the vertical distribution of tropospheric ozone over western India, J. Geophys. Res., 119, 10012-10026, 2014

Lelieveld, J et al. (2001), The Indian Ocean Experiment: widespread air pollution from south and Southeast Asia., Science, Vol. 291, p.1031-1036.

Lelieveld J, C. Barlas, D. Giannadaki, and A. Pozzer, (2013), Model calculated global, regional and megacity premature mortality due to air pollution, Atmos. Chem. Phys., 13, 7023–7037

Londhe, A. L., C. S. Bhosale, J. R. Kulkarni, B. P. Kumari, and D. B. Jadhav (2003), Space-time variability of ozone over the Indian region for the period 1981–1998, J. Geophys. Res., 108 (D24), 8781, doi:10.1029/2002JD002942

Mahapatra P. S. , J. Jena, S. Moharana, H. Srichandan, T. Das , G. Roy Chaudhury and S. N. Das, Surface ozone variation at Bhubaneswar and intra-co-relationship study with various parameters, *J. Earth Syst. Sci.* 121, No. 5, 2012, pp. 1163–1175

Mallik C. and S. Lal, (2011) Changing long-term trends in tropospheric temperature over two megacities in the Indo-Gangetic Plain, Current Science, Vol. 101, p. 637-644

Mallik, C., S. Lal, S. Venkataramani, M. Naja, and N. Ojha (2013), Variability in ozone and its precursors over the Bay of Bengal during post monsoon: Transport and emission effects, J. Geophys. Res. Atmos., 118, doi:10.1002/jgrd.50764.

Mallik C. and S. Lal (2014), Seasonal characteristics of SO2, NO_2, and CO emissions in and around the Indo-Gangetic Plain, Environ. Monit. Assess. 186, 1295-1310, DOI 10.1007/s10661-013-3458

Mallik C., N. Chandra, S. Venkataramani and S. Lal (2016), Variability of atmospheric carbonyl sulfide at a semi-arid urban site in western India, Sci. of the Total Env., 551-552, p.725-737

Mallik C., S. Lal and S. Venkataramani (2015), Trace gases at a semi-arid urban site in western India: variability and inter-correlations, J. Atm. Chem. Vol. 72, p. 143-164

Martin, R. V., D. J. Jacob, R. M. Yantosca, M. Chin, and P. Ginoux, Global and regional decreases in tropospheric oxidants from photochemical effects of aerosols, J. Geophys. Res., 108(D3), 4097, doi:10.1029/2002JD002622, 2003.

Monks, PS, Archibald, AT, Colette, A, Cooper, O, Coyle, M, Derwent, R, Fowler, D, Granier, C, Law, KS, Mills, GE, Stevenson, DS, Tarasova, O, Thouret, V, Von Schneidemesser, E., Sommariva, R, Wild, O & Williams, ML (2015), 'Tropospheric ozone and its precursors from the urban to the global scale from air quality to short-lived climate forcer' Atmos. Chem. and Phys., vol 15, no. 15, pp. 8889-8973, 10.5194/acp-15-8889-2015

Moorthy, K.K., Babu, S.S., Manoj, M.R., Satheesh, S.K., (2013). Buildup of aerosols over the Indian Region. Geophys. Res. Lett. 40, 1011e1014. http://dx.doi.10.1002/grl.50165.

Nair, P.R., D. Chand, S. Lal, K. S. Modh, M. Naja, K. Parameswaran, S. Ravindran, and S. Venkataramani, Temporal variations in surface ozone at Thumba (8.6°N, 77°E) a tropical coastal site in India, Atmospheric Environment, 36, 603-610, 2002.

Naja, M. and S. Lal, (1997), Solar eclipse induced changes in surface ozone at Ahmedabad. Indian J. Radio & Space Phys., 26, 312-318.

Naja, M. and S. Lal, (1996) Changes in surface ozone amount and its diurnal and seasonal patterns, from 1954-55 to 1991-93, measured at Ahmedabad (23N), India. Geophys. Res. Lett., 23, 81-84.

Naja, M. and Lal S. (2002), Surface ozone and precursor gases at Gadanki (13.5N, 79.2E), a tropical rural site in India, J. Geophys. Res. Vol. 107, p. 8-1 to 8-13.

Naja, M., S. Lal, and D. Chand, (2003) Diurnal and Seasonal variabilities in surface ozone at a high altitude site Mt Abu (24.6°N, 72.7°E, 1680 m asl) in India, Atmospheric Environment, 37, 4205-4215.

Naja M., C. Mallik, T. Sarangi, V. Sheel, S. Lal (2014), SO_2 measurements at a high altitude site in the central Himalayas: Role of regional transport, Atmos. Env., Vol. 99, p. 392-402

Naja, M., D. Chand, L. Sahu, and S. Lal, Trace gases over marine regions around India, Indian Journal of Marine Sciences, 33(1),95-106,IPCCode:Int.Cl7.C09K3/30, 2004.

Nishanth T, Praseed KM, Satheesh Kumar MK, Valsaraj KT (2012) Analysis of Ground Level O3 and No_x Measured at Kannur, India. J Earth Sci Climate Change 3:111. doi:10.4172/2157-7617.1000111

Ojha, N., M. Naja, K. P. Singh, T. Sarangi, R. Kumar, S. Lal, M. G. Lawrence, T. M. Butler, and H. C. Chandola (2012), Variabilities in ozone at a semi-urban site in the Indo-Gangetic Plain region: Association with the meteorology and regional processes, J. Geophys. Res., 117, D20301, doi:10.1029/2012JD017716.

Ojha, N., M. Naja, Sarangi T., R. Kumar, P. Bhardwaj, S. Lal, S. Venkataramani, R. Sagar, A. Kumar, and H. C. Chandola, On the processes influencing the vertical distribution of ozone over the central Himalayas: Analysis of yearlong ozonesonde observations, Atmospheric Environment, doi:10.1016/j.atmosenv.2014.01.031, 2014.

Parrish, D.D., Trainer, M., Holloway, J.S., Yee, J.E., Warshawsky, M.S., Fehsenfeld, F.C., 1998. Relationships between ozone and carbon monoxide at surface sites in the North Atlantic region. Journal of Geophysical Research 103, 13357–13376.

Patra, P.K., Lal S., Sheel, V., Subbaraya, B.H., C. Bruehl, R. B. Borchers, and Fabian, P. (2000), Chlorine partitioning in the stratosphere based on insitu measurements. Tellus, Vol. 52B, p.934-946

Peshin S.K and Singh Siddhartha, (2013), WMO/GAW Inter-comparison of Dobson Ozone Spectrophotometer in Tsukuba, Japan-2006 and Inter-comparison of National Standard 112 with Delhi Dobson 36., Mausam, 64, 317-322.

Rai R. M. Agrawal and SB Agrawal (2010), Threat to food security under current levels of ground level ozone: A case study for Indian cultivers of rice, Atm. Env. 44, 4272-82

Ram K., M. M Sarin, (2012), Carbonaceous Aerosols over Northern India: Sources and Spatio-temporal variability, Proceedings of the Indian National Science Academy 78 (3), 523-533

Ramachandran S. and T.A. Rajesh, Black carbon aerosol mass concentrations over Ahmedabad an urban location in western India: Comparison with urban sites in Asia, Europe, Canada and USA, *Journal of Geophysical Research, 112*, D06211, doi:10.1029/2006JD007488, 2007

Ramanathan V., Y. Feng, Air pollution, greenhouse gases and climate change: Global and regional perspectives Atmospheric Environment 43 (2009) 37–50

Randel, W. J., Park, M., Emmons, L., Kinnison, D., Bernath, P., Walker, K. A., Boone, C., and Pumphrey, H. (2011): Asian monsoon transport of pollution to the stratosphere, Science, 328, 611–613, doi:10.1126/science.1182274, 2010. 9943

Ravishankara A. R., Y. Rudich, and D. J. Wuebbles, Physical Chemistry of Climate Metrics, Chemical reviews, 2015

Renuka K., H. Gadhavi, A. Jayaraman, S. Lal, M. Naja, S. V. Bhaskara Rao, Study of Ozone and NO_2 over Gadanki – a rural site in South India, J Atmos Chem, DOI 10.1007/s10874-014-9284-y, 2014

Rosenfeld D., U. Lohmann, G. B. Raga, C. D. O'Dowd, M. Kulmala, S. Fuzzi, A. Reissell, M. O. Andreae (2008), Flood or Drought: How Do Aerosols Affect Precipitation? Science 321, 1309

Sahu, L. K., and S. Lal (2006), Distributions of C2-C5 NMHCs and related trace gases at a tropical urban site in India. Atmos. Env., Vol. 40, p.880-891

Saraf, N., and G. Beig (2004), Long-term trends in tropospheric ozone over the Indian tropical region, Geophys. Res. Lett., 31, L05101, doi:10.1029/ 2003GL018516.

Satsangi A., T. Pachauri, V. Singla, A. Lakhani K. M. Kumari, Water Soluble Ionic Species in Atmos pheric Aerosols: Concentrations and Sources at Agra in the Indo-Gangetic Plain (IGP), Aerosol and Air Quality Research, 13: 1877–1889, 2013

Sitch. S., et al., (2007), Indirect radiative forcing of climate change through ozone effects on the land-carbon sink, Nature, 448, 791-794.

Sahu, S.K., Beig, G., Parkhi, N.S., 2012. Emerging pattern of anthropogenic NO_x emission over Indian subcontinent during 1990s and 2000s. Atmospheric Pollution Research 3, 262–269.

Sinha B., K Singh Sangwan, Y Maurya, V Kumar, C Sarkar, BP Chandra, Assessment of crop yield losses in Punjab and Haryana using 2 years of continuous in situ ozone measurements, Atmospheric Chemistry and Physics 15 (16), 9555-9576, 2015

Sinha,V., Kumar,V., and Sarkar,C.: Chemical composition of pre-monsoon air in the Indo-Gangetic Plain measured using a new PTR-MS and air quality facility: high surface ozone and strong influence of biomass burning, Atmos. Chem. Phys., 14, 5921-5941, 2014.

Sarangi T., M. Naja, N. Ojha, R. Kumar, S. Lal, S. Venkataramani, A. Kumar, R. Sagar and H. C. Chandola (2014), First simultaneous measurements of ozone, CO and NO_y at a high altitude regional representative site in the central Himalayas, J. Geophys. Res. Atmos., 119, doi:10. 1002/2013JD020631.

Sarangi T., M. Naja, S. Lal, S. Venkataramani, P Bhardwaj, N. Ojha, R. Kumar and H. C. Chandola, First observations of light non-methane hydrocarbon (C2-C5) over a high altitude site in the central Himalayas (2016), 125, 450-460.

Singla, V., Pachauri, T., Satsangi, A., Kumari, K. M. and Lakhani A. (2012). Comparison of BTX Profiles and Their Mutagenicity Assessment at Two Sites of Agra, India. *ScientificWorldJournal* 2012: 272853: doi:10.1100/2012/272853.

Sharma Pc, J. C. Kuniyal, K. Chand, R. P. Guleria, P. P. Dhyani, C. Chauhan, Surface ozone concentration and its behaviour with aerosols in the northwestern Himalaya, India, Atmospheric Environment 71 (2013) 44e53

Sheel V., S. Lal, A. Richter and John P. Burrows (2010), Comparison of satellite observed tropospheric NO_2 over India with model simulations, Atmospheric Environment, vol. 44, 3314-3321, 2010

Subbaraya, B.H., Jayaraman, A., Lal S., Venkataramani, S. (1994), Variability in the vertical distribution of ozone measured over Thumba during the 1990 DYANA campaign. J. Atmos. and Terr. Physics, Vol. 56, p.1915.

Tandon A. and A. K. Attri, (2011) Trends in total ozone column over India: 1979-2008, Atmos. Env, 45, 1648-1654.

Tang, Y., et al. (2004), Three-dimensional simulations of inorganic aerosol distributions in east Asia during spring 2001, J. Geophys. Res., 109, D19S23, doi:10.1029/2003JD004201.

Ueda H., A. Iwai, K. Kuwako and M. E. Hori, 2006 Impact of anthropogenic forcing on the Asian summer monsoon as simulated by eight GCMs, GRL Volume 33, Issue 6

Venkanna R., GN Nikhil, TS Rao, PR Sinha, YV Swamy, (2014), Environmental monitoring of surface ozone and other trace gases over different time scales: chemistry, transport and modeling, International Journal of Environmental Science and Technology 12 (5), 1749-1758

Verma N., A. Satsangi, A. Lakhani, K. M. Kumari, Prediction of Ground level Ozone concentration in Ambient Air using Multiple Regression Analysis, JCBPS; Section A; 2015, Vol. 5, No. 4; 3685-3696.

Vogel B., G. Günther, R. Müller, J.-U. Grooß, and M. Riese (2015), Impact of different Asian source regions on the composition of the Asian monsoon anticyclone and on the extratropical lowermost stratosphere, Atmos. Chem. Phys. Discuss., 15, 9941–9995

Worden, J., et al. (2009), Observed vertical distribution of tropospheric ozone during the Asian summertime monsoon, J. Geophys. Res., 114, D13304, doi:10.1029/2008JD010560.

Chapter 15
Proxy Climatic Records of Past Monsoons

R. Ramesh, H. Boragaonkar, S. Band and M.G. Yadava

1 Introduction

Modeling of climate variability on interdecadal to century timescale is a major research area in the field of climatology (e.g., Masson-Delmotte et al. 2013). The shortness of instrumental climate records that cover the past ~ 100 years at the most renders this research rather difficult. In the Himalayan/Tibetan Plateau region, this problem is more severe due to lack of long instrumental records. For a better understanding of climate variability on different timescales, it is important to expand the geographical and temporal spans of climatic records by including historical records and climate proxies such as tree rings, ice-cores, speleothems, marine, and lake sediments. By understanding the past monsoonal variability and its relationship to various forcing factors, it is possible to improve our knowledge of the dynamics of the climate system. This will help develop more reliable predictive models of Asian monsoon variability.

Any well-dated, high-resolution paleoclimatic proxy offers a means of extending the climate record back in time. The Asian monsoon, one of the most important components of the climate system that impacts almost half of the world's population, has recently been the focus of international attention to decipher its driving mechanisms and feedbacks. A time series of Indian monsoon rainfall from instrumental records prepared from a dense network of meteorological stations covering a period of more than a century is available on regional scale as well as for India as a whole (Parthasarathy et al. 1995; Rajeevan et al. 2006; Guhathakurta et al. 2014). Optimizing the number of stations, appropriate statistical model has

R. Ramesh (✉) · S. Band · M.G. Yadava
Physical Research Laboratory, Navrangpura, Ahmedabad 380009, India
e-mail: rramesh@prl.ernet.in

H. Boragaonkar
Indian Institute of Tropical Meteorology, Earth System Science Organization, Pune, Maharashtra 410008, India

M.N. Rajeevan and S. Nayak (eds.), *Observed Climate Variability and Change Over the Indian Region*, Springer Geology, DOI 10.1007/978-981-10-2531-0_15

helped to extend the series further back to early nineteenth century (Sontakke et al. 1993). These indicate a highly variable, trend-less behavior of Indian summer monsoon rainfall with epochal changes. Climate variability studies over the Indian Himalayan region, Nepal, and Tibetan Plateau (Li and Tang 1986; Seko and Tahahashi 1991; Borgaonkar 1996) are mostly based on very specific localities and short data periods. Reasonably, long records of about a century from about 15 stations over the Indian Himalaya, however, do not show any long-term trend in rainfall in the instrumental period.

Climatic anomalies such as the occurrence of glacial phases, transitions of glacial and interglacial periods, and variability of monsoon as reflected in the frequency and intensity of droughts need to be examined with data sets much longer than the instrumental records. There are some records of paleomonsoon in India and other parts of Asia, which have been useful in bringing out some features of long-term climate change (see for a review, Tiwari et al. 2009; Ramesh et al. 2010; Sanyal and Sinha 2010). In this paper, we summarize our current knowledge about monsoons in the past from *quantifiable proxies with unambiguous dates only*. We mainly discuss results based on marine ($\delta^{18}O$ of planktic foraminifera) and terrestrial proxies such as tree rings and speleothems (cave deposits such as stalactites and stalagmites), because these are well dated and used for high-resolution monsoon reconstructions. Stable oxygen isotopic ($\delta^{18}O$) variations in these proxies help quantify monsoon changes with *uncertainties* of the order of $\sim$13 and 8 mm, respectively. These uncertainty estimates are based on (a) the 'amount effect' in tropical rainfall, i.e., a change of 1.5 ‰ per 100 mm change of monthly rain, and (b) the experimental uncertainty of $\delta^{18}O$ measurements, i.e., ± 0.2 and ± 0.1 ‰, respectively, for tree-ring cellulose and speleothem carbonates.

However, the $\delta^{18}O$ values recorded in continental proxies at some sites may not necessarily reflect local rainfall variability; instead, they could be useful in the reconstruction of large-scale convective systems or upstream rainfall (Lekshmy et al. 2014, 2015; Midhun and Ramesh 2015). Figure 1 shows the locations of marine and terrestrial records discussed below in detail.

2 Holocene Monsoon Variability

Climate changes over northwest India inferred from climate proxies indicate warm and humid climate with frequent floods between 10,000 and 4500 years ago (Pant and Maliekal 1987). Results from more precisely dated lake sediments (Enzel et al. 1999) from Lunkaransar, Rajasthan, however, showed that the monsoon fluctuated often in the early Holocene (the last $\sim$10,000 years) and then intensified abruptly around 6300 years ago. The lake completely desiccated around 4800 years ago. The major Harappan-Indus civilization began and flourished in this region 1000 years after desiccation of the lake during arid climate. Globally, six major climate events are recorded in the Holocene 9000–8000 ,6000–5000, 4200–3800, 3500–2500, 1200–1000, and 600–150 a (Mayewski et al. 2004).

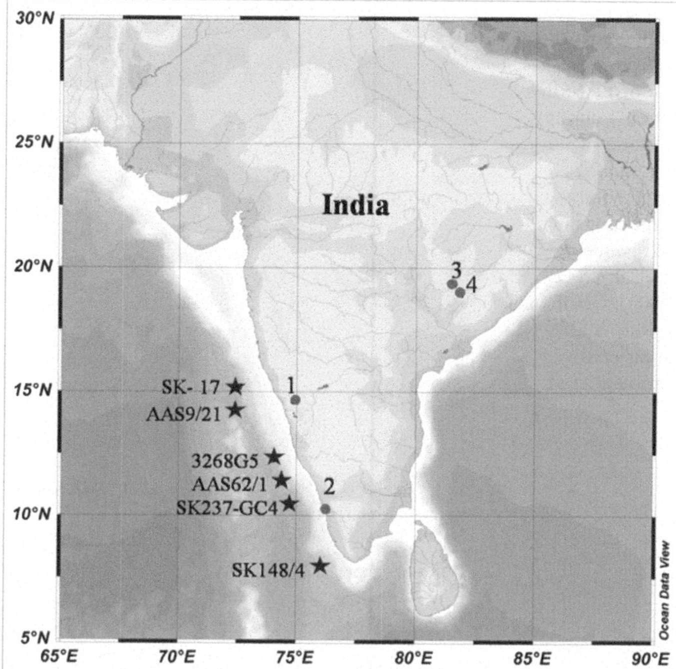

Fig. 1 Locations of marine and terrestrial proxies. Stars are sediment cores, Core SK-17 (Singh et al. 2006), Core AAS9/21 (Govil and Naidu 2010), Core 3268G5 (Sarkar et al. 2000), Core AAS62/1 (Kessarkar et al. 2013), Core SK237-GC4 (Saraswat et al. 2013), and Core SK148/4 (Rao et al. 2010). Terrestrial proxies are (*1*) Akalagavi cave stalagmite (Yadava et al. 2004), (*2*) Parambikulam tree cellulose (Managave et al. 2011), and (*3*) and (*4*) Dandak and Kotumsar caves (Yadava and Ramesh 2005)

3 Marine Records of Holocene Monsoon

Contrary to terrestrial deposits, marine deposits preserve complete and continuous monsoon record. Extensive studies have been carried out on the evolution of monsoon through the Holocene (Sirocko et al. 1993; Tiwari et al. 2011) using marine sediments. $\delta^{18}O$ variations recorded by marine sediment cores from the eastern Arabian Sea have shown a consistent increasing trend in the monsoon rainfall (Sarkar et al. 2000) during the Holocene. Abrupt increase in the monsoon was observed between 10000–9500 year BP (Overpeck et al. 1996). The increasing trend of monsoon rainfall during the Holocene is confirmed by $\delta^{18}O$ variations in speleothems from Central India (Yadava and Ramesh 2005) and several other marine sediment cores (Fig. 2) from the eastern Arabian Sea (Sarkar et al. 2000; Singh et al. 2006; Govil and Naidu 2010; Rao et al. 2010; Kessarkar et al. 2013; Saraswat et al. 2013). For a recent summary, refer to the study by kumar and Ramesh (2016).

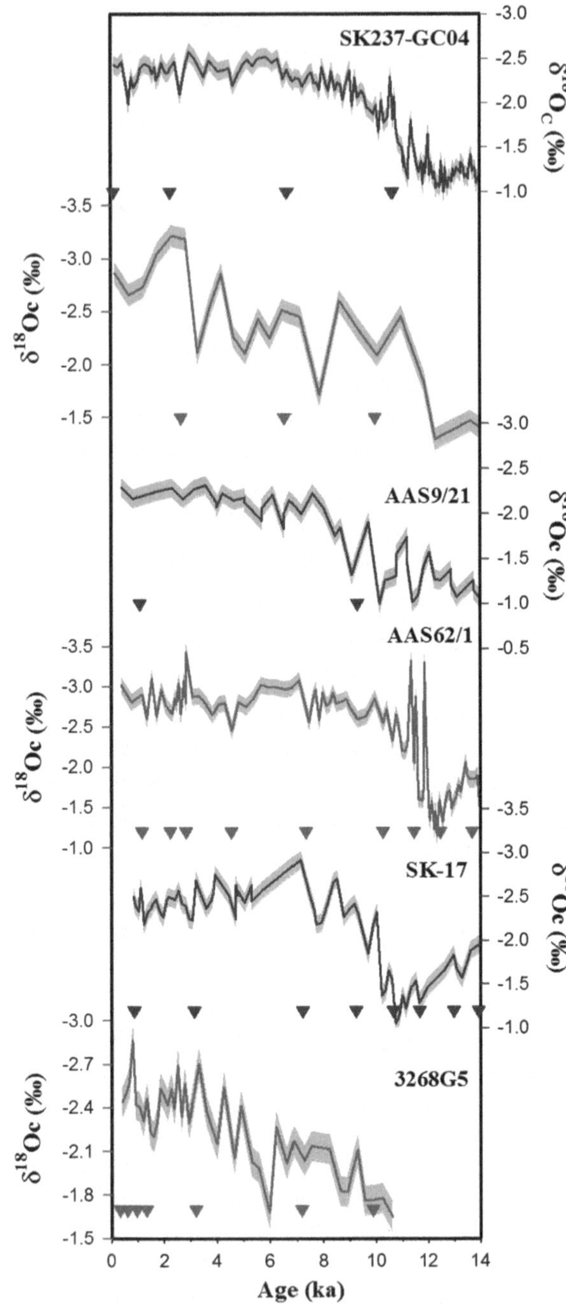

Fig. 2 $\delta^{18}O$ variations in planktic foraminifera from six different cores from the eastern Arabian Sea (see this figure for locations) during the Holocene, plotted with error bars (*gray shaded* regions). *Inverted triangles* represent ^{14}C ages for the respective core sediments. *Source* Sarkar et al. (2000), Singh et al. (2006), Govil and Naidu (2010), Rao et al. (2010), Kessarkar et al. (2013), and Saraswat et al. (2013)

4 Historical Climate Records

To put the contemporary climate change in perspective, historical, documentary evidences are very useful. Such evidences, however, are generally found scattered in many places and languages with different interpretations. Sometimes, the information is vague and fragmentary. Therefore, to extract valuable information on past climate, a systematic and logical approach is crucial in interpreting historical evidences. Relative to other parts of Asia, historical records in India and China are relatively longer. They provide informative evidences of past climate. In India, historical climate evidences are mostly related to monsoonal rain, for example, occurrence of drought or flood and no harvest due to lack of rain or good grain production. A systematic effort was made by Pant et al. (1993) to extract climate-related information, such as droughts and floods leading to famine, for the last millennium from historical records of various climatic zones of India (Fig. 3a). Relatively, fewer droughts were observed during CE 900–1600 due to non-availability of related historical information. Since CE 1600, droughts are more frequent and are more or less randomly distributed (Fig. 3b).

5 Terrestrial Records

Although the oceanic records are continuous, they are proxy to intensification of southwesterly winds. Terrestrial records on the other hand are direct measure of reconstructing monsoon rainfall. Also, unlike oceanic records, they have annual to decadal resolution and can be precisely dated. Two such proxies will be discussed in detail below.

6 Dendroclimatic Reconstruction

Information on climate during the last 1000 years, which is the period of man's greatest impact on the planet, is vital to understand the various external and internal forcing on the climate and thereby make reliable future projections. Significant contributions to climate science within the last decade have firmly established tree rings as valuable sources of proxy data for evaluating long-term climate variability/trends and as useful tools for developing long-term records of extreme climatic events (Mann et al. 1999).

6.1 New Tree-Ring Chronologies: Western Himalaya

Kinnor and Gangotri regions of the higher ranges of the Western Himalaya (altitudes varying between 2850 and 3200 m above the mean sea level (msl)) were

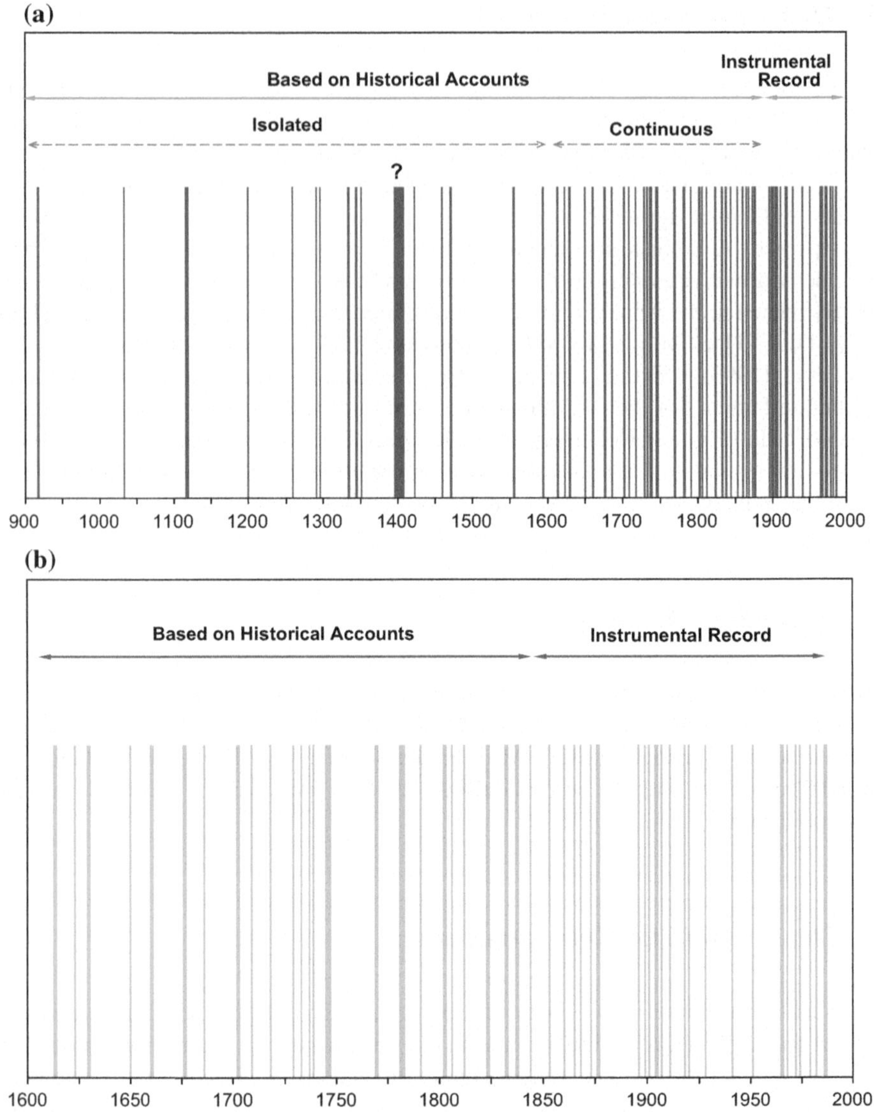

Fig. 3 **a** Occurrence of large-scale droughts over India during 900–1991 CE. Historical records were sporadic and isolated up to 1600 CE and continuous later. **b** Frequency of occurrence of drought events were relatively higher and randomly distributed after 1600 CE. *Source* Pant et al. (1993)

explored. The northern part of this region is classified as alpine, glacial climatic zone. Dry cold weather with severe and prolonged winter prevails here. The region behaves as upper tree line boundary; it is under snow cover during winter and spring. In five chronologies of some conifers (e.g., *Picea smithiana* Boiss and *Cedrus deodara* D. Don), a higher growth was seen in the last few decades. A strong positive relationship with the mean annual and winter (December–February) temperatures of the concurrent year is seen in these chronologies. Using these five chronologies, a 553-year-long master chronology was prepared (Borgaonkar et al. 2010a, b). This shows a few decadal and longer epochs of Little Ice Age (LIA) cooling during CE 1453–1590 and CE 1780–1930 (Fig. 4). These are relatable to the other proxy records (e.g., glacial fluctuations) (Mayewski et al. 1980; Thompson et al. 1995; Yao et al. 1995; Duan and Yao 2003).

Millennial-scale reconstruction of precipitation from tree-ring oxygen isotope records from high mountain ranges of north Pakistan indicates wet condition during recent centuries and dry conditions at the beginning of the past millennium and also during eighteenth and early nineteenth centuries (Treydte et al. 2006). However, isotope analysis of Dasuopu glacier from the central Himalayas indicated a major precipitation trend in the central Himalayas since last 300 years (Duan and Yao 2003). Monsoon rainfall in the central Himalayas has decreased over the past decades in the condition of global warming. This was also shown by Sano et al. through $\delta^{18}O$ measurements on tree rings from Nepal (Sano et al. 2010, 2011).

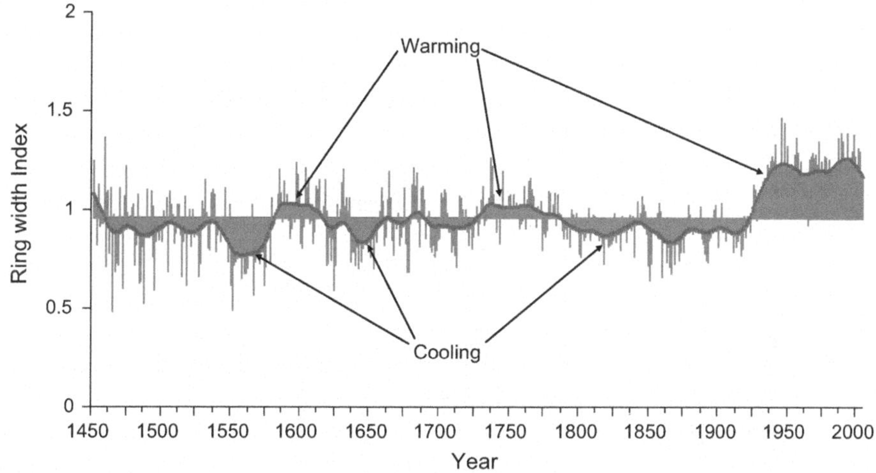

Fig. 4 A 553 years (1452–2004 CE) long tree-ring index chronology of high-altitude Himalayan conifers from Western Himalaya. *Smooth gray line* is 30-year cubic spline filter. Suppressed (cooling) and released (warming) growth patterns in tree-ring chronology have also been observed to be well related to the past glacial fluctuation records of the region. After Borgaonkar et al. (2010a) (color figure online)

6.2 Studies Based on Tropical Teak

To reconstruct past monsoon variations with a high resolution, good tree-ring data networks have been created in tropics of south and southwest Asia, during the last decade. Teak (*Tectona grandis*) from India, Java, Indonesia, and Thailand has been shown to be a potential source for past monsoon reconstruction of monsoon spatially (D'Arrigo et al. 1994; Buckley et al. 2007; Pant and Borgaonkar 1983; Murphy and Whetton 1989; Pumijumnong et al. 1995; Jacoby and D'Arrigo 1990; Bhattacharyya et al. 1992; Shah et al. 2007; Borgaonkar et al. 2010a, b). Other global parameters (e.g., El Nino Southern Oscillation, ENSO) related to monsoon could also be studied. Palmer Drought Sensitivity Index (PDSI) has been reconstructed for Java and Indonesia (D'Arrigo et al. 2006). This index reveals the drought history of 200 years and the influence of ENSO on monsoon. Teak chronologies are moisture sensitive. Such chronologies from South and Southeast Asia are capable of capturing ENSO-induced droughts. The variation of stable isotopic ratio of hydrogen (δD) in teak from Mumbai was related to the amount of rainfall and mean maximum temperature (Ramesh et al. 1989). The highest achievable resolution of ~20 days in teak was demonstrated by Managave et al. (2010a, b). Figure 5 (lower panel) shows the monsoon reconstruction of Parambikulam, Kerala, for more than two centuries in the past (Managave et al. 2011). This can be compared with the three-century-long monsoon reconstruction for northern Karnataka (top panel of the same Fig. 4) based on $\delta^{18}O$ variations in a stalagmite that had annual resolution.

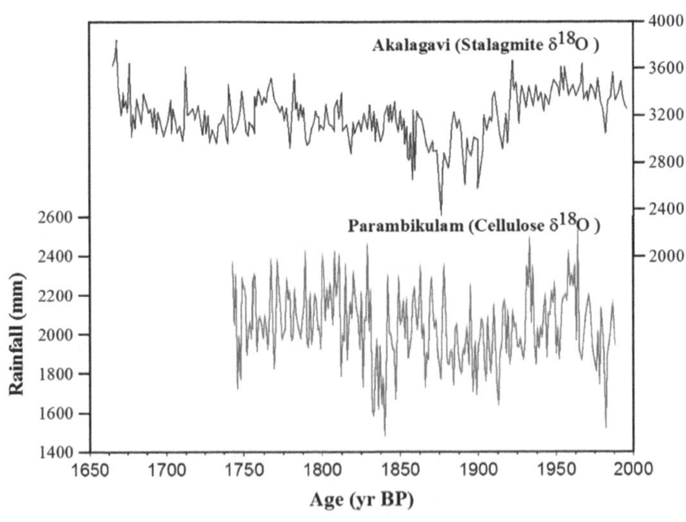

Fig. 5 Monsoon rainfall reconstructed from Akalagavi speleothem (*top*) and Parambikulam teak cellulose (*bottom*). *Solid curves* show the reconstructed values

6.3 Southern Indian Tree-Ring Chronologies from Kerala that Are Drought Sensitive

Borgaonkar et al. (2010a, b) created a 523-year (A.D. 1481–2003) tree-ring-width index chronology of Teak from Kerala from three forest sites. This indicates high sensitivity to Indian monsoon droughts. Annual rainfall in Kerala correlates significantly with monsoon (June–September) rainfall and also with all Indian rainfall. The availability of moisture for the growth of teak greatly depends on the excess/deficient rainfall. Whenever the monsoon was deficient in the past, associated with El Niño since the late eighteenth century (Fig. 6), it was observed that the frequency of occurrence of low tree growth was higher. Earlier, many low tree growth years were detected during known El Niño events, which are probably related to deficient Indian monsoon rainfall. It is known that the relationship between Indian summer monsoon rainfall (ISMR) and El Niño is negative in general. The spatial correlations between Kerala tree-ring chronology and sea surface temperatures (SSTs) over the Niño regions also follow similar patterns. Thus, ENSO-related monsoon signals are recorded in the tree-ring records with fidelity. Prior to the period of instrumental weather records, these tree-ring chronologies with a high degree of sensitivity to monsoon are quite useful to understand the vagaries of monsoon rainfall prior to the period of recorded data.

7 Speleothem Records from India

Speleothems are secondary carbonate deposits formed in limestone-dominated karst topography. Two morphotypes, stalactites and stalagmites, are preferentially used for paleoclimatic reconstructions. Since the speleothems have potential of recording annual to decadal monsoon variability and are precisely dated, the use of this proxy has increased in the last two decades. Fortunately in India, most of the limestone caves housing speleothems are located in the core monsoon region and a few in the Himalayan region where the influence of westerlies can also be studied in detail (Laskar et al. 2011, 2013; Lone et al. 2014; Kotlia et al. 2015; Allu et al. 2014; Yadava and Ramesh 1999, 2006).

Based on Gupteshwar stalactite and Dandak stalagmite, increasing trend in rainfall was recorded from 3400 to 2000 years BP (Yadava and Ramesh 2005). Monsoon was significantly stronger during 1200–800 years BP, the Medieval warm period (Laskar et al. 2013). The monsoon rainfall at Central India reconstructed from speleothem $\delta^{18}O$ (marked 3 in Fig. 1) (Sinha et al. 2007) shows the longest and severest droughts occurred in the fourteenth and mid-fifteenth centuries.

Annually resolved stalagmites from Akalagavi cave, India, and from Salalah region, Oman, point to increased rainfall during CE 1666 (Burns et al. 2002; Yadava et al. 2004). Low rainfall during CE 1880–1900 is also observed in both the records with increasing trend from CE 1991 to 1975. In a stalagmite from Panigarh cave, Uttar Pradesh, wetter conditions were observed during the LIA contrary to the

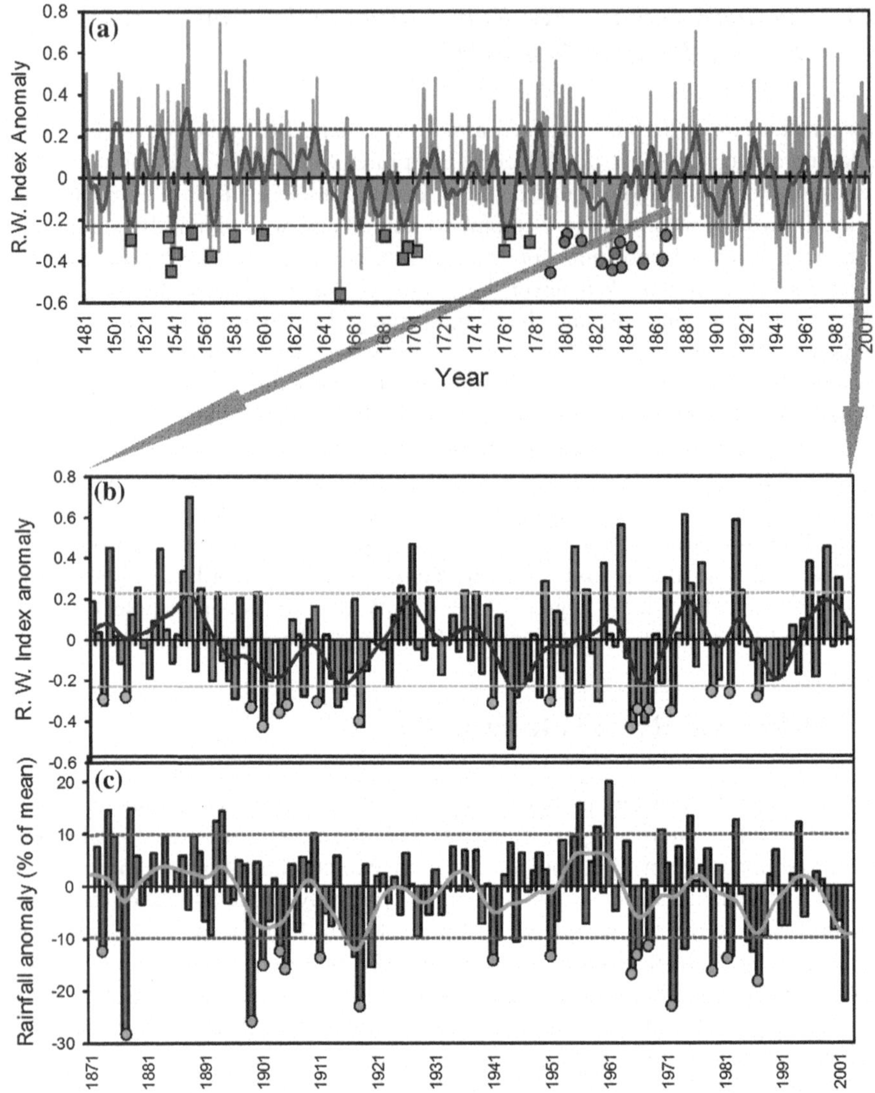

Fig. 6 **a** Tree-ring width index anomaly of Kerala Teak Ring Chronology (KTRC) in relation to long-term mean. *Smooth line* is 10-year cubic spline fit. *Dashed lines* in all the figures indicate "Mean ± Std. Dev." limit. *Filled circles* indicate low growth years occurred during the deficient rainfall (droughts) years associated with the El Nino. *Filled squares* are low growth years associated with El Nino years. **b**, **c** KTRC and ISMR anomalies, respectively, during the instrumental period CE 1871–2003. *Filled circles* in fig. **b** are low growth years and have one to one correspondence with deficient monsoon rainfall (drought) years associated with El Nino shown as *Red circles* in **c**. After Borgaonkar et al. (2010b) (color figure online)

drier phase in the core monsoon region of India (Liang et al. 2015). Kotlia et al. (2014) based on a stalagmite record from Sainji cave suggested that westerlies play an important role in governing the late Holocene climate of the Indian Himalaya. The wetter conditions during LIA may have been due to strengthening of westerlies between CE 1450 and 1700.

8 Conclusions

Paleomonsoon records show that cold periods of the climatic history had winter like circulation, whereas warm periods were characterized by a strong summer monsoon flow. General inferences on climate of the Holocene ($\sim$10 ka) drawn from various proxy sources indicate a clear increasing trend in the monsoon since 10 ka. In general, there was a warm humid climate with frequent floods during 10–4.5 ka over the northern regions of the subcontinent. The Little Ice Age over the Tibetan Plateau was not cold as in mid latitudes.

The Asian region has a vast potential of historical evidences. However, systematic and collective efforts are needed to elaborate the possibility of extracting paleoclimate data from many parts of the region to assess climate variability better.

Acknowledgments The authors are grateful to ISRO-GBP for funding.

References

Allu, N.C., Tiwari, M., Yadava, M.G., Chi, N., Shen, C.-C., Belgaonkar, S.P., Ramesh, R. and, Laskar, A.H., 2014. Stalagmite δ 18 O variations in southern India reveal divergent trends of Indian Summer Monsoon and East Asian Summer Monsoon during the last interglacial. Quat. Int. 10–15. doi:10.1016/j.quaint.2014.12.014

Bhattacharyya, A., Yadav, R.R., Borgaonkar, H.P. and, Pant, G.B., 1992. Growth-ring analysis of Indian tropical trees: dendroclimatic potential. Curr. Sci. 62, 736–741.

Borgaonkar, H.P., 1996. Tree Growth Climate Relationship and Long-Term Climate Change Over The Western Himalaya: A Dendroclimatic Approach,. Ph.D. Thesis, Univerisity of Pune, Pune-411008.

Borgaonkar H.P., Sikder A.B and, Somaru Ram, 2010. High altitude forest sensitivity to the recent warming: A tree-ring analysis of conifers from Western Himalaya, India. Quat. Int. 236, 158–166.

Borgaonkar, H.P., Sikder, A.B., Ram, S. and, Pant, G.B., 2010. El Nino and related monsoon drought signals in 523-year-long ring width records of teak (Tectona grandis L.F.) trees from south India. Palaeogeogr. Palaeoclimatol. Palaeoecol. 285, 74–84. doi:10.1016/j.palaeo.2009.10.026

Buckley B. M, Palakit, K., Duangsathaporn, K., Sanguantham, P. and, Prasomsin, P., 2007. Decadal scale droughts over northwestern Thailand over the past 448 years: Links to the tropical Pacific and Indian Ocean sectors. Clim. Dyn. 29, 63–71. doi:10.1007/s00382-007-0225-1

Burns, S.J., Fleitmann, D., Mudelsee, M., Neff, U., Matter, A., and, Mangini, A., 2002. A 780-year annually resolved record of Indian Ocean monsoon precipitation from a speleothem from south Oman. J. Geophys. Res. 107, 4434. doi:10.1029/2001JD001281

D'Arrigo, R.D., Jacoby Jr, G.. and, Krusic, P.J., 1994. Progress in dendroclimatic studies in Indonesia. Terr. Atmos. Ocean. Sci. 5.3, 349–363.

D'Arrigo, R.D., Wilson, R., Palmer, J., Krusic, P., Curtis, A., Sakulich, J., Bijaksana, S., Zulaikah, S. and, Ngkoimani, L.O., 2006. Monsoon drought over Java, Indonesia, during the past two centuries. Geophys. Res. Lett. 33. doi:10.1029/2005GL025465

Duan, K. and, Yao., T., 2003. Monsoon Variability in the Himalayas under the Condition of Global Waming. J. Meteorol. Soc. Japan 81, 251–257.

Enzel, Y., Ely, L.L., Mishra, S., Ramesh, R., Amit, R., Lazar, B., Rajaguru, S.N., Baker, V.R. and, Sandler, A., 1999. High-Resolution Holocene Environmental Changes in the Thar Desert, Northwestern India. Science (80-.). 284, 125–128. doi:10.1126/science.284.5411.125

Govil, P. and, Naidu, P.D., 2010. Evaporation-precipitation changes in the eastern Arabian Sea for the last 68 ka: Implications on monsoon variability. Paleoceanography 25, 1–11. doi:10.1029/2008PA001687

Guhathakurta, P., Rajeevan, M., Sikka, D.R., Tyagi, A., 2014. Observed changes in southwest monsoon rainfall over India during 1901–2011. Int. J. Climatol. 35, 1881–1898. doi:10.1002/joc.4095

Jacoby Jr, G.. and, D'Arrigo, R.D., 1990. Teak (Tectona grandis L.F.), A tropical species of large-scale dendroclimatic potential. Dendrochronologia 8, 83–98.

Kumar, P.K. and Ramesh, R., 2016. Revisiting reconstructed Indian monsoon rainfall variations during the last ~ 25 ka from planktonic foraminiferal $\delta^{18}O$ from the Eastern Arabian Sea. *Quaternary International.*

Kessarkar, P.M., Purnachadra Rao, V., Naqvi, S.W.A. and, Karapurkar, S.G., 2013. Variation in the Indian summer monsoon intensity during the Bølling-Ållerød and Holocene. Paleoceanography 28, 413–425. doi:10.1002/palo.20040

Kotlia, B.S., Singh, A.K., Joshi, L.M., Dhaila, B.S., 2015. Precipitation variability in the Indian Central Himalaya during last ca. 4,000 years inferred from a speleothem record: Impact of Indian Summer Monsoon (ISM) and Westerlies. Quat. Int. 371, 244–253. doi:10.1016/j.quaint.2014.10.066

Kotlia, B.S., Singh, A.K., Joshi, L.M., Dhaila, B.S., 2014. Precipitation variability in the Indian Central Himalaya during last ca. 4,000 years inferred from a speleothem record: Impact of Indian Summer Monsoon (ISM) and Westerlies. Quat. Int. doi:10.1016/j.quaint.2014.10.066

Laskar, A.H., Raghav, S., Yadava, M.G., Jani, R.A., Narayana, A.C. and, Ramesh, R., 2011. Potential of Stable Carbon and Oxygen Isotope Variations of Speleothems from Andaman Islands, India, for Paleomonsoon Reconstruction. J. Geol. Res. 2011, 1–7. doi:10.1155/2011/272971

Laskar, A.H., Yadava, M.G., Ramesh, R., Polyak, V.J., Asmerom, Y., 2013. A 4 kyr stalagmite oxygen isotopic record of the past Indian Summer Monsoon in the Andaman Islands. Geochemistry, Geophys. Geosystems 14, 3555–3566. doi:10.1002/ggge.20203

Lekshmy, P.R., Midhun, M., Ramesh, R., 2015. Spatial variation of amount effect over peninsular India and Sri Lanka: Role of seasonality. Geophys. Res. Lett. 42, 5500–5507. doi:10.1002/2015GL064517

Lekshmy, P.R., Midhun, M., Ramesh, R., Jani, R.A., 2014. 18O depletion in monsoon rain relates to large scale organized convection rather than the amount of rainfall. Sci. Rep. 4, 5661. doi:10.1038/srep05661

Li, C. and, Tang, M., 1986. Changes of air temperature of Qunghai-xizang plateau and its neighbourhood in the past 30 years. Plateau Meteorol. 5, 332–341.

Liang, F., Brook, G.A., Kotlia, B.S., Railsback, L.B., Hardt, B., Cheng, H., Edwards, R.L., Kandasamy, S., 2015. Panigarh cave stalagmite evidence of climate change in the Indian Central Himalaya since AD 1256: Monsoon breaks and winter southern jet depressions. Quat. Sci. Rev. 124, 145–161. doi:10.1016/j.quascirev.2015.07.017

Lone, M.A., Ahmad, S.M., Dung, N.C., Shen, C.C., Raza, W. and, Kumar, A., 2014. Speleothem based 1000-year high resolution record of Indian monsoon variability during the last deglaciation. Palaeogeogr. Palaeoclimatol. Palaeoecol. 395, 1–8. doi:10.1016/j.palaeo.2013.12.010

Managave, S.R., Sheshshayee, M.S., Bhattacharyya, A. and, Ramesh, R., 2011. Intra-annual variations of teak cellulose δ18O in Kerala , India : implications to the reconstruction of past summer and winter monsoon rains. Clim. Dyn. 37, 555–567. doi:10.1007/s00382-010-0917-9

Managave, S.R., Sheshshayee, M.S., Borgaonkar, H.P. and, Ramesh, R., 2010. Intra-annual oxygen isotope variations in Central Indian teak cellulose: possibility of improved resolution for past monsoon reconstruction. Curr. Sci. 98, 930–937.

Managave, S.R., Sheshshayee, M.S., Borgaonkar, H.P. and, Ramesh, R., 2010. Past break-monsoon conditions detectable by high resolution intra-annual d 18 O analysis of teak rings 37. doi:10.1029/2009GL041172

Mann, M.E., Bradley, R.S. and, Hughes, M.K., 1999. Northern hemisphere temperatures during the past millennium: Inferences, uncertainties, and limitations. Geophys. Res. Lett. doi:10.1029/1999GL900070

Masson-Delmotte, V., Schulz, M., Abe-Ouchi, A., Beer, J., Ganopolski, A., González Rouco, J.F., Jansen, E., Lambeck, K., Luterbacher, J., Naish, T. and Osborn, T., 2013. Information from paleoclimate archives. Climate change, in: IPCC Chapter 5. pp. 383–464.

Mayewski, P.A., Pregent, G.P., Jeschke., P.A. and, Ahmad, N., 1980. Himalayan and Trans-Himalayan Glacier Fluctuations and the South Asian Monsoon Record. Arct. Alp. Res. 12, 171–182.

Mayewski, P.A., Rohling, E.E., Stager, J.C., Karlén, W., Maasch, K.A., Meeker, L.D., Meyerson, E.A., Gasse, F., van Kreveld, S., Holmgren, K., Lee-Thorp, J., Rosqvist, G., Rack, F., Staubwasser, M., Schneider, R.R. and, Steig, E.J., 2004. Holocene climate variability. Quat. Res. 62, 243–255. doi:10.1016/j.yqres.2004.07.001

Midhun, M., Ramesh, R., 2015. Validation of δ18O as a proxy for past monsoon rain by multi-GCM simulations. Clim. Dyn. doi:10.1007/s00382-015-2652-8

Murphy J. O and, Whetton P. H, 1989. A re-analysis of tree-rings chronology from Java; Proce. koninkl. Nederl. Akad. Wetensch. B 92, 241–257.

Overpeck, J., Anderson, D., Trumbore, S. and, Prell, W., 1996. The southwest Indian Monsoon over the last 18 000 years. Clim. Dyn. 12, 213–225. doi:10.1007/BF00211619

Pant, G.B. and, Borgaonkar H.P., 1983. Growth rings of teak trees and regional climatology. Environ. Manag. Ed. by L.R Singh, R.C.Tiwari R.P.Srivastav, Allahabad Geogr. Soc. India 153–155.

Pant, G.B. and, Maliekal, J.A., 1987. Holocene climatic changes over northwest India: An appraisal. Clim. Change 10, 183–194. doi:10.1007/BF00140254

Pant, G.B., Rupa Kumar, K., Sontakke, N.A. and, Borgaonkar H.P., 1993. Climate variability over India on century and larger time scale In: "Advances in tropical meteorological" (Eds. Keshavamurthy, R.N. and Joshi, P.C.),. Tata McGraw-Hills Publ. Co. Ltd.New Delhi 149–158.

Parthasarathy, B., Munot, A.A. and, Kothawale, D.R., 1995. Monthly and seasonal rainfall series for all-India homogeneous regions and meteorological subdivions : 1871–1994, I.I.T.M. Res. Rep. NO. RR-65.

Pumijumnong, N., Eckstein, D. and, Sass, U., 1995. Tree-Ring Research on Tectona Grandis in Northern Thailand. IAWA J. 16, 385–392. doi:10.1163/22941932-90001428

Rajeevan, M., Bhate, J., Kale, J.D., Lal, B., 2006. High resolution daily gridded rainfall data for the Indian region: Analysis of break and active monsoon spells. Curr. Sci. 91, 296–306.

Ramesh, R., Bhattacharya, S.K. and, Pant, G.B., 1989. Climatic significance of δD variations in a tropical tree species from India. Nature. doi:10.1038/337149a0

Ramesh, R., Tiwari, M., Chakraborty, S., Managave, S.R., Yadava, M.G. and, Sinha, D.K., 2010. Retrieval of south Asian monsoon variation during the Holocene from natural climatic archives. Curr. Sci. 99, 1770–1786.

Rao, V.P., Kessarkar, P.M., Thamban, M. and, Patil, S.K., 2010. Paleoclimatic and diagenetic history of the late quaternary sediments in a core from the Southeastern Arabian Sea: Geochemical and magnetic signals. J. Oceanogr. 66, 133–146. doi:10.1007/s10872-010-0011-2

Sano, M., Ramesh, R., Sheshshayee, M. S., and Sukumar, R., 2011. Increasing aridity over the past 223 years in the Nepal Himalaya inferred from a tree-ring δ18O chronology. The Holocene. doi:0.1177/0959683611430338

Sano, M., Sheshshayee, M.S., Managave, S., Ramesh, R., Sukumar, R., Sweda, T., 2010. Climatic potential of δ18O of Abies spectabilis from the Nepal Himalaya. Dendrochronologia 28, 93–98. doi:10.1016/j.dendro.2009.05.005

Sanyal, P. and, Sinha, R., 2010. Evolution of the Indian summer monsoon : synthesis of continental records. Geol. Soc. London, Spec. Publ. 342, 153–183.

Saraswat, R., Lea, D.W., Nigam, R., Mackensen, A. and, Naik, D.K., 2013. Deglaciation in the tropical Indian Ocean driven by interplay between the regional monsoon and global teleconnections. Earth Planet. Sci. Lett. 375, 166–175. doi:10.1016/j.epsl.2013.05.022

Sarkar, A., Ramesh, R., Somayajulu, B.L.K., Agnihotri, R., Jull, A.J.T. and, Burr, G.S., 2000. High resolution Holocene monsoon record from the eastern Arabian Sea. Earth Planet. Sci. Lett. 177, 209–218. doi:10.1016/S0012-821X(00)00053-4

Seko, K. and, Tahahashi, S., 1991. Characteristics of winter precipitation and its effect on glaciers in the Nepal Himalaya. Bull. Glacial Res. 9, 9–16.

Shah, S.K., Bhattacharyya, A., Chaudhary, V., 2007. Reconstruction of June – September precipitation based on tree-ring data of teak (Tectona grandis L .) from Hoshangabad , Madhya Pradesh , India 25, 57–64. doi:10.1016/j.dendro.2007.02.001

Singh, A.D., Kroon, D. and, Ganeshram, R.S., 2006. Millennial scale variations in productivity and OMZ intensity in the Eastern Arabian Sea. J. Geol. Soc. India 68, 369–377.

Sinha, A., Cannariato, K.G., Stott, L.D., Cheng, H., Edwards, R.L., Yadava, M.G., Ramesh, R. and, Singh, I.B., 2007. A 900-year (600 to 1500 A.D.) record of the Indian summer monsoon precipitation from the core monsoon zone of India. Geophys. Res. Lett. 34. doi:10.1029/2007GL030431

Sirocko, F., Sarnthein, M., Erlenkeuser, H., Lange, H., Arnold, M. and, Duplessy, J.C., 1993. Century-scale events in monsoonal climate over the past 24,000 years. Nature 364, 322–324. doi:10.1038/364322a0

Sontakke, N.A., Pant, G.B. and, Singh, N., 1993. Optimization of the raingauges for a representative all Indian and Sub-divisional rainfall series. Theor. Appli. Clim. 47, 159–173.

Thompson, L.G., Mosley-Thompson, E., Davis, M.E., Lin, P.N., Dai, J., Bolzan J. F and, Yao, T., 1995. A 1000 year ice core climate record from the Guliya Ice Cap, China : its relationship to global climate variability. Ann. Glaciol. 21, 175–181.

Tiwari, M., Managave, S.R., Yadava, M.G. and, Ramesh, R., 2009. Spatial and temporal coherence of paleomonsoon records from marine and land proxies in the Indian region during the past 30 ka. Spec. Platin. Jubil. Publ. Indian Acad. Sci. Bangalore, (ed. N. Mukunda) 517–535.

Tiwari, M., Singh, A.K. and, Ramesh, R., 2011. High-Resolution Monsoon Records Since Last Glacial Maximum: A Comparison of Marine and Terrestrial Paleoarchives from South Asia. J. Geol. Res. 2011, 1–12. doi:10.1155/2011/765248

Treydte, K.S., Schleser, G.H., Helle, G., Frank, D.C., Winiger, M., Haug, G.H. and, Esper, J., 2006. The twentieth century was the wettest period in northern Pakistan over the past millennium. Nature 440, 1179–82. doi:10.1038/nature04743

Yadava, M.G. and, Ramesh, R., 2006. Stable Oxygen and Carbon Isotope Variations as Monsoon Proxies : A Comparative Study of Speleothems from Four Different Locations in India. J. Geol. Soc. India 68, 461–475.

Yadava, M.G. and, Ramesh, R., 2005. Monsoon reconstruction from radiocarbon dated tropical Indian speleothems. The Holocene, 15(1), pp.48-59.

Yadava, M.G. and, Ramesh, R., 1999. Speleothems-Useful proxies for past monsoon rainfall. J. Sci. Ind. Res. (India). 58, 339–348.

Yadava, M.G., Ramesh, R., Pant, G.B., 2004. Past monsoon rainfall variations in peninsular India recorded in a 331-year-old speleothem. The Holocene 14, 517–524. doi:10.1191/0959683604hl728rp

Yao, T., Jiao, K., Tian, L., Li, Z., Li, Y., Liu, J., Huang, C., Xie, C. and, Thompson, L.G., 1995. Climatic and Environmental records in Guliya Ice Cap. Sci. China (Series B) 38, 228–237.

Chapter 16
Regional Climate Change Scenarios

J. Sanjay, M.V.S. Ramarao, M. Mujumdar and R. Krishnan

1 Introduction

The information about long-term (century-scale) climate change is needed to develop national or regional adaptation and mitigation policies. The reliable climate information will be useful and actionable only when they are available at the sub-continental to regional scales. The coarser spatial resolution and systematic error (called bias) of global climate models (GCMs) limits the examination of possible impacts of climate change and adaptation strategies on a smaller scale. The grid interval of atmosphere-ocean coupled GCMs (AOGCMs) used in the Coupled Model Intercomparison Project phase 5 (CMIP5; Taylor et al. 2012) ranges from 1.0° to 3.8°. The CMIP5 models indicate large bias in the monthly cycle of near surface air temperature and precipitation climate over the Indian region, particularly at elevations over the Himalayas (Flato et al. 2013). Recently, the stand-alone atmospheric GCMs run at higher resolution than AOGCMs have been made possible with high-performance computer systems to provide complementary regional-scale climate information. The advantages of a 20-km grid atmospheric GCM in simulating the Indian climate have been identified, especially the complex features of the Indian summer monsoon, including improved regional precipitation (Rajendran and Kitoh 2008; Krishnan et al. 2013). Another approach is the use of an atmospheric GCM with variable grid-point resolution zooming over the region of interest. The global simulation using telescopic zooming to a high resolution of more than 35 km over India was shown to provide improved representation of the organized convective activity over the South Asian monsoon region and in realistically capturing the regional details of the precipitation variability and their links to monsoon circulation (Sabin et al. 2013). However, as large computer resources are needed, the model integration period and number of ensemble members is

J. Sanjay (✉) · M.V.S. Ramarao · M. Mujumdar · R. Krishnan
Indian Institute of Tropical Meteorology, Earth System Science Organization, Pune, India
e-mail: sanjay@tropmet.res.in

M.N. Rajeevan and S. Nayak (eds.), *Observed Climate Variability and Change Over the Indian Region*, Springer Geology, DOI 10.1007/978-981-10-2531-0_16

limited for such high-resolution atmospheric GCMs. Therefore, modellers rely on dynamical and/or statistical downscaling methods to make available higher-resolution, long-term climate data for impact and adaptation studies in specific regions.

The dynamical downscaling method using high resolution limited area regional climate models (RCMs) utilizes the outputs provided by AOGCMs as lateral boundary condition to provide physically consistent spatiotemporal variations of climatic parameters at spatial scales much smaller than the AOGCMs' grid. The RCMs by resolving the topographical details, coastlines, and land-surface heterogeneities allow the reproduction of small-scale processes and information that are most useful for impact assessment and in decision making for adaptation (Flato et al. 2013). Past research have studied the ability of individual RCMs to capture the general features of the Indian climate, particularly the summer monsoon climate (e.g. Krishnakumar et al. 2011; Dash et al. 2013 and the references therein). These studies showed that in general RCMs reproduce the precipitation seasonal mean and annual cycle quite accurately, although individual models can reveal significant biases in some sub-regions and seasons. In addition to the RCM-related bias, more errors and uncertainties in the downscaled climate can be inherited from the driving AOGCMs through the lateral boundary conditions. Parts of these uncertainties for the present climate can be addressed by generating a collection of simulations by driving the same RCM with different AOGCMs. The other sources of uncertainty in our understanding of the future climate change need to be addressed using the RCMs in a similar manner were addressed in CMIP5 using AOGCMs. Several different emission scenarios such as the Representation Concentration Pathways (RCPs) need to be used to force the RCMs so as to properly sample the uncertainty linked to future changes of the external forcing of the climate system. Similarly, by using several RCMs or a group of simulations with one RCM perturbed in its representation of the physics, parts of the uncertainties associated with how changes in external forcing factors control the climate need to be assessed. Finally, several simulations with each RCM driven with the same RCP scenario but with different initial conditions need to be generated to understand to what extent is the future climate change signal masked/amplified by natural variations of the climate system. The detailed evaluation of RCM-produced projections and a full classification of these underlying uncertainties are necessary for providing valuable information for assessing regional climate change vulnerability and impact application studies. The computer resources needed to generate such large ensembles of RCM outputs vary, depending on a number of factors (e.g. time step; horizontal and vertical resolutions; domain size; time period of simulation; and coupling system with additional models for ocean, land surface or terrestrial vegetation), but are quite demanding.

The World Climate Research Programme CORDEX (Coordinated Regional climate Downscaling Experiment; Giorgi et al. 2009; http://wcrp-cordex.ipsl.jussieu.fr/) initiative aims to foster international partnership in order to produce an ensemble of high-resolution past and future climate projections at regional scale, by downscaling several AOGCMs participating in the CMIP5. This ensemble can be used to demonstrate uncertainties on the regional scale or to obtain probabilistic climate change information in a region. The CORDEX South Asia component led by the Centre for Climate Change Research (CCCR) at the Indian Institute of Tropical Meteorology (IITM) aims to develop such multi-model ensemble of high-resolution (50-km) past and future climate projections over the South Asia region. The goal of this collaborative initiative is to generate robust national climate change information through dissemination of CORDEX South Asia datasets for regional climate change impact assessments and for developing adaptation strategies. The available RCM outputs under the CORDEX South Asia initiative are archived and published on the CCCR-IITM climate data portal (http://cccr.tropmet.res.in/cordex/files/downloads.jsp). Table 1 provides the basic references for these RCMs, their driving AOGCMs and the contributing partner modelling institutes. These datasets will shortly be accessible from the CCCR-IITM Earth System Grid Federation (ESGF) data node, which was the climate model data dissemination mechanism used for CMIP5. Some of these simulations have been analysed in terms of Indian precipitation extreme indices, its associated intermodal variability,

Table 1 List of CORDEX South Asia regional climate model (RCM) experiments

Experiment name	RCM description	Driving GCM	Contributing institute
LMDZ4 (IPSL)	Institut Pierre-Simon Laplace (IPSL) Laboratoire de Me´te´orologie Dynamique Zoomed version 4 (LMDZ4) atmospheric general circulation model (Sabin et al. 2013)	IPSL Coupled Model version 5 (IPSL-CM5-LR; Dufresne et al. 2013)	Centre for Climate Change Research (CCCR), Indian Institute of Tropical Meteorology (IITM), India
CCLM4 (MPI)	COnsortium for Small-scale MOdelling (COSMO) model in CLimate Mode version 4.8 (CCLM; Dobler and Ahrens 2008)	Max Planck Institute for Meteorology, Germany, Earth System Model (MPI-ESM-LR; Giorgetta et al. 2013)	Institute for Atmospheric and Environmental Sciences (IAES), Goethe University, Frankfurt am Main (GUF), Germany
RCA4 (ICHEC)	Rossby Centre regional atmospheric model version 4 (RCA4; Samuelsson et al. 2011)	Irish Centre for High-End Computing (ICHEC), European Consortium ESM (EC-EARTH; Hazeleger et al. 2012)	Rosssy Centre, Swedish Meteorological and Hydrological Institute (SMHI), Sweden
RegCM411 (GFDL) RegCM445 (GFDL)	The Abdus Salam International Centre for Theoretical Physics (ICTP) Regional Climatic Model version 4 (RegCM4; Giorgi et al. 2012)	Geophysical Fluid Dynamics Laboratory, USA, Earth System Model (GFDL-ESM2 M-LR; Dunne et al. 2012)	CCCR, IITM
REMO (MPI)	MPI Regional model 2009 (REMO2009; Teichmann et al. 2013)	MPI-ESM-LR (Giorgetta et al. 2013)	Climate Service Center, Hamburg, Germany

and up to 100-year return periods (Mishra et al. 2014). Here, we compare the outputs of the downscaled RCMs to those of the driving AOGCMs over the recent climate. It is important first to generally assess the ability of the CORDEX RCMs to simulate the general characteristics of the Indian climate (e.g. seasonal distribution of temperature and precipitation, and summer monsoon season climatology) and, second, to examine whether the downscaled simulations add value to those by the driving GCMs. Therefore, we focus not only on the main climate statistics, but we inspect also the ability of the RCMs to reproduce the temperature and precipitation seasonal cycle in sub-regions over India. Further, we analyse the changes in seasonal mean 2-m air temperature and precipitation to show the spread in average climate conditions by the middle of the century (2031–2060) for the Indian sub-continent.

2 Models and Experiments

2.1 CORDEX South Asia Models

The models used for CORDEX South Asia (listed in Table 1) describe the atmosphere and its coupling with the land surface. This ensemble includes four RCMs and a variable grid atmospheric GCM. The dynamics and physics set up for each model differ, but the regional domain at 50-km horizontal resolution is common to all RCMs as specified by the CORDEX experiment protocol (http://wcrp-cordex.ipsl.jussieu.fr/index.php/experiment-guidelines/cordex-experiment-protocol), covering West and South Asia region, and we present results within the interior model domain over the Indian sub-continent (e.g. Fig. 1). This ensemble consists of five 140-yr transient downscaled regional climate change simulations during the time period 1950–2100 and taking 6-hourly lateral and monthly ocean surface boundary conditions from four AOGCMs that participated in the CMIP5 RCP4.5 scenario experiments (listed in Table 1). It may be noted that two RCM experiments viz. CCLM4 (MPI) and REMO (MPI) were driven with the same MPI-ESM-LR AOGCM. Also RegCM411 RCM outputs are only available for the historical 1950–2005 period. The latest version of this RCM (RegCM445) includes an updated physics, in particular a more comprehensive representation of the land surface processes. The temporal evolution of the greenhouse gas concentrations is prescribed in the RCMs similar to that used by the CMIP5 AOGCMs, based on the observed values for the historical period and based on the RCP4.5 scenario for the future projection period. The remaining anthropogenic and natural forcings such as ozone and aerosols are kept constant. The regional land cover and land use changes are also not included in these downscaled climate simulations over South Asia.

The comparison of model results to the reference climate in the period 1976–2005 is presented to demonstrate how large biases the state-of-the-art RCMs shows when forced by lateral boundary conditions from AOGCMs. The quality of the simulations over India is assessed using the monthly mean 2-m air temperature and

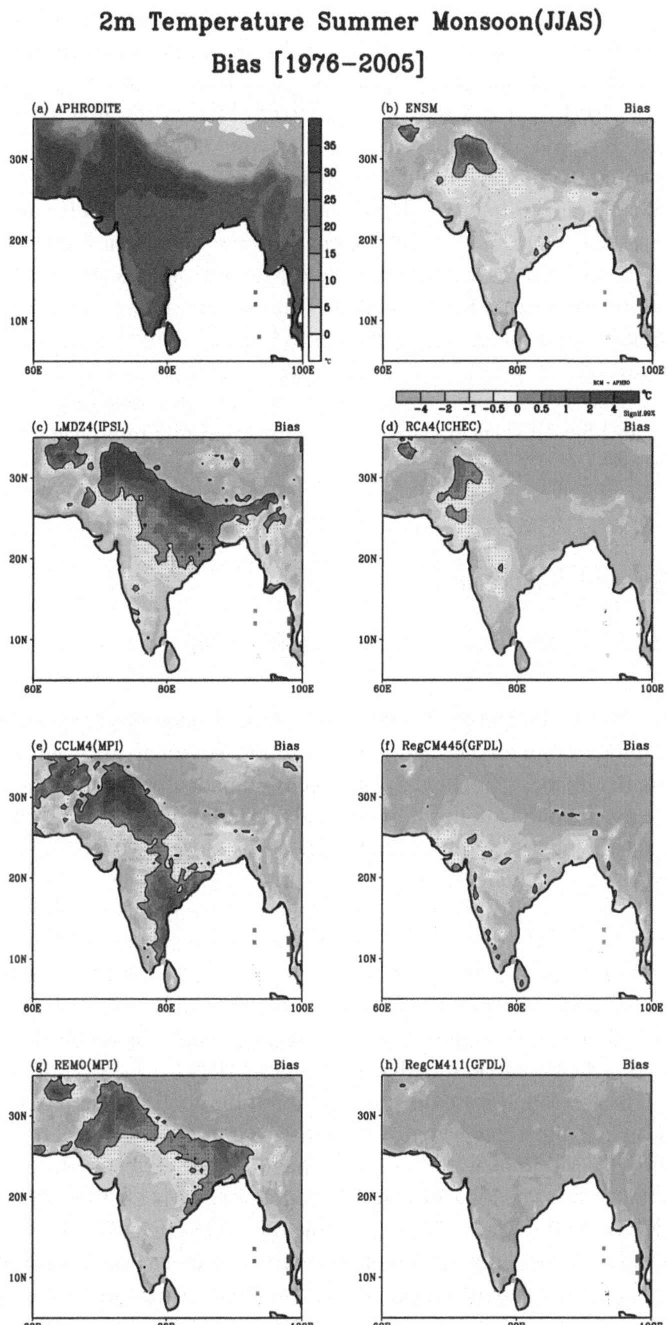

Fig. 1 **a** Summer monsoon (JJAS) season mean 2-m air temperature (°C; APHRODITE) for 1976–2005 and biases of 2-m air temperature (°C) in the CORDEX South Asia simulations driven by CMIP5 AOGCM historical experiments: **b** multi-model ensemble mean (ENSM) and **c**–**h** six different CORDEX RCMs listed in Table 1. Stippling denotes areas where the 30-year mean differences are not statistically significant at the 1 % level using Student's *t* test

the rain gauge-based global land precipitation dataset available at 0.5° spatial and monthly temporal resolution from the Asian Precipitation—Highly-Resolved Observational Data Integration Towards Evaluation of Water Resources (APHRODITE 1951–2005; Yatagai et al. 2012). This dataset is constructed to represent grid box average observations, allowing direct comparison to model results. The monthly model outputs from the CORDEX RCMs and the driving CMIP5 AOGCMs are bi-linearly interpolated to the APHRODITE spatial resolution. The seasonal averages for the Indian summer monsoon (June–September; JJAS) and post-monsoon (October–December; OND) months are computed. The annual averages (ANN) are also calculated. These seasonal and annual averages are computed for each year in 30-year periods both for the baseline period (1976–2005) and for a future time period (2031–2060). The Student's t test at the 1 % significance level is used to determine the robustness for the climate change signal in the future period and for differences between the simulated and reference climate in the baseline period.

3 Results and Discussion

3.1 Simulated Baseline Climate (1976–2005)

The simulated 2-m air temperature climate over India is in relatively good agreement with the reference climate. The summer monsoon seasonal mean differences are generally smaller than 1 °C (Fig. 1), although statistically significant differences above 2°C exist in parts of north India and the adjoining Himalayan mountain region. The six-member ensemble mean (Fig. 1b) is closer to observed temperature over the central India than the individual simulations for the summer monsoon season (Fig. 1c–f). This is due to the cancellation of the biases of opposite signs in the models when added over this region. It may be noted that the two RCMs viz. CCLM4(MPI) and REMO (MPI) driven with the same CMIP5 AOGCM show different temperature bias patterns, particularly over the western parts of India. The REMO (MPI) was reported to have a larger annual mean cold bias over the entire west Asia when compared with the driving CMIP5 MPI-ESM-LR (Teichmann et al. 2013). It is also found from the comparison of the spatial pattern of temperature in two versions of the RegCM4 driven with the GFDL-ESM2 M that the latest version of this RCM (RegCM445; Fig. 1f) has to some extent reduced the larger cold bias in the earlier version of the same model (RegCM411; Fig. 1h). Compared to the APHRODITE dataset, precipitation in CORDEX South Asia RCMs six-member ensemble mean is slightly underestimated during summer in the central and Indo-Gangetic plains over India, while significant overestimation is found over parts of south peninsula (Fig. 2b). However, the ensemble mean is closer to reference precipitation than most individual simulations for the summer monsoon season (Fig. 2c–f). This is because the biases of opposite signs in the models over parts of

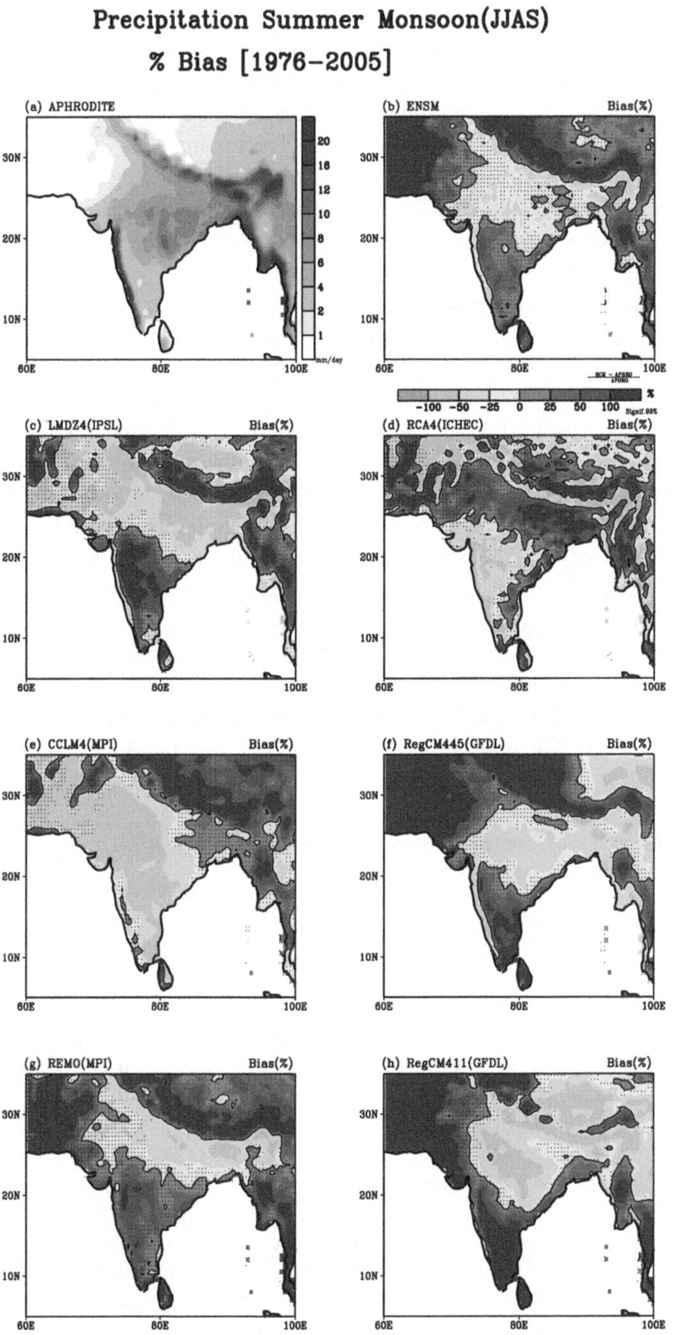

Fig. 2 **a** Summer monsoon (JJAS) season mean precipitation (mm day^{-1}; APHRODITE) for 1976–2005 and biases of precipitation (%) in the CORDEX South Asia simulations driven by CMIP5 AOGCM historical experiments: **b** multi-model ensemble mean (ENSM) and **c**–**h** six different CORDEX RCMs listed in Table 1. Stippling denotes areas where the 30-year mean differences are not statistically significant at the 1 % level using Student's *t* test

south India tend to cancel each other when added, cf. LMDZ4 (IPSL) and RCA4 (ICHEC) in Fig. 2. The summer monsoon season wet bias seen in REMO (MPI) over the central and southern parts of India (Fig. 2g) was also found for the simulated annual mean precipitation with this RCM, while the driving MPI-ESM-LR had a dry bias over this region (Teichmann et al. 2013). The relatively lesser wet biases over south India in the latest version of RegCM4 (Fig. 2f) compared to the earlier version of this model (Fig. 2h) have also to some extent contributed to the improved spatial pattern of precipitation in the six-member ensemble mean.

Table 2 summarizes the climatological skill of 2-m temperature in the CORDEX historical experiments and the corresponding driving CMIP5 AOGCMs in simulating summer monsoon (JJAS), post-monsoon (OND) seasonal and annual means for the 30-year period 1976–2005 over the South Asia land areas (60–100°E, 5–35° N). In general, it is found that the RCMs and their driving AOGCMs tend to underestimate the spatial mean and overestimate the standard deviation over this large domain. The pattern correlation with APHRODITE is found to be relatively improved for all the RCMs in comparison with their driving AOGCMs in both seasons and annually. The spatial variability simulated by individual models relative to APHRODITE is assessed using a Taylor diagram (Taylor 2001; Fig. 3a). It is seen that the individual RCMs and their driving AOGCMs consistently yield higher Taylor skill in simulating the annual mean 2-m temperature distribution over land areas in South Asia. These models tend to overestimate the spatial mean precipitation for both seasons and annually in this 30-year period over this region (Table 3). The RCMs consistently show higher magnitude of spatial standard

Table 2 Performance of 2-m temperature (°C) climatology (1976–2005) averaged over land grid points in South Asia (60–100°E, 5–35°N)

Experiment name	Spatial mean			Standard deviation			Pattern correlation		
	JJAS	OND	Annual	JJAS	OND	Annual	JJAS	OND	Annual
APHRODITE	23.4	14.8	18.4	8.7	11.8	10.9	–	–	–
LMDZ4 (IPSL)	22.0 (22.0)	12.1 (11.0)	16.1 (15.2)	10.3 (10.8)	13.2 (13.9)	12.5 (13.1)	**0.98** (0.96)	**0.99** (0.98)	**0.99** (0.98)
RCA4 (ICHEC)	20.1 (18.8)	9.2 (10.7)	13.7 (14.3)	10.8 (9.1)	14.2 (12.4)	13.3 (11.6)	**0.99** (0.97)	**0.99** (0.98)	**0.99** (0.98)
CCLM4 (MPI)	21.9 (22.5)	12.8 (11.4)	16.6 (16.6)	10.3 (10.0)	13.7 (11.8)	13.0 (11.6)	**0.98** (0.94)	**0.99** (0.97)	**0.99** (0.96)
RegCM445 (GFDL)	19.0 (23.2)	9.8 (13.4)	13.1 (17.5)	11.2 (10.6)	13.4 (12.6)	12.9 (12.2)	**0.97** (0.96)	**0.98** (0.97)	**0.98** (0.97)
REMO (MPI)	21.5 (22.5)	10.2 (11.4)	15.1 (16.6)	10.4 (10.0)	13.2 (11.8)	12.6 (11.6)	**0.99** (0.94)	**0.99** (0.97)	**0.99** (0.96)
RegCM411 (GFDL)	18.5 (23.2)	8.2 (13.4)	12.0 (17.5)	10.2 (10.6)	13.1 (12.6)	12.5 (12.2)	**0.97** (0.96)	**0.99** (0.97)	**0.98** (0.97)

The spatial skill for the six different CORDEX RCMs listed in Table 1 are compared with (in parenthesis) the corresponding CMIP5 AOGCM historical experiment used to drive the RCMs. The bold text shows the improved performance of the RCM relative to its driving AOGCM

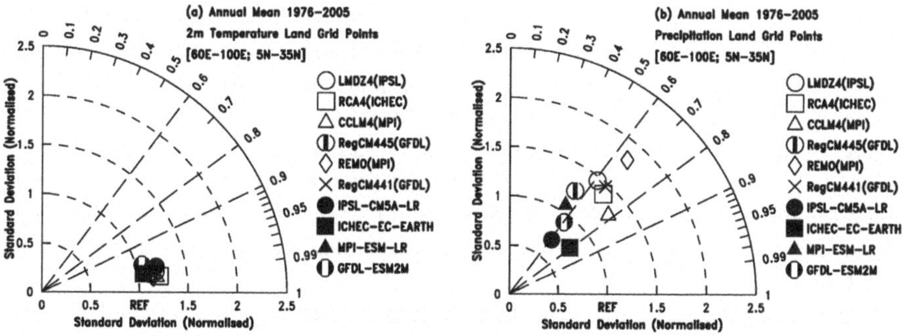

Fig. 3 Taylor diagram for the annual mean **a** 2-m air temperature (°C) and **b** precipitation (mm day^{-1}) climatology (1976–2005) averaged over land grid points in South Asia (60–100°E, 5–35°N). The radial coordinate shows the standard deviation of the spatial pattern, normalized by the observed standard deviation. The azimuthal variable shows the correlation of the modelled spatial pattern with the observed spatial pattern. The distance between the reference (REF) dataset (APHRODITE) and individual points corresponds to root-mean-square error (RMSE). The diagram shows the skill for the six different CORDEX RCMs listed in Table 1 and for the four CMIP5 model historical experiments used to drive the CORDEX South Asia RCMs

Table 3 Performance of precipitation (mm d^{-1}) climatology (1976–2005) averaged over land grid points in South Asia (60–100°E, 5–35°N)

Experiment name	Spatial mean			Standard deviation			Pattern correlation		
	JJAS	OND	Annual	JJAS	OND	Annual	JJAS	OND	Annual
APHRODITE	4.7	1.0	2.2	4.2	1.3	1.8	–	–	–
LMDZ4 (IPSL)	5.2 (3.1)	1.6 (1.4)	2.7 (1.7)	5.6 (2.8)	1.9 (1.2)	2.7 (1.3)	0.61 (0.62)	**0.78** (0.71)	**0.61** (0.61)
RCA4 (ICHEC)	5.2 (5.7)	1.5 (1.4)	2.7 (2.8)	5.4 (3.0)	1.8 (1.4)	2.6 (1.4)	0.66 (0.76)	0.81 (0.91)	0.68 (0.79)
CCLM4 (MPI)	4.7 (5.5)	1.4 (1.6)	2.5 (2.6)	4.8 (3.7)	1.3 (1.4)	2.4 (1.7)	**0.80** (0.63)	0.72 (0.73)	**0.78** (0.60)
RegCM445 (GFDL)	5.2 (5.0)	1.8 (1.2)	3.2 (2.4)	3.1 (4.2)	2.0 (1.1)	2.3 (2.0)	0.54 (0.56)	0.07 (0.70)	**0.54** (0.53)
REMO (MPI)	6.0 (5.5)	2.0 (1.6)	3.2 (2.6)	6.5 (3.7)	2.7 (1.4)	3.3 (1.7)	**0.65** (0.63)	**0.74** (0.73)	**0.66** (0.60)
RegCM411 (GFDL)	5.1 (5.0)	1.8 (1.2)	3.0 (2.4)	6.0 (4.2)	1.8 (1.1)	2.7 (2.0)	0.46 (0.56)	0.69 (0.70)	**0.67** (0.53)

The spatial skill for the six different CORDEX RCMs listed in Table 1 are compared with (in parenthesis) the corresponding CMIP5 AOGCM historical experiment used to drive the RCMs. The bold text shows the improved performance of the RCM relative to its driving AOGCM deviation for precipitation than APHRODITE observations suggesting that the high-resolution downscaling has overestimated the spatial variability of precipitation over this region. The simulated spatial patterns relative to APHRODITE (Table 3) vary among the RCMs and are found to be better than the driving AOGCMs in both seasons only for few individual RCMs viz. LMDZ4 (IPSL), CCLM4 (MPI) and REMO (MPI). The Taylor diagram for the annual mean precipitation distribution over land areas in South Asia (Fig. 3b) shows the large spread in the Taylor skill between the individual RCMs and their driving AOGCMs.

Figure 4 presents the monthly 2-m air temperature (left panels) and the precipitation (right panels) annual cycle for the period 1976–2005 simulated in the individual RCMs and their driving AOGCMs for the central, south-west and

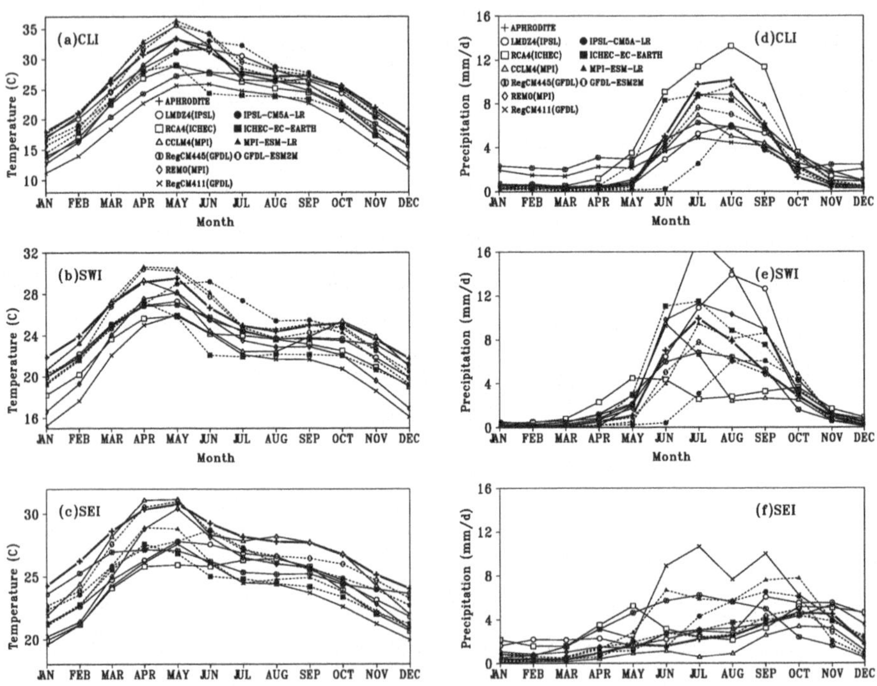

Fig. 4 Mean seasonal cycle for the period 1976–2005 of (*left panels*) 2-m air temperature (°C) and (*right panels*) precipitation rate (mm day^{-1}) for the six different CORDEX RCMs (*thin lines*) listed in Table 1 and for the four CMIP5 models (*dashed lines*) used to drive the RCMs used in the CORDEX South Asia historical experiments. The observed values based on APHRODITE (*thick line*) are used as reference. The analysis used the land grid points in the sub-regions **a**, **d** Central India (CLI; 20–25°N, 78–82°E), **b**, **e** South-West India (SWI; 20–25°N, 78–82°E), and **c**, **f** South-East India (SEI; 20–25°N, 78–82°E)

south-east sub-regions over India. All the models simulate the phase of the seasonality in temperature well than the amplitude relative to APHRODITE observations in the three sub-regions. The individual model skill in simulating the seasonal cycle of temperature is summarized for the three sub-regions in Table 4. The root-mean-square error (RMSE) normalized with the APHRODITE annual range in temperature reveals that three RCMs viz. LMDZ4(IPSL), RCA4(ICHEC) and CCLM4(MPI) are able to outperform their driving AOGCMs in simulating the amplitude of seasonality in temperature over central India. The correlation coefficient between the model simulated and APHRODITE annual cycle of temperature further confirms that these RCMs improve not only the amplitude but also the phase of the temperature seasonality compared to their driving AOGCMs over central India. However, the RCMs are in general not able to improve the amplitude or the phase of the annual cycle of temperature over the hilly regions in south-west India and the drier regions in south-east India. Despite large inter-model variations found in the simulated precipitation seasonality, some RCMs appear to agree relatively closer with APHRODITE than their driving AOGCMs at least in capturing the phase of the seasonality over the three sub-regions (Fig. 4; right panels). However, the individual model skill summarized in Table 5 shows that only LMDZ4 (IPSL) is able to show an added value compared to its driving AOGCM in simulating the amplitude and phase of the seasonality in precipitation for all three sub-regions over India.

Table 4 Performance of 2-m temperature (°C) monthly annual cycle climatology (1976–2005) averaged over land grid points in three sub-regions: Central India (CLI; 20–25°N, 78–82°E), South-West India (SWI; 20–25°N, 78–82°E) and, (c) South-East India (SEI; 20–25°N, 78–82°E)

Experiment name	Normalized RMSE			Correlation coefficient		
	CLI	SWI	SEI	CLI	SWI	SEI
LMDZ4 (IPSL)	**0.11** (0.16)	0.22 (0.21)	0.39 (0.31)	**0.97** (0.93)	**0.97** (0.83)	0.91 (0.92)
RCA4 (ICHEC)	**0.23** (0.25)	**0.36** (0.37)	0.53 (0.49)	**0.99** (0.98)	0.91 (0.95)	0.86 (0.98)
CCLM4 (MPI)	**0.08** (0.13)	0.18 (0.16)	**0.17** (0.44)	**0.99** (0.97)	0.93 (0.98)	**0.98** (0.98)
RegCM445 (GFDL)	0.26 (0.11)	0.22 (0.15)	0.35 (0.14)	0.92 (0.99)	0.95 (0.99)	0.94 (0.97)
REMO (MPI)	0.21 (0.13)	0.41 (0.16)	0.38 (0.44)	**0.99** (0.97)	0.96 (0.98)	0.96 (0.98)
RegCM411 (GFDL)	0.40 (0.11)	0.58 (0.15)	0.58 (0.14)	0.95 (0.99)	0.93 (0.99)	**0.98** (0.97)

The root-mean-square error (RMSE) normalized with the APHRODITE annual range and the correlation coefficient between the simulated and APHRODITE annual cycle for the six different CORDEX RCMs listed in Table 1 are compared with (in parenthesis) the corresponding CMIP5 model historical experiment used to drive the RCMs. The bold text shows the improved performance of the RCM relative to its driving AOGCM

Table 5 Performance of precipitation (mm day^{-1}) monthly annual cycle climatology (1976–2005) averaged over land grid points in three sub-regions: Central India (CLI; 20–25°N, 78–82°E), South-West India (SWI; 20–25°N, 78–82°E) and, (c) South-East India (SEI; 20–25°N, 78–82°E)

Experiment name	Normalized RMSE			Correlation coefficient		
	CLI	SWI	SEI	CLI	SWI	SEI
LMDZ4 (IPSL)	**0.81** (1.09)	**0.66** (0.99)	0.72 (0.71)	**0.93** (0.82)	**0.90** (0.63)	**0.86** (0.74)
RCA4 (ICHEC)	0.65 (0.37)	1.23 (0.37)	0.99 (0.29)	**0.96** (0.94)	0.53 (0.98)	0.63 (0.93)
CCLM4 (MPI)	0.66 (0.30)	0.71 (0.29)	**0.51** (1.10)	0.96 (0.97)	0.82 (0.91)	**0.96** (0.76)
RegCM445 (GFDL)	0.84 (0.44)	0.54 (0.35)	1.22 (0.22)	0.95 (0.99)	0.96 (0.99)	0.25 (0.90)
REMO (MPI)	0.35 (0.30)	0.33 (0.29)	**0.45** (1.10)	**0.98** (0.97)	**0.98** (0.91)	**0.92** (0.76)
RegCM411 (GFDL)	0.97 (0.44)	1.21 (0.35)	1.88 (0.22)	0.96 (0.99)	0.98 (0.99)	0.54 (0.90)

The root-mean-square error (RMSE) normalized with the APHRODITE annual mean value and the correlation coefficient between the simulated and APHRODITE annual cycle for the six different CORDEX RCMs listed in Table 1 are compared with (in parenthesis) the corresponding CMIP5 model historical experiment used to drive the RCMs. The bold text shows the improved performance of the RCM relative to its driving AOGCM

3.2 *Simulated Climate Change (2031–2060)*

The simulated summer monsoon 2-m air temperature patterns are changing between 1976–2005 and 2031–2060 for the RCP4.5 scenario (Fig. 5). The ensemble mean of five-members indicate large increase of above 1.5 °C over the central and northern parts of India (Fig. 5a). The individual model simulations show very different response over India (Fig. 5b-f), with LMDZ4 (IPSL) indicating a warming that is above 2.0 °C higher than in RegCM445(GFDL) by middle of the twenty-first century for most parts of the country. The ensemble mean summer monsoon precipitation change for the same period indicates less than 25 % drying in the central and eastern parts of India and moistening of similar magnitude over the rest of India (Fig. 6a). However, for most parts of India, this ensemble mean change is not found to be statistically significant. Also some of the individual models indicate precipitation changes of similar magnitude with opposite sign in these regions (Fig. 6b–f). Further, these precipitation changes are also not statistically significant, implying that the CORDEX RCMs simulated change in summer monsoon precipitation over India are uncertain not only in magnitude but also in sign.

The analysis of the CORDEX multi-RCM temperature and precipitation projections for South Asia land areas for the period 1950–2100 (Fig. 7) shows that for RCP4.5 scenario the annual mean warming is likely to be in the range 1.0–2.0 °C by 2030, 1.8–3.0 °C by 2060 and 2.0–3.1 °C by 2090 relative to the 1976–2005 period (Fig. 7b). The summer monsoon precipitation projections for this region

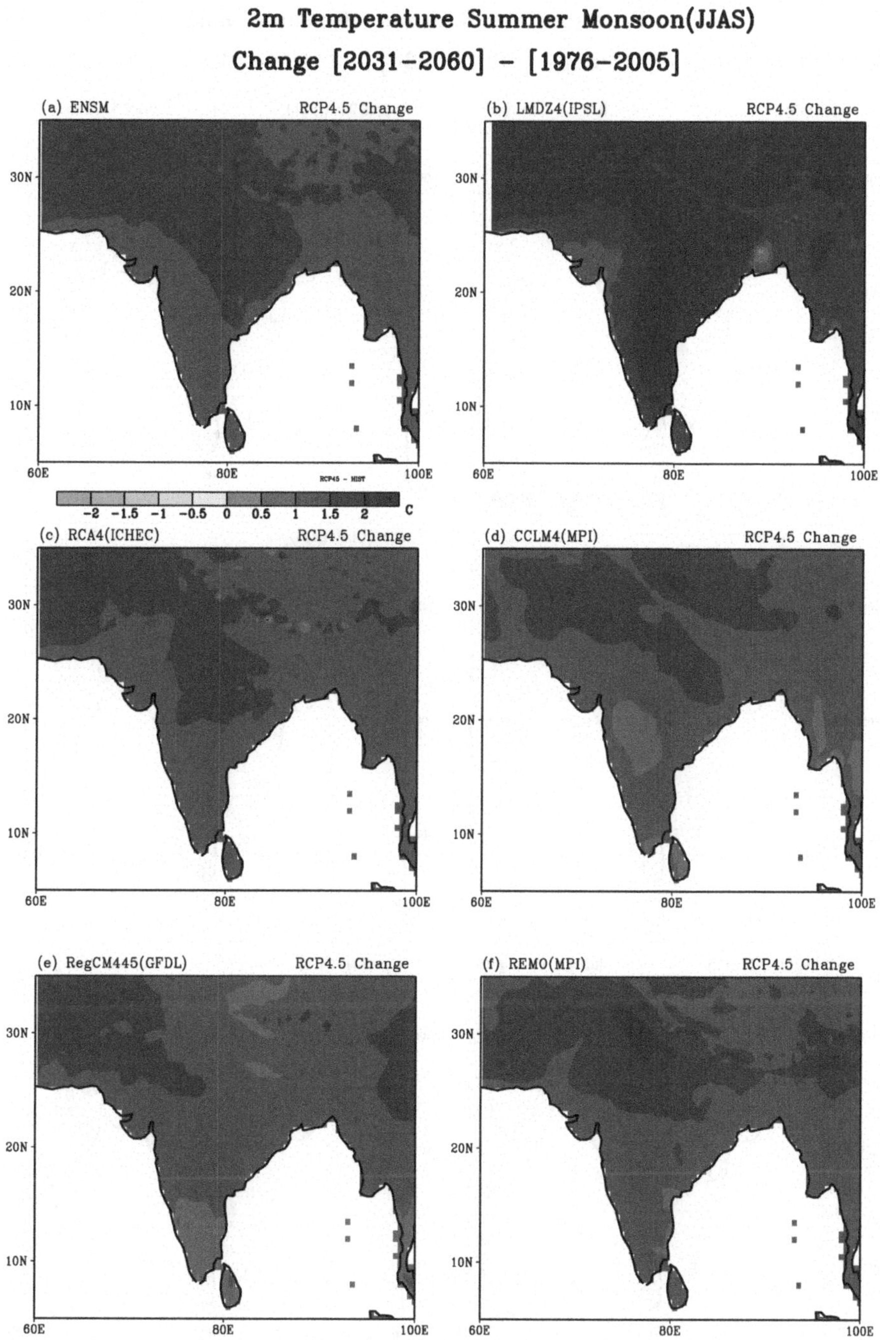

Fig. 5 Summer monsoon (JJAS) season mean 2-m air temperature (°C) changes in 2031–2060 with respect to 1976–2005 for the CORDEX South Asia simulations driven by CMIP5 AOGCM RCP4.5 scenario experiments **a** multi-model ensemble mean (ENSM) and **b**–**f** five different CORDEX RCMs listed in Table 1. Stippling denotes areas where the 30-year mean changes are not statistically significant at the 1 % level using Student's *t* test

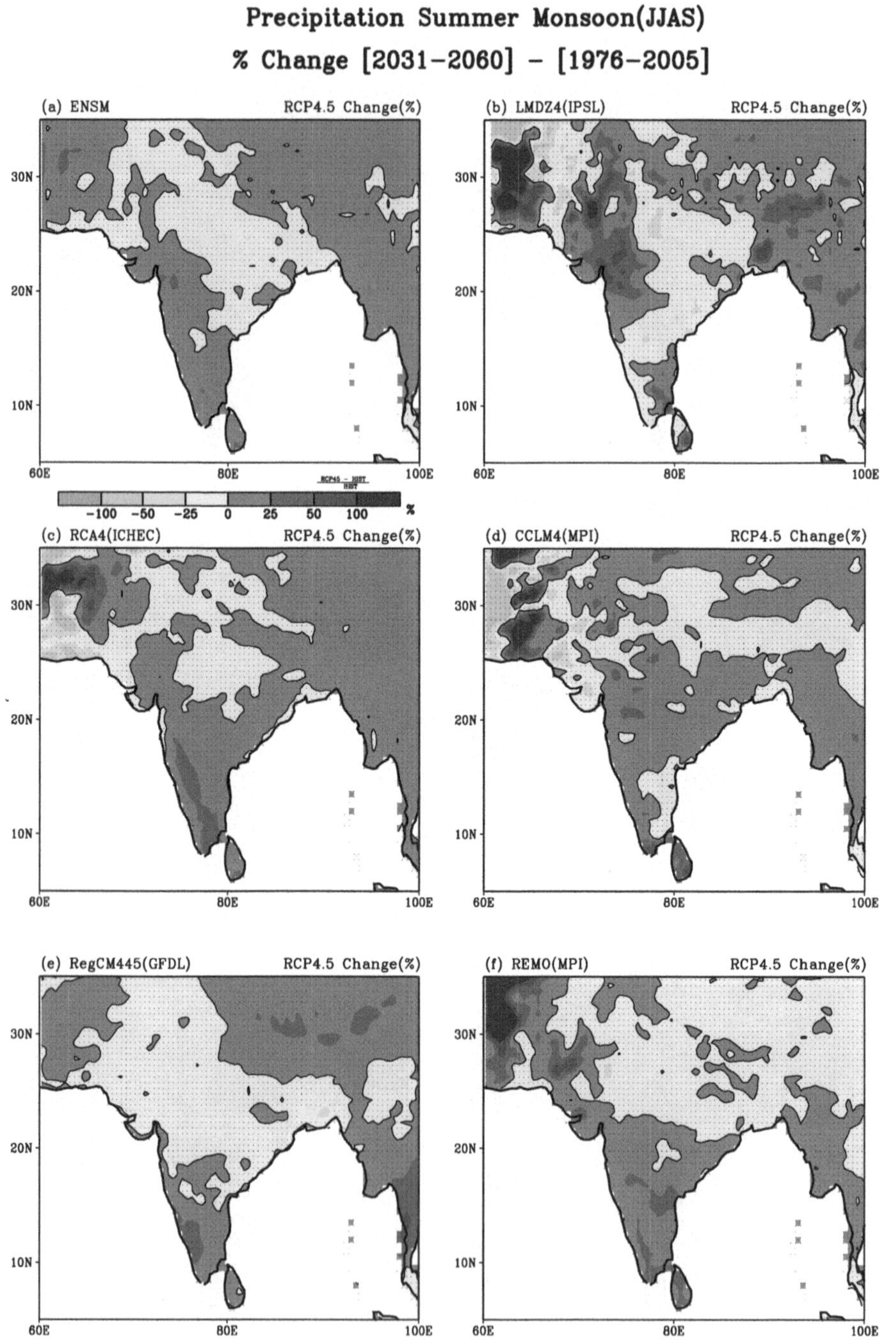

Fig. 6 Summer monsoon (JJAS) season mean precipitation changes (%) in 2031–2060 with respect to 1976–2005 for the CORDEX South Asia simulations driven by CMIP5 AOGCM RCP4.5 scenario experiments **a** multi-model ensemble mean (ENSM) and **b–f** five different CORDEX RCMs listed in Table 1. Stippling denotes areas where the 30-year mean changes are not statistically significant at the 1 % level using Student's *t* test

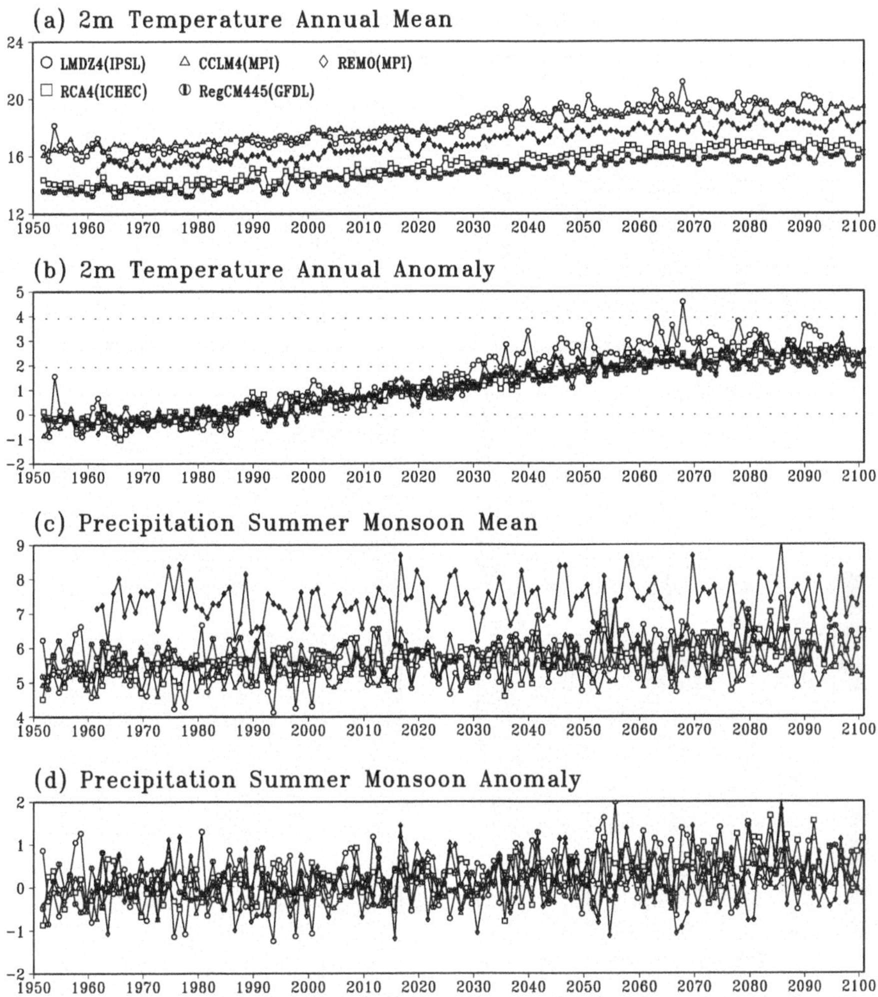

Fig. 7 Time series for the period 1950–2100 of: 2-m temperature (°C) annual **a** mean and **b** change relative to 1976–2005; and precipitation (mm day^{-1}) summer monsoon (JJAS) season **c** mean and **d** change relative to 1976–2005, averaged over land grid points in South Asia (60–100°E, 5–35°N) for the five different CORDEX RCMs listed in Table 1

show large spread among the individual models for the entire analysis period (Fig. 7d) suggesting that these downscaled precipitation projections are not reliable.

The percentile distribution of near-surface air temperature or precipitation gives insight into the spatial and temporal patterns of their extremes (e.g. 90th percentile). The mid-term (2031–60) projections in the RCP4.5 scenario experiments for South Asia during summer monsoon months (June-September) are shown in Fig. 8 and Fig. 9, displaying mid-term changes in extreme temperature (Fig. 8) and precipitation (Fig. 9) for the CORDEX South Asia RCMs (right-hand panels) and their

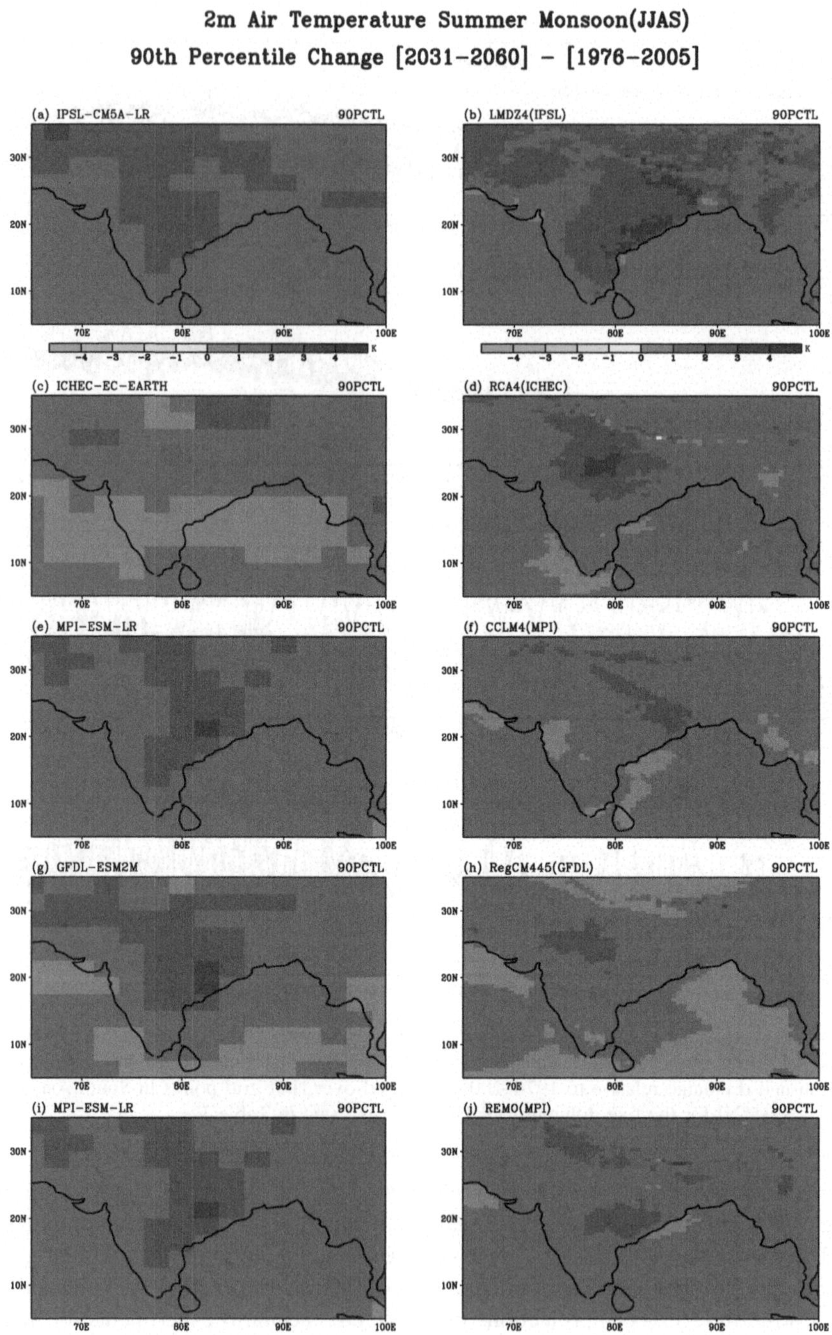

Fig. 8 Changes in the 90th percentile of the daily distribution of 2-m air temperature (°C) during summer monsoon months (June–September) in the 30-year period 2031–2060 with respect to 1976–2005 for the five CORDEX South Asia simulations (*right panels*) driven by CMIP5 AOGCM RCP4.5 scenario experiments (*left panels*) listed in Table 1

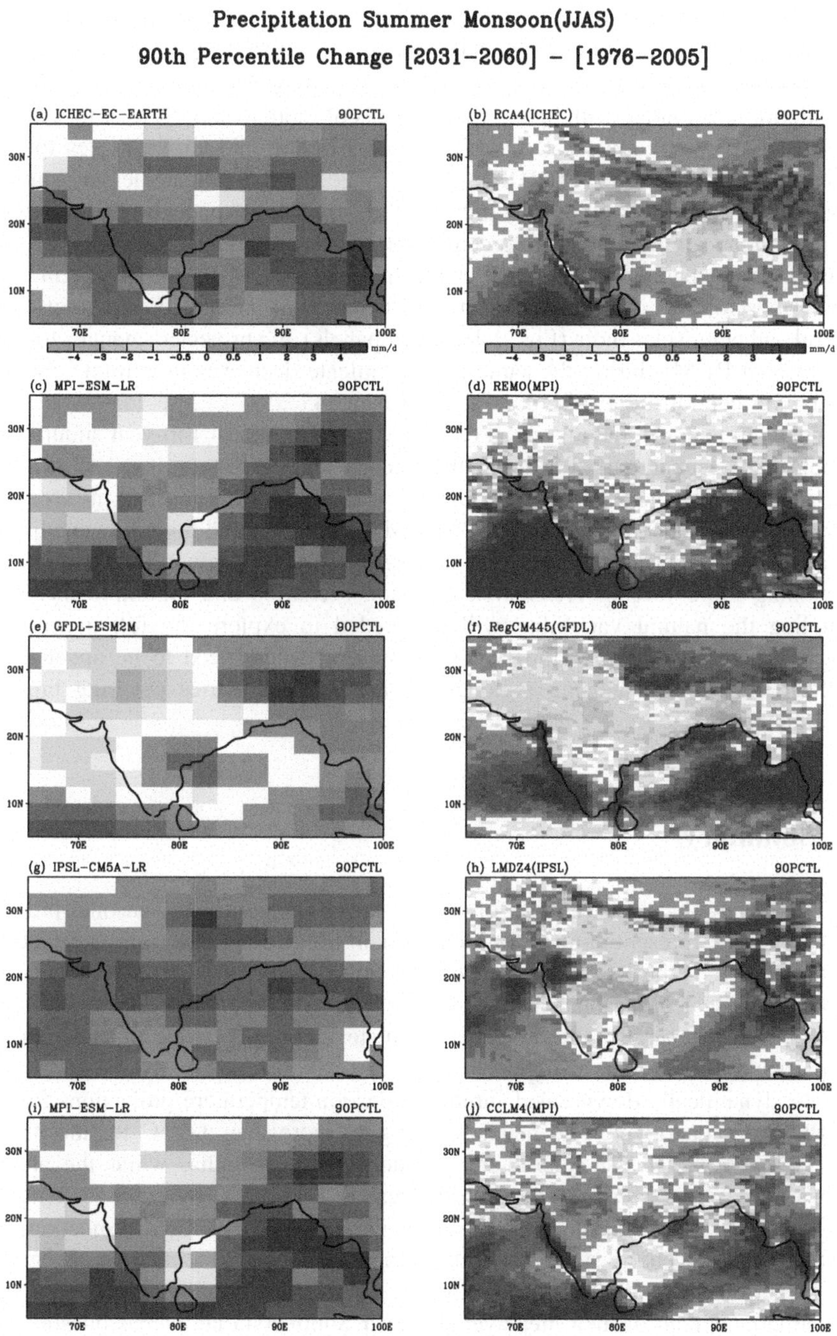

Fig. 9 Changes in the 90th percentile of the daily distribution of precipitation (mm day^{-1}) during summer monsoon months (June–September) in the 30-year period 2031–2060 with respect to 1976–2005 for the five CORDEX South Asia simulations (*right panels*) driven by CMIP5 AOGCM RCP4.5 scenario experiments (*left panels*) listed in Table 1

corresponding driving coarser resolution CMIP5 AOGCMs (left-hand panels) relative to the reference period 1976–2005. Most of the AOGCM projections (Fig. 8, left-hand panels) show a warming of 2–3 °C extending from the eastern region to the interior north India, with the highest changes of more than 3 °C over the eastern coast of India. The downscaled spatial patterns of the mid-term changes for the extreme temperatures in CORDEX South Asia RCMs indicate lesser summer monsoon seasonal warming over India (Fig. 8, right-hand panels). The mid-term changes in the summer monsoon precipitation for the downscaled CORDEX RCMs show different spatial patterns than that for their driving CMIP5 AOGCMs (Fig. 9). While most of the AOGCMs indicate an increase in extreme precipitation over central and peninsular India (Fig. 9, left-hand panels), the projected changes for few downscaled RCMs during the same period indicate decreases in extreme precipitation over central India (Fig. 9, right-hand panels).

Finally, it is noted that this small ensemble of five transient 140-year simulations samples only a small part of the total uncertainty range. However, it was shown that this small ensemble of climate change simulations were useful in order to demonstrate the uncertainties related to RCM formulation and boundary conditions in a physically consistent manner at the regional scale. A much larger ensemble containing more forcing AOGCMs, emission scenarios and ensemble members sampling the natural variability will be needed to explore the full uncertainty ranges. Therefore, the preliminary results presented here need to be updated for CORDEX South Asia using regional climate change metrics covering broader uncertainty ranges.

4 Summary

- The geographical distribution of surface air temperature and seasonal precipitation in the present climate for land areas in South Asia is strongly affected by the choice of the RCM and boundary conditions (i.e. driving AOGCMs), and the downscaled seasonal averages are not always improved. However, some RCMs are generally able to better simulate the annual cycle of temperature, particularly over central India.
- The dynamically downscaled summer monsoon temperature projections for the RCP4.5 scenario indicate mean warming of more than 1.5 °C for the period 2031–2060 over the central and northern parts of India, while the annual warming range over South Asia land areas is 1.8–3.0 °C by 2060.
- The results based on the available small ensemble of five RCM projections for the RCP4.5 scenario suggest that the summer monsoon precipitation change for the period 2031–2060 are uncertain not only in magnitude but also in sign, and the large spread of their area averages over South Asia land areas question the reliability of these downscaled precipitation projections for applying in regional climate change impact assessment studies.

Acknowledgments The World Climate Research Programme's Working Group on Regional Climate, and the Working Group on Coupled Modelling, former coordinating body of CORDEX and responsible panel for CMIP5 are gratefully acknowledged. The climate modelling groups (listed in Table 1 of this chapter) are sincerely thanked for producing and making available their model output. The Earth System Grid Federation infrastructure (ESGF; http://esgf.llnl.gov/index.html) is also acknowledged. We are thankful to the IITM Prithvi and Aaditya High Power Computing (HPC) support for providing the computer resources for in-house model runs. The Climate Data Operators software (CDO; https://code.zmaw.de/projects/cdo/) and the Grid Analysis and Display System (GrADS; http://iges.org/grads/) were extensively used throughout this analysis. We thank Sandeep Ingle for managing CCCR climate data portal and for technical support. The Indian Institute for Human Settlements (IIHS) through the Adaptation at Scale in Semi-Arid Regions (ASSAR) consortia of the Collaborative Adaptation Research initiative in Africa and Asia (CARIAA) financially supported M. V. S. Ramarao for carrying out this scientific research.

References

Dash SK, Mamgain A, Pattnayak KC, Giorgi F (2013) Spatial and temporal variations in Indian summer monsoon rainfall and temperature: an analysis based on RegCM3 simulations. Pure appl Geophys 170:655–674

Dobler, A., and B. Ahrens (2008) Precipitation by a regional climate model and bias correction in Europe and South Asia, Meteorol. Z., 17, 499–509.

Dufresne, J. L. and Coauthors (2013) Climate change projections using the IPSL-CM5 Earth System Model: from CMIP3 to CMIP5. Clim Dyn (2013) 40:2123–2165. DOI 10.1007/s00382-012-1636-1

Dunne, J.P. and Coauthors (2012) GFDL's ESM2 Global Coupled Climate–Carbon Earth System Models. Part I: Physical Formulation and Baseline Simulation Characteristics. J. Clim, 25, 6646-6665.

Flato, G., J. and Coauthors (2013) Evaluation of Climate Models. In: Climate Change 2013: The Physical Science Basis. Contribution of Working Group I to the Fifth Assessment Report of the Intergovernmental Panel on Climate Change [Stocker, T.F., D. Qin, G.-K. Plattner, M. Tignor, S.K. Allen, J. Boschung, A. Nauels, Y. Xia, V. Bex and P.M. Midgley (eds.)]. Cambridge University Press, Cambridge, United Kingdom and New York, NY, USA.

Giorgi F, Jones C, Asrar G (2009) Addressing climate information needs at the regional level: the CORDEX framework. World Meteorol Organ (WMO) Bull 58(July):175–183

Giorgi, F. and Coauthors (2012) RegCM4: model description and preliminary tests over multiple CORDEX domains. Clim Res 52:7–29.

Giorgetta, M.A. and Coauthors (2013) Climate and carbon cycle changes from 1850 to 2100 in MPI-ESM simulations for the coupled model intercomparison project phase 5. Journal of Advances in Modeling Earth Systems, 5, 572-597. doi:10.1002/jame.20038

Hazeleger, W. and Coauthors (2012) EC-Earth V2.2: description and validation of a new seamless earth system prediction model. Clim Dyn, 39:2611–2629. DOI 10.1007/s00382-011-1228-5

Krishnan, R., Sabin, T. P., Ayantika, D. C., Kitoh, A., Sugi, M., Murakami, H., Turner, A. G., Slingo, J. M. and Rajendran, K.: Will the South Asian monsoon overturning circulation stabilize any further?, Clim. Dyn., 40, 187–211, doi:10.1007/s00382-012-1317-0, 2013.

Krishnakumar K, S. K. Patwardhan, A. Kulkarni, K. Kamala, K. Koteswara Rao and R. Jones (2011) Simulated projections for summer monsoon climate over India by a high-resolution regional climate model (PRECIS). *CURRENT SCIENCE*, 101:312-326.

Mishra, V., D. Kumar, A. R. Ganguly, J. Sanjay, M. Mujumdar, R. Krishnan, and R. D. Shah (2014), Reliability of regional and global climate models to simulate precipitation extremes over India, J. Geophys. Res. Atmos., 119, 9301–9323, doi:10.1002/2014JD021636.

Rajendran, K and A. Kitoh (2008) Indian summer monsoon in future climate projection by a super high-resolution global model. Current Science, 95:1560-1569.

Sabin, T. P. and Coauthors (2013) High resolution simulation of the South Asian monsoon using a variable resolution global climate model. Clim Dyn 41:173–194. DOI 10.1007/s00382-012-1658-8.

Samuelsson, P. and Coauthors (2011) The Rossby Centre regional climate model RCA3: Model description and performance. Tellus, 63A, 4–23.

Taylor KE. 2001. Summarizing multiple aspects of model performance in a single diagram. J. Geophys. Res. 106: 7183–7192.

Taylor KE, Stouffer RJ, Meehl GA (2012) An overview of CMIP5 and the experiment design. Bull Am Meteorol Soc 93(4):485–498. doi:10.1175/BAMS-D-11-00094.1

Teichmann, C. and Coauthors (2013) How Does a Regional Climate Model Modify the Projected Climate Change Signal of the Driving GCM: A Study over Different CORDEX Regions Using REMO. *Atmosphere*, 4:214-236. doi:10.3390/atmos4020214

Yatagai, A., Kamiguchi, K., Arakawa, O., Hamada, A., Yasutomi, N. and Kitoh, A. (2012) APHRODITE: Constructing a long-term daily gridded precipitation dataset for Asia based on a dense network of rain gauges, Bull. Am. Met. Soc., 93, 1401-1415, doi:10.1175/BAMS-D-11-00122.1.

Author Index

M.N. Rajeevan and S. Nayak (eds.), *Observed Climate Variability and Change Over the Indian Region*, Springer Geology, DOI 10.1007/978-981-10-2531-0

The manufacturer's authorised representative in the EU is Springer Nature Customer Service Centre GmbH, Europaplatz 3, 69115 Heidelberg, Germany. If you have any concerns regarding our products, please contact ProductSafety@springernature.com

Printed and bound by CPI Group (UK) Ltd, Croydon, CR0 4YY
25/09/2026
02222896-0009